Results and Problems in Cell Differentiation

A Series of Topical Volumes in Developmental Biology

11

Differentiation and Neoplasia

Edited by
R.G. McKinnell
M.A. DiBerardino, M. Blumenfeld
and R.D. Bergad

With 77 Figures

Springer-Verlag Berlin Heidelberg GmbH 1980

Professor ROBERT G. MCKINNELL
University of Minnesota, Department of Genetics and Cell Biology
250 Biological Sciences Center, St. Paul, MN 55108, USA

Dr. MARIE A. DIBERARDINO
Department of Anatomy, Medical College of Pennsylvania
Philadelphia, PA 19129, USA

Dr. MARTIN BLUMENFELD/Dr. ROBERT D. BERGAD
University of Minnesota, Department of Genetics and Cell Biology
250 Biological Sciences Center, St. Paul, MN 55108, USA

ISBN 978-3-662-11561-9   ISBN 978-3-540-38267-6 (eBook)
DOI 10.1007/978-3-540-38267-6

Library of Congress Cataloging in Publication Data. Main entry under title: Differentiation and neoplasia. (Results and problems in cell differentiation; v. 11) Bibliography: p. Includes index. 1. Carcinogenesis. 2. Cell differentiation. 3. Cancer cells. I. McKinnell, Robert Gilmore. II. Series. QH607.R4 vol. 11 [RC268.5] 616.99'407 80-20018

Preface

There is no commonly accepted mechanism to explain differentiation of either normal or neoplastic cells. Despite this fact, the organizers of the 3rd International Conference on Differentiation recognized that there is much emerging evidence which supports the view that both normal cells and many cancer cells share common differentiative processes. Accordingly, the organizers perceived that clinical scientists and developmental biologists would greatly benefit by together considering differentiation. In that way, developmental biologists would be apprised of recent insights in cancer cell biology and the physician scientist would be updated on events in developmental biology and both would gain new understanding of the cell biology of neoplasia.

A specific example may reveal the potential value of developmental biologists interacting with cancer physicians. An example chosen at random suggests that probably any paper included in the symposium volume would serve the purpose. Dr. Stephen Subtelny reviewed recent studies by his laboratory concerning germ cell migration and replication in frog embryos. How might those results interest the cancer scientist?

Dr. Subtelny showed that primordial germ cells of a fertile graft will reverse their migratory direction and move into a sterile host. Perhaps in this context it would not be inappropriate to state that the germ cells of the graft metastasized into the host. Germ cells from grafts of a different species will populate the previously sterile host gonad. Surely, study of the mechanisms of a useful migration of cells will provide valuable information for those who consider that most malignant aspect of neoplasia: metastasis.

A subtle control of primordial germ cell proliferation was noted by Subtelny. Triploid embryos had fewer (but larger) germ cells than did their diploid siblings. Somehow, the polyploid organism limited cell replication so that a normal-sized embryo with normal-sized gonads resulted – structured of larger but fewer cells. The cell cycle inhibition that adjusted cell number to fit the larger cell size was obviously not cytotoxic because the triploid embryos were no less vigorous than their diploid compatriots.

Surely, clinical scientists seek insights concerning noncytotoxic factors which inhibit mitosis. The genetic material of primordial germ cells does not ordain that the cells become gametes. Rather, it is a nonnuclear germ plasm determinant, sensitive to ultraviolet radiation during first cleavage, that controls differentiation of primordial germ cells. If differentiative processes are important in understanding neoplasia, as many of us believe, then continued study of this ultraviolet-sensitive material and the cells whose differentiation it controls will be of value to both cell biologists and those who study neoplasia.

The example cited above represents the rationale of the symposium *Differentiation and Neoplasia* held in Minneapolis 28 August, 1978 through 1 September, 1978. The symposium participants submitted the papers which comprise this volume. I believe, and many of those who attended the Minneapolis meeting believe, that the rationale for the meeting was fulfilled.

The symposium could not have taken place had it not been for generous contributions of funds. The organizers of the meeting express their grateful appreciation to: Fogarty International Center and the National Cancer Institute, National Institutes of Health, Washington, D.C.; American Cancer Society, Minnesota Division, Inc.; Grotto Foundation, Inc.; Hoffmann-LaRoche, Inc.; Association pour le Développement de la Recherche sur le Cancer; Ministère des Affaires Etrangères de France, Sous-Direction des Affaires Scientifiques; The Upjohn Company; Merck Sharp and Dohme; College of Biological Sciences, Office of International Programs, and the Department of Concerts and Lectures, University of Minnesota; SYNTEX Research; and Honeywell Systems and Research Center.

Summer 1980   R.G. McKinnell

Contents

Differential Histone Phosphorylation During *Drosophila* Development

By M. BLUMENFELD, P. C. BILLINGS, J. W. ORF, C. G. PAN, D. K. PALMER, and
L. A. SNYDER (With 3 Figures)

Nonhistone Proteins and Chromosome Structure

By D. E. COMINGS and T. A. OKADA (With 2 Figures) 49

The Current Status of Cloning and Nuclear Reprograming in Amphibian Eggs

By M. A. DiBERARDINO and N. J. HOFFNER

Control of Early Embryonic Development:
An Analysis of a Cytoplasmic Component and Its Mode of Action

By A. J. BROTHERS (With 2 Figures)

Surface Antigens in Early Embryonic Development

By R. KEMLER, D. MORELLO, CH. BABINET, and F. JACOB (With 1 Figure)

The Role of Fibronectin in Cellular Behavior

By R. O. HYNES (With 2 Figures)

Desmin and Intermediate Filaments in Muscle Cells

By E. LAZARIDES (With 2 Figures)

The Microtubule Cytoskeleton in Normal and Transformed Cells in Vitro

By B. R. BRINKLEY, L. J. WIBLE, B. B. ASCH, D. MEDINA, M. M. MACE,
P. T. BEALL, and R. M. CAILLEAU (With 12 Figures)

**Cytoplasmic Zone Analysis in the Study of the Polysomes
of Differentiated Cells**

By J.-E. EDSTRÖM and U. LÖNN (With 4 Figures)

**Mechanism of Morphogenetic Tissue Interactions:
The Message of Transfilter Experiments**

By L. SAXÉN

In Memoriam NELSON TRACY SPRATT, JR.

Migration and Replication of the Germ Cell Line in Rana pipiens

By S. SUBTELNY (With 2 Figures)

**The Effects of Temperature-Sensitive Rous Sarcoma Virus
and Phorbol Diester Tumor Promoters on Cell Lineages**

By H. HOLTZER, J. BIEHL, M. PACIFICI, D. BOETTIGER, R. PAYETTE, and
C. WEST

Control of Genome Integrity in Terminally Differentiating and Postmitotic Aging Cells

By S. P. MODAK and C. UNGER-ULLMANN (With 3 Figures)

On RNA Action in Differentiation: Induction and Differentiation of Somites in Chick Embryo

By S. RANZI (With 3 Figures)

RNA Viruses, Cancer and Development

By H. M. TEMIN (With 4 Figures)

The Regulation of Differentiation in Murine Virus-Induced Erythroleukemic Cells

By CH. FRIEND

Activation of Normal Differentiation Genes and the Origin and Development of Myeloid Leukemia

By L. SACHS

Cellular Heterogeneities in Acute Myeloblastic Leukemia

By E. A. McCULLOCH, R. N. BUICK, M. D. MINDEN, and C. A. IZAGUIRRE
(With 3 Figures)

Growth Regulations of Human Malignant Cells

By A. M. MAUER, S. B. MURPHY, and F. A. HAYES

The Control of Tumor Metastasis

By E. GORELIK, M. FOGEL, S. SEGAL, and M. FELDMAN

Teratocarcinoma Cells as Agents for Producing Mutant Mice

By M. J. Dewey and B. Mintz (With 4 Figures)

Development of Embryo-Derived Teratomas in Vitro

By N. Škreb and V. Crnek (With 3 Figures)

Loss of Tumorigenicity and Gain of Differentiated Function by Embryonal Carcinoma Cells

By E. D. Adamson and Ch. F. Graham

Clinical Oncology and Cell Differentiation

By B. J. Kennedy

Contributors

ADAMSON, E.D.
ASCH, B.B.
BABINET, CH.
BEALL, P.T.
BEYER, A.L.
BIEHL, J.
BILLINGS, P.C.
BLUMENFELD, M.
BOETTIGER, D.
BRACHET, J.
BRINKLEY, B.R.
BROTHERS, A.J.
BUICK, R.N.
CAILLEAU, R.M.
CERVENKA, J.
COMINGS, D.E.
CORTESE, R.
CRNEK, V.
DEROBERTIS, E.M.
DEWEY, M.J.
DIBERARDINO, M.A.
EDSTRÖM, J.-E.
FELDMAN, M.
FOGEL, M.
FRANKE, W.
FRIEND, CH.
GOLDSTEIN, M.N.
GORELIK, E.
GRAHAM, CH.F.

GURDON, J.B.
HAYES, F.A.
HICKS, R.M.
HOFFNER, N.J.
HOLTZER, H.
HYNES, R.O.
ILLMENSEE, K.
IZAGUIRRE, C.A.
JACOB, F.
KEMLER, R.
KENNEDY, B.J.
KROHNE, G.
LAZARIDES, E.
LÖNN, U.
MACE, M.M.
MARTIN, K.
MAUER, A.M.
MCCULLOCH, E.A.
MEDINA, D.
MELTON, D.A.
MILLER, O.L., JR.
MINDEN, M.D.
MINTZ, B.
MODAK, S.P.
MORELLO, D.
MÜLLER, U.
MURPHY, S.B.
NOWELL, P.C.
OKADA, T.A.

ORF, J.W.
OSHEIM, Y.N.
PACIFICI, M.
PALMER, D.K.
PAN, C.G.
PAYETTE, R.
PITOT, H.C.
PLURAD, S.
RANZI, S.
SACHS, L.
SAXÉN, L.
SCHEER, U.
SEGAL, S.
SHOGER, R.L.
SIGNORET, J.
SIRICA, A.E.
ŠKREB, N.
SNYDER, L.A.
SPRING, H.
STEVENS, L.C.
SUBTELNY, S.
TEMIN, H.M.
TRENDELENBURG, M.F.
UNGER-ULLMANN, C.
WEISS, M.C.
WEST, C.
WIBLE, L.J.
ZENTGRAF, H.

Cell Differentiation Yesterday and Today

J. BRACHET[1]

(Presidental address to the International Society of Differentiation)
Département of Biologie Moléculaire, Université Libre de Bruxelles, Belgium,
and Laboratorio di Embriologia Molecolare, CNR, Arco Felice, Italy

Not really yesterday: this story starts half a century ago, when I was a medical student in the University of Brussels. We had, among other things, lectures in histology, embryology and pathology. I remember very well A. Dalcq's lectures on morphogenetic movements, neural plate induction, germ layers formation, and organogenesis. I have not forgotten P. Gerard's description of tissues and organs. Perhaps still more vivid in my mind remains the discussion by Albert Dustin (who discovered the inhibition of mitosis by colchicine) of the mysterious causes of cancerous metaplasia: is the metaplasia due to chemicals (tar in the case of chimney sweepers) or to physical agents (UV for sailors)? Is it contagious and due to a virus? Or hereditary? An enormous amount of work has been undertaken since 1930 in order to answer these questions; yet we do not know for certain whether cancerous transformation is due to a multiplicity of causes or to a single one as mutation of "cancer genes".

Yes, all these things were told us, 50 years ago, by excellent teachers; but I do not remember a single word said about cell differentiation! I have not kept my student's notebooks; however, I have little doubt that differentiation was indeed discussed in the histology and embryology courses; but the presentation of this topic must have been so cursory that it left no trace in my mind.

In order to refresh my memory, I had a look at the old treatises of cytology and histology that I had been using in the late twenties. "A tout seigneur, tout honneur". I looked first at E.B. Wilson's admirable *The cell in development and heredity* (3rd ed., 1925). Not a word about differentiation, except for Frank Lillie's differentiation without cleavage in *Chaetopterus* eggs. This is a fascinating phenomenon (in which I am still personally interested), where eggs treated with parthenogenetic agents develop into unicellular ciliated larvae. *Chaetopterus* eggs still remain a good model for the study of cell differentiation in the absence of cytokinesis; yet they only represent an isolated and exceptional case for the student of cell differentiation.

I had better luck with Pol Bouin's monumental *Eléments d'histologie* (1929). Among thousands of pages, the great French histophysiologist (who discovered,

---

1 Due to illness, Dr. Brachet was unfortunately unable to attend the Minneapolis Conference

among other things, that testosterone is synthesized by interstitial cells and not by the germ cells in the testis) devoted only 18 pages to cell differentiation, and half of them deal with experimental embryology (mosaic and regulatory eggs, primary induction, etc. Bouin rightly stressed that cells exert "morphogenetic influences" between each other. This was demonstrated by an analysis of lens induction and of hormonal stimulation of target cells. Today, cell recognition and inter-communication, positional information, transfilter inductions, etc. are very important topics, which have been the subject of several papers at the present Conference. We know that the microtubules and microfilaments associated with the cell surface are responsible for cell shape and locomotion. In some cells, especially lymphocytes, signals received at the cell surface level induce DNA replication and mitosis. How this happens remains, however, a matter for speculation and further work. Whether DNA replication and mitosis are, as many believe, an absolute prerequisite for cell differentiation is not yet certain. There are undoubtedly exceptions, since an *anucleate* fragment of the unicellular alga *Acetabularia* can differentiate a complex and elegant umbrella.

Going back to Bouin, he was well aware half a century ago of the still much discussed problem of *dedifferentiation*. He already pointed out that in plants and in many lower animals (Sponges, Hydra, Planarians, Ascidians) apparently differentiated cells remain totipotent. This was suggested by the pioneering work of Paul Brien (in Brussels) on regeneration and asexual reproduction in Hydra and Ascidians. However, on the other hand, Etienne Wolff and his school were showing that lower organisms such as the Planarians contain a reserve of undifferentiated "embryonic" totipotent cells and that the regeneration blastoma is due to the migration and differentiation of these "neoblasts". Bouin's conclusion was that cell differentiation is labile in lower organisms; but in higher organisms, it is dictated by the specificity of the three germ layers, according to Bard's old principle (1885): *omnis cellula e cellula ejusdem naturae*. Albert Brachet, in his *Traité d' Embryologie des Vertébrés* (1930), considered this principle too dogmatic on the basis of morphological observations and experimental studies. He proposed that germ layers, like blastomeres, have "real" and "total" potentialities; the former are expressed in normal development, the latter under various experimental condi-tions. Pol Bouin's final conclusion was not very different: all cells have a variety of potentialities; one of them becomes dominant, while the others remain latent. This "dominant" differentiation is what is often called today "terminal" differentia-tion, where the adjective terminal conveys a feeling of irreversibility which was missing in Bouin's definition. There is a biological system, which already interested P. Bouin, which deserves mention in this context: lens regeneration in Urodeles, which was discovered by Charles Bonnet in 1780 and was studied cytologically by G. Wolff (1895, 1904). Lens regeneration and induction are now analyzed with an array of modern methods by T. Yamada, J. Piatigorsky and many others. This is easy because the lens contains specific proteins, the crystallins, whose synthesis results from the production of the corresponding messengers (mRNA's). T. Yama-da's conclusion that the cells of the upper rim of the iris lose their pigment—and thus dedifferentiate—and, after undergoing mitoses, differentiate, according to a new genetic program, into lens cells, is based on strong experimental evidence. It was nevertheless the subject of warm discussion at the Copenhagen Conference on Differentiation between those who for theoretical reasons, either reject or accept the possibility of cell dedifferentiation and redifferentiation in another direction. Due to the extreme diversity of the living world, this is the kind of argument where one should keep an open mind and avoid dogmatic positions. One can expect that the dedifferentiation-redifferentiation controversy will go on until many of the systems where dedifferentiation is believed to take place have been analyzed with the modern methodology of molecular biology (detection of specific proteins by immunofluorescence isolation, characterization and quantitation of the mRNA's, etc.).

Also of interest for the history of cell differentiation is the *Trattato di Istologia* by Giuseppe Levi (1927), who was one of the pioneers of in vitro cell culture. He thinks that loss of differentiation (anaplasia) occurs during liver regeneration and in some, but not all, cell lines in culture. Cell proliferation is stimulated by the addition of embryo extracts (Carrel's trephones), which inhibit cell differentiation. Numerous papers are still devoted today to the antagonism between cell proliferation and cell differentiation (see, for instance, the vast literature concerning the opposite effects of cAMP and cGMP on these two processes) and to the isolation and purification of growth stimulating factors (for instance, the epidermal growth factor, EGF). There is also nowadays a fair amount of work on substances which, on the contrary, inhibit specifically the growth of a given tissue (the chalones). Of particular interest in this field of research, because of its high tissue specificity and of our advanced knowledge on its chemical structure and its mode of action, is the nerve growth factor (NGF) discovered by one of G. Levi's pupils, Rita Levi-Montalcini.

Thus, in the 1925–1930 period, the cytologists who were interested in cell differentiation were mainly influenced by progress in experimental embryology. The progress culminated in the discovery of the "organizer" by Hans Spemann. However, it is good to remember, as has been pointed out by A. Dalcq, that the primary result of neural induction is organogenesis (formation of a nervous system), not differentiation of nerve cells. But nobody cared about the role that genes might play in differentiation processes; indeed, at that time, Morgan's work on *Drosophila* was still the subject of skeptical remarks among french-speaking biologists (until my father, Albert Brachet, visited many United States laboratories in 1929). Thomas Hunt Morgan was both an embryologist and a geneticist.

It was a great day for biology when he linked together, in the 1930's, his two fields of interest into a general theory which remains at the root of our present thinking. Morgan knew, from his own work, that the cytoplasm of most eggs is heterogeneous and that the genes are located in the chromosomes. He proposed that, during egg cleavage, genetically identical totipotent nuclei would be distributed, as the result of mitotic activity, in the mosaic of the chemically different territories (germinal localizations) which build up the egg cytoplasm. The germinal localizations would affect (some would now say modulate) gene activity in the nuclei. Specific genes would be activated (or derepressed, if one prefers) in certain germinal localizations, but not in others. These genes would remain silent in other parts of the egg. Gene activation would lead to modifications of the surrounding cytoplasm; this would increase its initial heterogeneity and, as a consequence, lead to further gene activation. Repeated gene-cytoplasm interactions would ultimately lead to cell differentiation.

Morgan's theory struck me very deeply as soon as I read it, since it was the first valid attempt to link together embryology and genetics. This theory remains at the root of our present models on the control of development. The most popular of these models is that of "selective gene activation" (see E. Davidson 1971), which is not basically different from Morgan's ideas. Thanks to molecular biology, our understanding of the control of gene activity in eucaryotes is making steady progress and it is not surprising that three sessions of the present Conference are devoted to this subject.

In eggs, we now know that blastula nuclei are genetically identical to the zygote nucleus and are thus totipotent. This has been demonstrated by the nuclear transplantation experiments of R. Briggs and T. King (1952) and of J. Gurdon (1964). Whether nuclei taken from adult cells are still totipotent is not so certain. In J. Gurdon's best experiments with adult nuclei, only tadpoles, but no adults were obtained, and the percentage of successful experiments was much lower than with blastula nuclei. These experiments show that adult nuclei still possess the genetic information necessary for differentiation until the tadpole stage, but that they have more restricted potentialities than younger nuclei. This conclusion would fit with the possibility, first proposed by E. Scarano (1969), that DNA might undergo modifications (base methylation, in particular) during development. This restriction of potentialities for development in old nuclei is one of the arguments presented for the model recently put forward by A. Caplan and C. Ordahl (1978). They propose a scheme of "progressive gene repression" instead of "selective gene activation". During development some of the genes would undergo irreversible repression, while others would remain unrepressed and be responsible for the differentiated phenotype.

Be it as it may, there is no doubt that the major problem we are still facing is the control of gene expresssion. In very early stages of development (from the oocyte to the end of cleavage), there is no doubt that the cytoplasm plays an essential role in this control. For instance, protein synthesis is stimulated when progesterone is added to *enucleate Xenopus* oocytes. In such a case, stimulation of protein synthesis is entirely controlled at a post-transcriptional level. The very important work described in the present book by A.J. Brothers, M.A. Di Berardino and especially by J. Gurdon and his colleagues emphasizes the role played in development by products of genetic activity which have accumulated in the cytoplasm during oogenesis. For example, the cytoplasm of full-grown oocytes contains factors capable of reprogramming old nuclei. Two-dimensional gel electrophoresis of the neo-synthesized proteins has demonstrated (Gurdon and De Robertis 1977) that when old nuclei are injected into an oocyte, some of the adult proteins are no longer synthesized; simultaneously, some of the oocyte proteins are synthesized again by the adult nuclei. But, at later stages of development, terminal differentiation, which is characterized by the extensive synthesis of one (or a few) specific proteins (hemoglobin, myosin, chondrine, etc.) is obviously the result of controls exerted primarily at the level of gene transcription. It seems to be a general rule that the amount of specific protein synthesized by a terminally differentiating cell is closely related to the amount of the corresponding mRNA. The appearance of a specific protein is an easy biochemical marker for cell differentiation; it implies that the corresponding structural gene has been activated and transcribed. The existence, in addition, of post-transcriptional mechanisms for gene expression should not, however, be ruled out.

When, for the preparation of the present address, I reread the books of E.B. Wilson, P. Bouin, and G. Levi, I thought: "Now be courageous and read what you wrote in your first book about differentiation". I do not like to read my own books after a few years, since I know too well that they are full of factual errors and of wild ideas which often turned out to be wrong. My first book, *Embryologie chimique* (1944) was written under the roar of bombers and missiles, and the shrieks of three yelling kids. We had, of course, no access to American and English scientific literature after 1940. Thus, I had a ready excuse if I had forgotten to say something about differentiation. Luckily, the magic word (differentiation) was in the index and I read, with a good deal of apprehension, p. 470 of the French edition. To quote from the English translation by L.G. Barth (1950): "One might predict that the formation of specific proteins plays an important role ("rôle capital", in the French edition) in the "chemodifferentiation" (Huxley) of organs and the molecular rearrangement of proteins are probably amongst the essential agents of organogenesis". I then suggested the use of immunological methods for the identification of these differentiation specific proteins and proposed the differentiation of somites into muscles as a suitable experimental system. This was not too bad, but there was not a single word about gene control, although I was very well aware of the pioneer work of Ephrussi and Beadle on eye color in *Drosophila*.

In the present book, one can find (thanks to an excellent selection of speakers by Prof. R.G. McKinnell and his colleagues) original and important contributions to all the aspects of the fundamental problem of the genetic control of differentiation: isolation and replication of pure genes, respective role of histones and nonhistone proteins in the structure and genetic activity of chromatin, genetic manipulation in amphibian and mammalian embryos, analysis of gene transcription under the E.M., etc. One also finds, in this book, important information about the role played by the cell surface and the cytoskeleton in cell differentiation, information linking changes in locomotion activity and cell shape to the assembly and disassembly of actin and tubulin molecules. Such changes are important not only for in vitro cultured cells, but also for the coordinated cell movements required for gastrulation and neurulation (morphogenetic movements).

There is a considerable change, almost a mutation, between the Minneapolis Conference and the two preceeding ones (Nice and Copenhagen). In Minneapolis a much greater emphasis is placed on cancer, on neoplasia. Embryologists and cytologists who study cell proliferation and differentiation have always been interested in cancer cells, where the mechanisms which control these two fundamental attributes of all cells are clearly no longer normal. To say that cancer cells are malignant because they cannot differentiate would be a gross oversimplification. In the present Conference, a good deal will be said about the induction of differentiation in three kinds of malignant cells: Friend's erythroleukemic cells, teratocarcinoma, and neuroblastoma cells.

Friend's cells are of great interest because they synthesize, after induction with various chemicals, several of the proteins (hemoglobin, spectrin, in particular) characteristic of adult red blood cells. The synthesis of these proteins can easily be followed quantitatively and it has even been possible to quantify the synthesis of hemoglobin mRNA. In view of our impressive knowledge of hemoglobin synthesis in normal reticulocytes, Friend's cells are an exceptional material for a comparison between virus-induced malignant cells and their normal counterpart.

Teratocarcinoma has become, thanks to B. Ephrussi, a fascinating subject for both embryologists and cancer biologists; it is now under study in a number of outstanding laboratories.

Ingenious experiments by B. Mintz, C. Graham, and R. Gardner have shown that a combination of malignant, totipotent teratocarcinoma cells with normal embryonic cells can lead to a loss of the malignant character. F. Jacob has demonstrated that teratoma malignant cells share surface antigens with sperm and with early embryonic stages (morulae). Why then is a teratoma a monster and not a normal embryo? A possible answer to that question might be that teratoma lacks the gradient organization which is needed for proper morphogenesis in vertebrate embryos. Is a pluripotent teratoma cell equivalent to an egg? B. Mintz, at a recent discussion on differentiation in Brussels (Francqui Symposium) gave a negative answer to that question. Limited evidence from molecular biology seems to support her. Totipotent teratocarcinoma express 7,700 mRNA sequences, as compared with 13,200 for myoblasts and 6,200 for Friend's cells; 6,000 of these sequences are in common to the three cell types (N.A. Appara et al., Cell 12, 509-520, 1977);— but in sea urchin eggs, the complexity of maternal mRNA is so high that it could code for as many as 20,000 proteins (B. Hough-Evans et al., Dev. Biol. 60, 258-277, 1977). Thus the informational content of sea urchin egg (and probably of a mouse egg) is about three times that of a totipotent teratocarcinoma cell. Storage of an exceptionally high amount of information in the cytoplasm should be considered as a molecular marker for differentiation in the egg, the totipotent cell par excellence.

Suppression of cell differentiation by treatment with bromodeoxyuridine (BrdUR) remains an important tool for the analysis of the molecular and genetic bases of differentiative processes; however, the mode of action of BrdUR on differentiation remains poorly understood. There is a newcomer at the present Conference, TPA (a derivative of phorbol, and a well known tumor promoter), which also suppressed in vitro differentiation. While BrdUR is believed to directly modify the genome (as a result of incorporation into DNA molecules), TPA is thought to act mainly at the cell membrane level. It will be interesting to see whether these two inhibitors, which are chemically unrelated, act on a single metabolic event which would be of crucial importance for differentiation. It will also be worthwhile to study the effects of TPA on embryonic differentiation which, for unknown reasons (possibly the gradient organization of the embryo?) is curiously resistant to BrdUR.

Cell fusion (somatic hybridization) is a very powerful tool for the analysis of somatic cell genetics, cell differentiation and malignancy. One can expect, in the near future, marked progresses in developmental genetics. The search for developmental mutants, especially in *Drosophila*, Amphibians and the mouse, is progressing. It will be highly rewarding when the molecular defects of the mutants are identified. It is also to be expected that a more direct approach, the injection of purified genes into eggs (as has been done by D. Brown and J. Gurdon for the 5S RNA genes) will yield results of exceptional interest.

Further progress in our comprehension of differentiation should come from endocrinology, which has been almost left out of the present Conference. We are understanding much better now how hormones stimulate target cells and induce the synthesis of specific proteins such as ovalbumin or vitellogenin, or the progression of meiosis during oocyte maturation.

I see that I am no longer talking about the present, but about the future, about the next Conference organized by the International Society of Differentiation. This Society was founded by Dr. D. Viza at the Nice Conference in 1971. Membership has largely increased, thanks to the efforts of the present Secretary, Dr. Robert McKinnell. The new President of the Society will have an easy life, thanks to the enthusiasm and devotion of the members of the Board. I had a very easy life, also thanks to them, during my long Chairmanship of the Society. I am sure that the Society will, like a fertilized egg which has a good genetic background, develop quickly and harmoniously. Differentiation will long remain a fascinating subject, and will attract an increasing number of gifted young scientists. I wish them good luck in their research, research which requires not only intelligence, skill, and imagination, but also adequate equipment (still nonexistent in most of the world), adequate financial and economical conditions and, above all, peace among all the nations of the world.

Gene Injections into Amphibian Oocytes

D.A. Melton[1], R. Cortese[1,2], E.M. de Robertis[1]
M.F. Trendelenburg[1], and J.B. Gurdon[1]

[1]MCR Laboratory of Molecular Biology, Hills Road, Cambridge, England
[2]Istituto di Chimica Biologica, Facolta di Medicina e Chirurgia
Universita di Napoli, Napoli, Italy

I. Introduction

This article reviews initial results from an experimental system designed to identify molecules which regulate gene activity. In particular, the experiments described are concerned with the transcription of cloned DNA's as studied by microinjection into *Xenopus* oocytes. We first discuss the rationale behind this experimental approach and then summarize evidence which demonstrates that cloned tRNA genes direct the synthesis of genuine tRNA's when injected into oocytes. Other experiments described focus on the idea that oocyte injections can be used to identify regions of DNA which contain the information necessary for the accurate transcription of genes. Finally, the prospects for reisolating injected DNA's from oocytes to look for hypothetical controlling molecules are discussed.

II. Experimental Strategy

The general strategy behind injecting genes into frog oocytes derives from the following logic. One conclusion which can be drawn from nuclear transplantation experiments is that eggs and oocytes contain cytoplasmic components which regulate gene expression (Gurdon 1977). For example, when the nucleus of a fully differentiated skin cell is transplanted into an enucleated frog egg, the egg divides and develops into a tadpole containing nerve, muscle, lens and other differentiated cells. In this instance, a nucleus specialized for the synthesis of keratin gives rise, by mitosis, to nuclei which express other specialized genes (Gurdon et al. 1975). Similarly, if *Xenopus* tissue culture nuclei are injected into fully grown oocytes, these nuclei are reprogrammed to express *oocyte* specific genes, genes which were previously not expressed in the tissue culture cells (de Robertis and Gurdon 1977). Such experiments demonstrate that nuclear gene activity conforms to the cytoplasm's biosynthetic program, that is to say, the cytoplasm of eggs and oocytes must contain components which regulate gene expression.

Given this assumption, one might try to identify these cytoplasmic components by reisolating injected nuclei after these nuclei have changed their biosynthetic

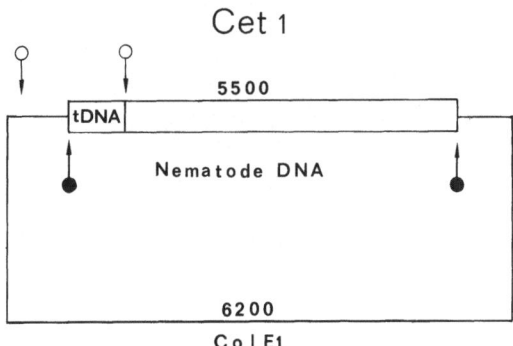

Fig. 1. A nematode recombinant DNA plasmid, Cet 1. Twelve recombinant DNA plasmids were isolated by inserting nematode DNA into the Eco Rl site (●) of the plasmid Col El. The DNA containing the tRNA gene (tDNA) in Cet 1 is 300 base pairs in length and can be isolated by a double enzyme digestion with Eco Rl and Taq (♀). Lengths of DNA are expressed as nucleotide base pairs

program. Alternatively, a more precisely defined genetic unit, such as a cloned gene, can be used for oocyte injections. If the oocyte contains regulatory molecules which are (1) gene specific and (2) act by binding to injected DNA, then these molecules could be identified by injecting and reextracting a purified gene. The strategy then is to first identify regions on the DNA, other than the structural gene itself, required for correct gene transcription, i.e., to identify promoters and possible regulatory regions. Secondly, one will try to find molecules presumed to be present in oocytes, which bind specifically to these regions of the DNA and thereby regulate gene activity.

A first step towards such an analysis is to demonstrate that genes injected into oocytes are in fact accurately transcribed. The accurate transcription and expression of DNA's injected into the oocyte's nucleus has been demonstrated for SV40 DNA (de Robertis and Mertz 1977; Mertz and Gurdon 1977), *Xenopus* tDNA (Kressmann et al. 1978) and most convincingly for *Xenopus* 5S DNA (Brown and Gurdon 1977, 1978; Gurdon and Brown 1978). We summarize evidence below which demonstrates that tRNA genes from a heterologous source, the nematode *Caenorhabditis elegans*, direct the synthesis of genuine tRNA's when injected into oocytes.

A. Nematode tDNA Is Transcribed in Oocytes

Nematode transfer RNA genes direct the synthesis of distinct 4S RNA's when injected into oocytes (Cortese et al. 1978). The nematode tDNA's used for oocyte injections were purified by cloning in the plasmid Col El (Fig. 1). The result of a DNA injection experiment using a tDNA recombinant plasmid, called Cet 1, is

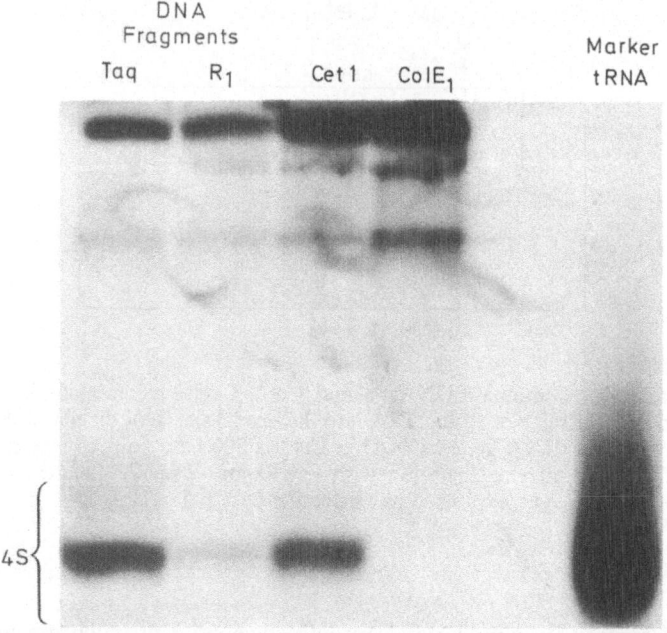

Fig. 2. Polyacrylamide gel electrophoresis of ^{32}P-RNA synthesized by oocytes injected with Cet 1 DNA. Autoradiograms of total RNA synthesized in oocytes injected with DNA and ^{32}P-α-GTP. Linear DNA fragments injected are 300 base pairs (Taq) and 5500 base pairs (Rl) in length (see Fig. 1). Note the absence of a distinct 4S band in control oocytes injected with Col EI DNA

shown in Fig. 2. The 4S RNA synthesized in oocytes injected with Cet 1 DNA is a distinct 4S RNA species which comigrates with tRNA markers. This 4S RNA also comigrates in a high resolution acrylamide gel with one species of tRNA isolated from nematode RNA by hybridization to Cet 1 DNA. Furthermore, this 4S RNA has been shown to contain several modified bases, such as pseudouridine, which are characteristic of tRNA's (Cortese et al. 1978). On the basis of these data we conclude that heterologous tRNA genes are transcribed and their transcripts matured when injected into *Xenopus* oocytes.

B. Nematode tDNA Has Its Own Promoter

Two kinds of experiment demonstrate that the nematode DNA portion of the Cet 1 plasmid contains all the information necessary for the correct transcription of the tRNA gene. In one series of experiments Cet 1 DNA is mixed prior to injection with a concentration of α-amanitin (1 μg/ml) which is known to inhibit the transcription of Col El DNA. In this case, there are no detectable Col El transcripts synthesized in injected oocytes as measured by hybridization to filter bound Col El DNA. Yet, at the same time, the transcription of the nematode tDNA gene, as measured by counting the 4S band in an acrylamide gel, is not impaired. The tDNA transcription is inhibited at higher α-amanitin concentrations (10–100 μg/ml). These results show that Col El DNA and nematode tDNA are transcribed in oocytes by different RNA polymerases, polymerases II and III respectively. Assuming that the different RNA polymerases recognize different promoters, one can conclude that the nematode DNA insert of Cet 1 has its own (RNA polymerase III) promoter.

A more precise definition of the region of DNA which contains this tDNA promoter is provided by injecting purified linear DNA restriction fragments. The injection of the entire nematode DNA insert of Cet 1, a 5500 base pair Eco RI fragment (Fig. 1), directs the synthesis of the same 4S RNA as is synthesized in oocytes injected with whole Cet 1 DNA (Fig. 2). The Cet 1 plasmid can be further dissected using a restriction enzyme called Taq which cuts the DNA into 15 fragments, only one of which hybridizes to ^{32}P-labeled tRNA. The injection into oocytes of this 750 base pair Taq DNA fragment also supports the synthesis of the Cet 1 tRNA (Fig. 2). Other Taq DNA fragments of Cet 1 injected into oocytes as controls do not support the synthesis of a tRNA. Using this method of injecting DNA restriction fragments we have narrowed down the functional tRNA gene to a 300 base pair DNA segment (Fig. 1; Cortese et al. 1978). Neither an Hae III nor an Hinf II restriction enzyme digest of this DNA fragment supports tRNA synthesis when injected into oocytes.

These results confirm those from the α-amanitin experiment and demonstrate that a relatively small DNA segment (300 base pairs) contains both a structural tRNA gene (about 100 base pairs) and the region(s) of DNA required for its accurate transcription. It is interesting to note that for this tRNA gene the promoter seems to be closely associated to the structural gene. More generally, these experiments show how DNA injections into oocytes can be used to dissect cloned DNA's to find transcriptionally active genes.

C. Oocyte Injections as a Promoter Assay

The results from nematode tDNA (Cortese et al. 1978) and frog 5S DNA (Brown and Gurdon 1978) injection experiments suggest that the transcription of DNA's injected into oocytes is initiated at specific regions of the DNA, i.e., the oocyte seems to recognize eukaryotic promoters on injected DNA's (Gurdon et al. 1978). This conclusion is strengthened by the fact that the injection of a *prokaryotic* tRNA gene, a tyrosine tRNA suppressor, does *not* support the synthesis of a tRNA in oocytes. This suggests then that the oocyte recognizes eukaryotic but not prokaryotic promoters, at least for RNA polymerase III type genes. Furthermore, recently developed methods using the electron microscope to visualize the transcription of injected DNA's have shown that RNA polymerase I promoters are recognized specifically in oocytes (Fig. 3; Trendelenburg and Gurdon 1978). These various data collectively suggest that the oocyte may selectively recognize promoters and encourage one to explore the possibility of using oocyte injections as a general assay for eukaryotic promoters.

In principle, DNA injections into oocytes could be used to identify regulatory regions, promoters, etc., even on short pieces of DNA which do not contain

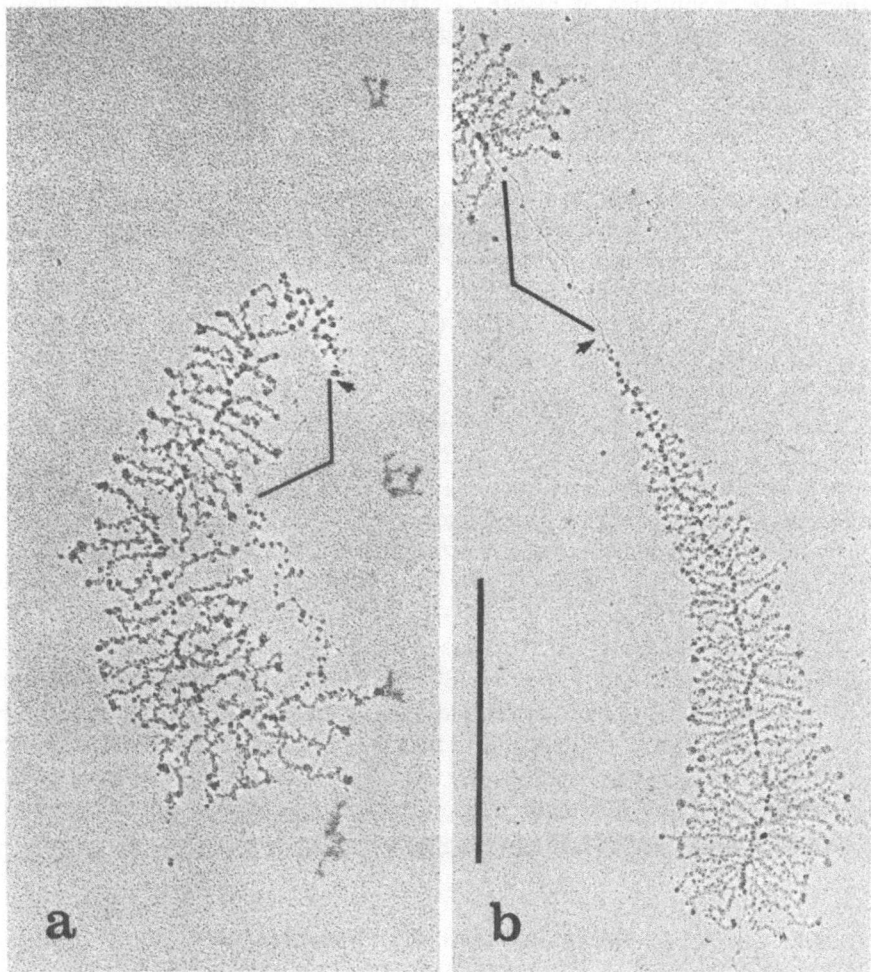

Fig. 3a,b. Electron micrographs of spread *Xenopus laevis* ribosomal DNA transcription units. This preparation was made from a *Xenopus* oocyte nucleus which had been injected with circular recombinant plasmid DNA which contains a single *Xenopus* 40S rDNA repeat unit. The initiation of transcription *(small arrows)* is at similar regions of the rDNA for injected plasmid molecules **(a)** and endogenous amplified rDNA **(b)**. The position of the nontranscribed rDNA spacer region is denoted by *brackets*. (Scale indicates 1 μm)

structural genes. For instance, one might attach a putative promoter sequence, by means of enzymatic ligation, to another DNA segment known not to contain its own promoter. Satellite DNA's are not normally transcribed and indeed they direct the synthesis of virtually no RNA when injected into oocytes. Thus, the joining of a promoter DNA segment to satellite DNA should cause the satellite DNA to be transcribed. Preliminary results using a RNA-polymerase I promoter and calf satellite DNA indicate that this is a valid design of experiment for identifying promoters. Further studies of this type should address the question of whether RNA polymerase II type promoters can be identified and characterized with this functional assay.

III. Prospectus

The experiments summarized above are concerned with identifying regions of DNA which are important for eukaryotic gene regulation. If promoters and regulatory regions for genes can in fact be characterized by oocyte injections, one would next like to search for molecules which bind specifically to these regions. This might be explored by the methods described by Weideli et al. (1978) wherein a purified *Drosophila* DNA is used to fish out specific proteins from the *Drosophila* soluble protein pool. Similarly, one might inject cloned DNA into oocytes and reextract the DNA to look for RNA's and/or proteins which bind specifically to some regions of the DNA.

Further studies with this injection system might focus on the molecular configuration of the transcriptionally active genes. It is known that DNA's injected into the oocyte nuclues become associated with endogenous nucleoproteins to form stable chromatin (Wyllie et al. 1978). As yet, the relationship between this chromatin and transcriptional activity is obscure. However, two kinds of experiments may help clarify this issue. Experiments using the electron microscope to examine injected genes show that this method allows one to visualize the transcriptionally active DNA's (Trendelenburg et al. 1978). This method should make it possible to obtain useful information about the structure and transcription of cloned DNA's injected into oocytes. Secondly, as it is possible to assemble DNA into chromatin in vitro (Laskey et al. 1977) one can in principle use labeled proteins to assemble chromatin and then injected these DNA-protein complexes. This combination of in vitro chromatin assembly and an in vivo transcriptional assay may provide a way forward on the problem of gene regulation.

References

Brown, D.D., Gurdon, J.B.: High fidelity transcription of 5S DNA injected into *Xenopus* oocytes. Proc. Natl. Acad. Sci. USA 74, 2064-2068 (1977)

Brown, D.D., Gurdon, J.B.: Cloned single repeating units of 5S DNA direct accurate transcription of 5S RNA when injected into *Xenopus* oocytes. Proc. Natl. Acad. Sci. USA 75, 2849-2853 (1978)

Cortese, R., Melton, D., Tranquilla, T., Smith, J.D.: Cloning of nematode tRNA genes and their expression in the frog oocyte. Nucl. Acid Res. 5, 4593-4611 (1978)

de Robertis, E.M., Gurdon, J.B.: Gene activation in somatic nuclei after injection into amphibian oocytes. Proc. Natl. Acad. Sci. USA 74, 2470-2474 (1977)

de Robertis, E.M., Mertz, J.E.: Coupled transcription-translation of DNA injected into *Xenopus* oocytes. Cell 12, 175-182 (1977)

Gurdon, J.B.: Egg cytoplasm and gene control in development. Proc. R. Soc. London Ser. B 198, 211-247 (1977)

Gurdon, J.B., Brown, D.D.: The transcription of 5S DNA injected into *Xenopus* oocytes. Dev. Biol. 67, 346-356 (1978)

D. A. Melton et al.

Gurdon, J.B., Laskey, R.A., Reeves, O.R.: The development capacity of nuclei transplanted from keratinized skin cells of adult frogs. J. Embryol. Exp. Morphol. 34, 93-112 (1975)

Gurdon, J.B., Melton, D.A., de Robertis, E.M.: Genetics in an oocyte. Ciba Symp., in press (1978)

Kressmann, A., Clarkson, S.G., Pirotta, V., Birnstiel, M.L.: Transcription of cloned tRNA gene fragments and subfragments injected into the oocyte nucleus of *Xenopus laevis*. Proc. Natl. Acad. Sci. USA 75, 1176-1180 (1978)

Laskey, R.A., Honda, B.M., Mills, A.D., Morris, N.R., Wyllie, A.H., Mertz, J.E., de Robertis, E.M., Gurdon, J.B.: Chromatin assembly and transcription in eggs and oocytes of *Xenopus laevis*. Cold Spring Harbor Symp. Quant. Biol. 42, 171-178 (1977)

Mertz, J.E., Gurdon, J.B.: Purified DNAs are transcribed after microinjection into *Xenopus* oocytes. Proc. Natl. Acad. Sci. USA 74, 1502-1506 (1977)

Trendelenburg, M.F., Gurdon, J.B.: Transcription of cloned *Xenopus* ribosomal genes visualized after injection into oocyte nuclei. Nature 276, 292-294 (1978)

Trendelenburg, M.F., Zentgraf, H., Franke, W.W., Gurdon, J.B.: Transcription patterns of amplified *Dytiscus* genes coding for ribosomal RNA after injection into *Xenopus* oocyte nuclei. Proc. Natl. Acad. Sci. USA 75, 3791-3795 (1978)

Weideli, H., Schedl, P., Artaranis-Tsakonas, S., Steward, R., Yuan, R., Gehring, W.J.: Purification of a protein from unfertilized eggs of *Drosophila* with specific affinity for a defined DNA sequence and the cloning of this DNA sequence in bacterial plasmas. Cold Spring Harbor Symp. Quant. Biol. 42, 693-700 (1977)

Wyllie, A.H., Laskey, R.A., Finch, J., Gurdon, J.B.: Selective DNA conservation and chromatin assembly after injection of SV40 DNA into *Xenopus* oocytes. Dev. Biol. 64, 178-188 (1978)

Organization of Transcribed and Nontranscribed Chromatin

W. W. Franke[1], U. Scheer[1], H. Zentgraf[2]
M. F. Trendelenburg[1], U. Müller[3], G. Krohne[1]
and H. Spring[1]

Abteilung für Membranbiologie und Biochemie
Institut für experimentelle Pathologie, Deutsches Krebsforschungszentrum
[2] Institut für Virologie, Deutsches Krebsforschungszentrum
[3] Institut für Mikrobiologie der Universität
6900 Heidelberg, FRG

I. Introduction

In most eukaryotic nuclei two extreme forms of chromatin packing can be distinguished, condensed and dispersed chromatin. This does not exclude that, in addition, transitional or intermediate forms may exist. Ever since the pioneering morphological studies of Heitz (1929, 1932, 1956) cytologists have associated condensed chromatin, including the "heterochromatin" sensu Heitz (cf. also Brown 1966), with transcriptional inactivity, as opposed to the various levels of transcriptional activity thought to exist in regions of dispersed ("extended") chromatin, including the puffs and Balbiani rings described in polytene chromosomes of various dipteran insects (Pelling 1964; for refs. see Beermann 1972) and the loops of lampbrush chromosomes (Gall and Callan 1962). The discovery of the nucleosome (v-body) as a subunit of chromatin organization (Olins and Olins 1973, 1974; Woodcock 1973; Kornberg 1974; for recent reviews see Chambon 1978; Felsenfeld 1978), which is the predominant structure of first order packing in transcriptionally inactive chromatin, has led to the question whether the packing of DNA is different in transcriptionally active chromatin. In the context of the present study we define the nucleosome as a particle of about 10–13 nm overall diameter which contains a defined octamer of histones and a total of about 200 base pairs of DNA (for details see Kornberg 1974, 1977; Oudet et al. 1975; Felsenfeld 1978). Inherent in this definition is a formula of packing of DNA, resulting in a linear foreshortening, relative to the DNA molecule in B-conformation, of about 5.3 (for electron microscopic demonstrations of this packing of DNA in isolated nucleosomes see, e.g., also Woodcock et al. 1976). In order to avoid a misunderstanding frequent in the literature we want to emphasize that we do not use the term "nucleosome", in the context of this article, for any unit of 140–200 base pairs of DNA packed with proteins in a way that protects it from digestion with certain endonucleases such as micrococcal nuclease (see also below). We designate "nucleosome" only the particulate unit ("soma", i.e., body) as defined above.

II. Biochemical Evidence for Structural Differences in Transcriptionally Active Chromatin

Studies with deoxyribonuclease I have shown that transcribed chromatin of nucleolar as well as non-nucleolar origin is more sensitive to digestion than nontranscribed chromatin (Weintraub and Groudine 1976; see also Garel and Axel 1976, 1978; Flint and Weintraub 1977; Bellard et al. 1978; Levy et al. 1978; Mathis and Gorovsky 1978; Palmiter et al. 1978; Paul et al. 1978; Suzuki and Ohshima 1978). Differential kinetics of digestion of DNA present in transcriptionally active chromatin has also been demonstrated with deoxyribonuclease II (Gottesfeld and Butler 1977) and with micrococcal nuclease, both in non-nucleolar (e.g., Bellard et al. 1978; Bloom and Anderson 1978) and nucleolar (Reeves and Jones 1976; Reeves 1978; cf. Allfrey et al. 1978) genes. Transcriptionally engaged subunits of nucleolar chromatin obtained by nuclease digestion in *Physarum polycephalum* have been reported to have sedimentation properties and protein composition different from those obtained from typical nucleosomes (e.g., Allfrey et al. 1978). A series of compositional differences, including histone modifications, have been reported in isolated transcriptionally active chromatin from various sources (Levy-Wilson et al. 1977; Allfrey et al. 1978; Bonner et al. 1978; Gottesfeld 1978; for further refs. see Chambon 1978). Structural differences between "active" and "inactive" chromatin fractions have also been described by Moudrianakis et al. (1977) who concluded from their experiments that "nontranscribable chromatin is in the form of a string of contiguous, about 100 Å-diameter beads, whereas transcribable chromatin is in the form of relatively smooth fiber ca. 35 Å thick" (see also remark by E.N. Moudrianakis in the discussion of the paper by V.E. Foe 1978).

Digestion experiments with micrococcal nuclease have shown that the DNA of transcriptionally active chromatin is arranged in a protection pattern similar to that described in inactive chromatin (for refs. see Bonner et al. 1978; Garel and Axel 1978; Gottesfeld and Melton 1978; Mathis and Gorovsky 1978). However, the demonstration of a periodical arrangement of protective entities, most probably histones, does not allow the discrimination between the extended and the compacted state of the nucleosome-equivalent nucleo-histone unit (e.g., Jackson and Chalkley 1975; Oudet et al. 1978a,b; Woodcock and Frado 1978; cf. also Yaneva and Dessev 1976).

III. Nucleosomal Structure of Inactive Chromatin as Revealed by Electron Microscopy

When transcriptionally inactive chromatin, which in most cell types constitutes the bulk of the nuclear chromatin, is dispersed in buffer solutions of low ionic strength and examined in the electron microscope, for example by the use of the "spreading technique" developed by Miller and co-workers (Miller and Beatty 1969; Miller and Bakken 1972), the typical "beads-on-string" arrays are observed (Fig. 1; cf. Figs. 4d, 7c, 8a and e; cf. Olins and Olins 1973, 1974; Woodcock 1973). In such preparations, however, the individual nucleosomal particles are not always closely spaced, as it is probably the case in situ, but can show various degrees of greater nucleosome spacing. In different preparations, or in different strands of the same preparation, or even in different regions of the same chromatin strand, beads are arranged in close spacing, often arranged in a typical zig-zag pattern, or may show more or less extended internucleosomal "linker" regions (Fig. 1; see refs. quoted above). Consequently, in such preparations the apparent DNA packing ratio does not seem to be constant but varies from about 5 as in close spacing (see above; such a value is also caused by preparations made from solutions of nearly physiological ionic strength; cf. Griffith 1975; cf. Griffith and Christiansen 1978; Keller et al. 1978; Zentgraf et al. 1978) to an average of about 2 which is usually found after prolonged incubation in media of very low salt concentrations (e.g., below 1 mM alkali salt; Fig. 1; cf. Foe et al. 1976; Foe 1978; McKnight et al. 1978; Reeder et al. 1978a). Similar mean contraction ratios of 2.0 to 2.5 has also been oberserved in spread preparations of chromatin containing well defined circular DNA molecules such as SV40 DNA (Fig. 8a; for refs. see Griffith 1975; Griffith and Christiansen 1978; Keller et al. 1978; Müller et al. 1978; Oudet et al. 1978a,b; Zentgraf et al. 1978), nontranscribed amplified rDNA of oocytes of *Dytiscus marginalis* (Fig. 4; for details see Scheer and Zentgraf 1978), as well as nontranscribed SV 40 DNA (Laskey et al. 1978), plasmid DNA (Trendelenburg and Gurdon 1978), and mitochondrial DNA (Fig. 8e; H. Zentgraf, M.F. Trendelenburg, and W.W. Franke, unpublished results) that have been injected into nuclei of oocytes of *Xenopus laevis* (for possible mechanisms of nucleosome packing of microinjected DNA see Laskey et al. 1978). The reason for the observed variations of nucleosome packing in chromatin that has been dispersed and spread in low salt buffers is not clear. Experiments suggest that the specific packing density of nucleosomal particles depends on the ionic strength and the pH of the solutions used, the time of incubation, the mode of preparation, and the type of chromatin examined (see also below).

IV. Ultrastructure of Transcriptionally Active Chromatin

In view of the predominance, in most nuclei, of inactive or only weakly active chromatin we want to begin this discussion, for the sake of clarity, with a description of the one extreme, i.e., fully transcribed chromatin as is found, for example, in nucleoli containing actively transcribed genes coding for pre-rRNA (Fig. 2) or in lampbrush chromosome loops (Fig. 3). In electron micrographs of spread preparations such highly active chromatin is characterized by the predominance of transcriptional complexes which appear as lateral ribo-nucleoprotein (RNP) fibrils containing the forming RNA growing from the RNA-polymerase particles attached to the chromatin axis (cf. Miller and Beatty 1969; Franke et al. 1976a, 1978a,b; Franke and Scheer 1978). The following situations can be distinguished:

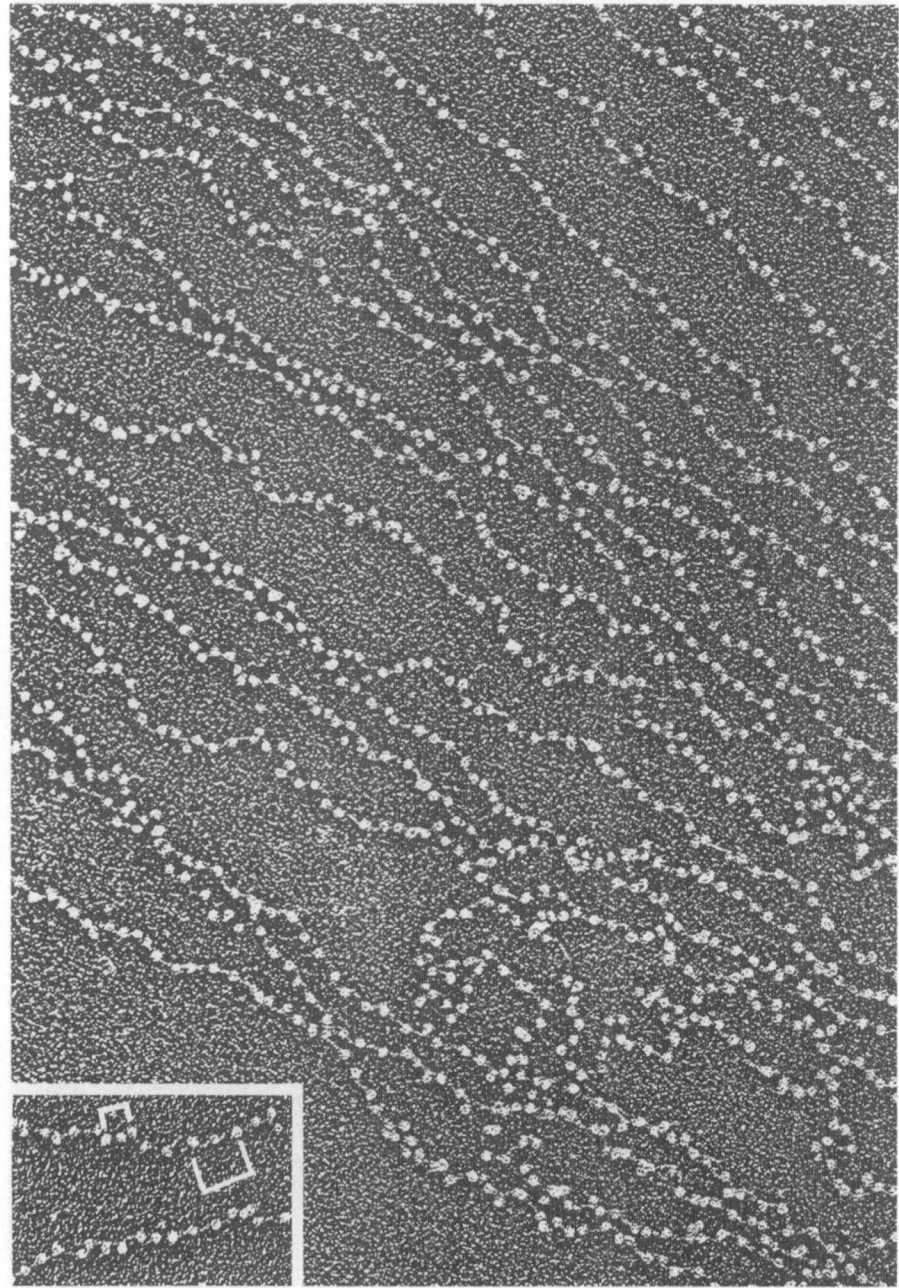

Fig. 1. Survey electron micrograph showing the appearance of hen erythrocyte chromatin when dispersed in low salt buffer, spread and metal shadowed (for methodical details see Franke et al. 1976a). Note the characteristic nucleosomal "beads-on-a-string" organization of the chromatin strands. Some variation of center-to-center distances of nucleosomes can be observed, sometimes even on the same chromatin strand. The insert shows one example *(bracket in the left)* of long spacing, with a mean distance of about 38 nm (corresponding to a

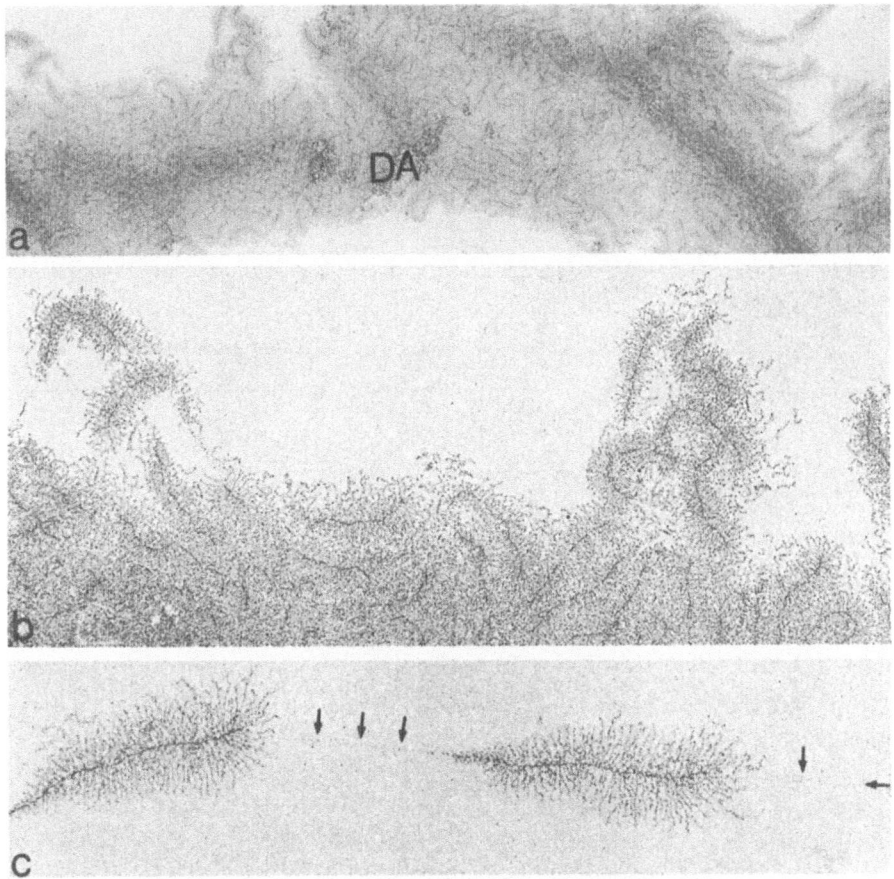

Fig. 2a-c. Nucleolar chromatin of the primary nucleus of the green algae *Acetabularia mediterranea* (Fig. 2a and b) and *Acetabularia cliftonii* (Fig. 2c). **a** shows a large mass of aggregated, fully active pre-rRNA genes, reflecting the aggregated state of this material in situ (*DA*, dense aggregate structure, which is often seen in central portions of such nucleolar spread preparations. **b** is a higher magnification of the previous figure, showing the individual transcriptional units in the margin of the spread nucleolar mass. **c** shows a typical example of a well-spread strand of transcriptionally active nucleolar chromatin, characterized by tandemly arranged pre-rRNA genes and interspersed spacer regions *(arrows)*. Note that the axis of the spacer is relatively thin and does not show nucleosome-sized particles. The mean length of the transcriptional units (e.g., ca. 2.2 μm in the *strand* shown in **c**), recognized by the associated lateral RNP-fibrils containing the nascent pre-rRNA, is in fair correspondence to the size of the pre-rRNA isolated from such nuclei (Spring et al. 1974, 1976). **a:** ×4,300; **b:** ×11,600; **c:** ×15,500

DNA packing ratio of about 1.7); the *smaller bracket* indicates an example of relatively close spacing, here with a mean distance of 22 nm (corresponding to a foreshortening ratio, relative to B-conformation DNA, of about 2.9). These estimations are based on a value of 200 base pairs of DNA per nucleosome and nucleosome-equivalent unit, respectively, and the electron microscopic DNA reference values given by Stüber and Bujard (1977). The heterogeneity in the spacing of nucleosome particles may have been artificially induced during the preparation. ×78,000; insert, ×80,000

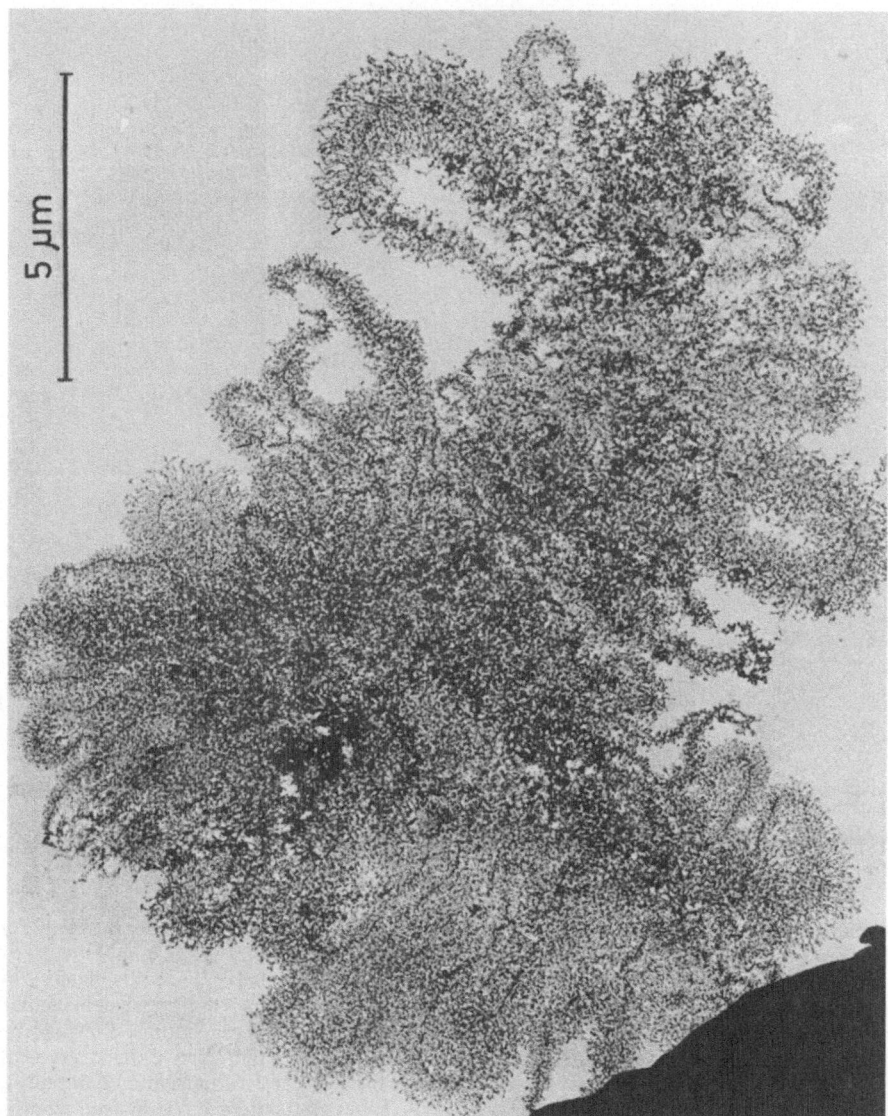

Fig. 3. Actively transcribed non-nucleolar chromatin as seen in a spread and positively stained whole lampbrush chromosome from a primary nucleus of the green alga *Acetabularia mediterranea* (for preparation see Spring et al. 1975). Note the numerous loops of various lengths, each containing transcriptional units. × 12,500

A. Nucleolar Chromatin

1. Fully active transcriptional units are recognized as "matrix units" showing close spacing of transcriptional complexes (Figs. 2, 4c, 5). In such matrix units one does not reveal additional nucleosome-sized particles that are free of lateral RNP fibrils; nor does one regularly see extended nonbeaded regions of chromatin axis as this would have to be expected if nucleosomes would exist in such regions but would unfold during the preparation. In many, though not in all (cf. Trendelenburg et al. 1973, 1976) organisms, the length and the DNA content of the matrix units corresponds to the sizes of the specific pre-rRNA's isolated from the same cells which are assumed to represent the primary—or at least only slightly processed—products of transcription (Scheer et al. 1973; Spring et al. 1974, 1976; Foe et al. 1976; Franke et al. 1976a, 1978a,b; Laird et al. 1976; Scheer et al. 1977; Foe 1978; McKnight et al. 1978; Reeder et al. 1978b; see these for further references).

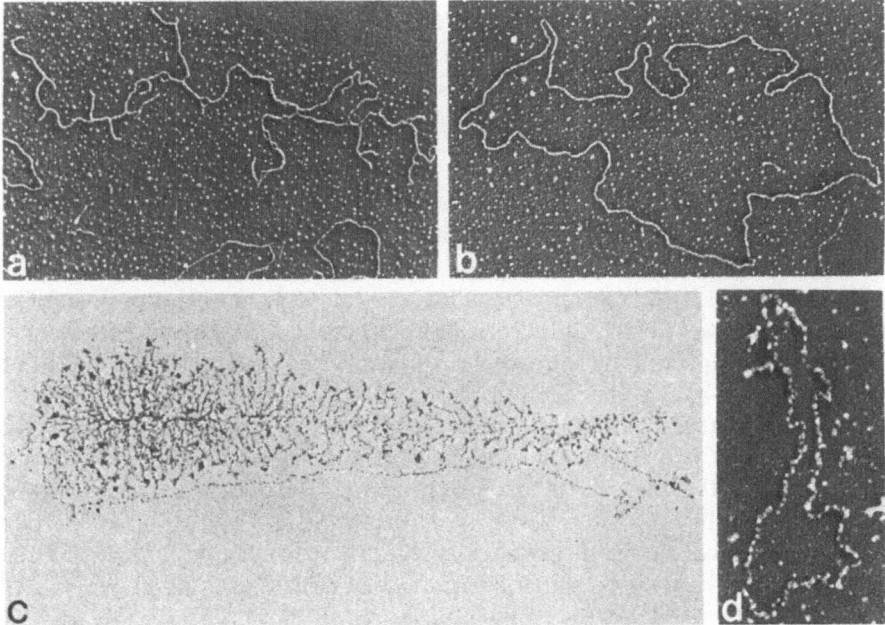

Fig. 4a–d. Electron microscopy of purified amplified rDNA and of nucleolar chromatin from the water beetle *Dytiscus marginalis*. Isolated circular rDNA appears in spread preparations either as supercoiled (**a**) or relaxed molecules (**b**); the contour length of this one pre-rRNA gene containing molecule is 8.6 μm. In spread preparations of transcriptionally active nucleolar chromatin the rings of the smallest size class, i.e., those containing one transcriptional unit and a spacer unit, reveal the typical arrangement of fibril-covered matrix units and fibril-free, thin spacer intercepts (**c**); contour length of the circle shown here is ca. 8.7 μm. By contrast, when rings of the smallest size class are isolated from transcriptionally inactive, rDNA-containing chromatin, e.g., from previtellogenic oocytes, they show the beaded nucleosomal organization and the expected foreshortening that results from rDNA packing into nucleosomes (**d**); contour length of the chromatin circle shown here is 3.35 μm (for details see Scheer and Zentgraf 1978). This illustrates that rDNA is extended in transcriptionally active stages and is packed into nucleosomes when inactive. **a–c:** × 24,000; **d:** × 42,000

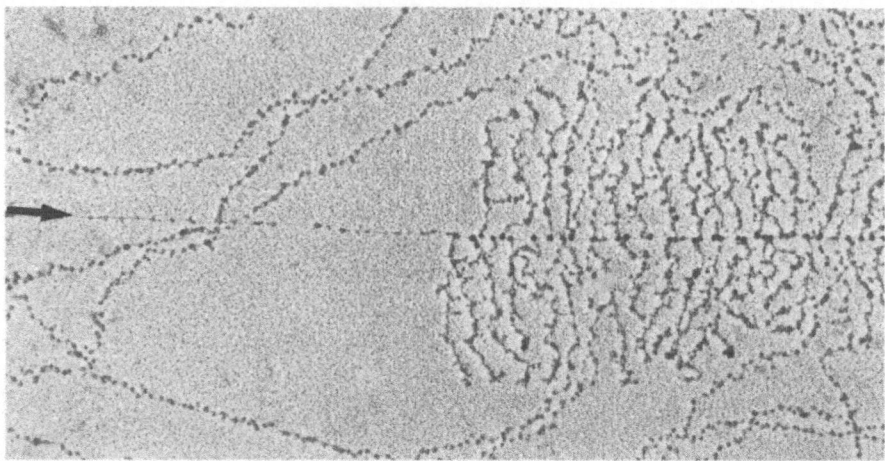

Fig. 5. Nucleolar chromatin of reduced transcriptional activity (preparation from a nearly mature oocyte of the Alpine newt, *Triturus alpestris*) often shows a "mixed appearance in spread preparations (cf. Scheer et al. 1976a; Franke et al. 1978a,b; Scheer 1978). Inactive chromatin fibrils show the typical nucleosomal ("beaded") organization and can be observed next to transcriptionally active rDNA regions characterized by fibril-covered matrix units and adjacent thin, nearly smooth ("nonbeaded") spacer intercepts. × 42,000

2. Apparent spacer regions (for definitions see Franke et al. 1976a,b, 1978a,b) interspersed between matrix units show only few, frequently none, nucleosome-sized particles (e.g., Foe et al. 1976; Franke et al. 1976a,b, 1978a,b; Foe 1978; Villard and Fakan 1978). Although it has been suggested by some authors that these particles represent nucleosomes (Foe et al. 1976; Woodcock et al. 1976; Foe 1978) several observations tend to indicate that such particles are not of nucleosomal nature (for definition see above) and that spacer regions located between fully active matrix units are not packed into nucleosomes, at least not in a mode similar to that observed in inactive chromatin from the same nucleus or even the same nucleolus:

a) Some spacer-associated particles of 10–13 nm diameter are preserved after treatment with Sarkosyl NL-30 at concentrations which result in the removal of histones (Franke et al. 1976a, 1978a,b; Scheer et al. 1977).

b) Some spacer-associated particles are associated with short lateral fibrils similar to those seen in matrix units (Scheer et al. 1973, 1977; Franke et al. 1976b). This, together with the observation mentioned under (a), suggests that at least some of the spacer-associated particles represent RNA-polymerase molecules and transcriptional complexes, respectively (for the special situation for 5-fluorouridine-treated nucleoli see also Rungger et al. 1978).

c) The DNA present in spacer regions is not considerably contracted (estimated contraction ratios ranging from 1.0 to 1.24 have been reported; Franke et al. 1976a, 1978a,b; Trendelenburg et al. 1976, 1978; Scheer et al. 1977; Franke and Scheer 1978; Reeder et al. 1978a; McKnight et al. 1978; for a comparison of mammalian rDNA repeating units and spacer units see also Puvion-Dutilleul and May 1978).

d) An especially clear demonstration of the altered configuration in spacer chromatin is the finding (see Melton et al., this vol.) that plasmids containing both spacer and pre-rRNA gene regions which are transcribed upon injection into germinal vesicles of *Xenopus laevis* show nucleosome-sized particles only in the nontranscribed plasmid DNA region but not in the extended transcribed gene region and the adjacent rDNA spacer region.

From this, we conclude, in harmony with other authors (Reeder et al. 1978a), that the DNA of spacer regions is also only slightly, if at all, contracted and thus cannot be packed in nucleosomes. We cannot decide, however, whether the obvious structural difference of rDNA spacer chromatin in active nucleoli reflects the true absence of nucleosomes in situ or a greater selective susceptibility of these nucleosomes, if they exist, to unfolding forces exerted during the preparation. Moreover, we want to mention various other alternatives for explaining the occasional observations of beadlike particles of nucleosomal size in spacer regions: these particles could represent histone complexes such as nucleosome-derived octamers, reaggregated histones, other protein aggregates, regular stain deposits, et cetera.

3. Regions in pre-rRNA matrix units of reduced transcriptional activity, which are recognized by reduced packing densities of transcriptional complexes (i.e., "gaps" within matrix units; cf. Scheer et al. 1975, 1976a), have also been shown to display a "smooth", i.e., nonnucleosomal appearance (Franke et al. 1976a, 1978a,b; Foe 1978; Scheer 1978). This observed absence of nucleosomal particles in partly inactive pre-rRNA genes is consistent with the finding that the lengths of such matrix units of reduced lateral fibril densities are not considerably decreased, i.e., the rDNA apparently remains in an extended, non-nucleosomal form (Franke et al. 1976a; Scheer et al. 1976a; cf. Foe 1978).

4. Regions of nucleolar chromatin that correspond to pre-rRNA genes but are not associated with identifiable transcriptional complexes sometimes also show a smooth, i.e., non-nucleosomal appearance and seem to contain extended DNA (Franke et al. 1976a, 1978a,b; Scheer et al. 1976a, 1977; Foe 1978). Such situations have been described during embryogenesis of the milkweed bug, *Oncopeltus fasciatus* (Foe et al. 1976; Foe 1978) during oogenesis of various amphibians[1] (Franke et al. 1976a; Scheer et al. 1976a; Scheer 1978) and during drug-induced inhibition of nucleolar transcription (cf. Scheer et al. 1975; Franke et al. 1976a).

---

1   While lampbrush chromosome loop retraction as well as centripetal accumulation and condensation of nucleoli are striking during late oogenesis in several urodelan species (for refs. see Scheer et al. 1976a), concomitant with progressively reduced transcriptional activity, such phenomena are much less pronounced in the oogenesis of the anuran species, *Xenopus laevis*. Late stages of oogenesis of *Xenopus laevis*, often collectively classified as stage 6 (cf. Scheer 1973), show much less reduced transcriptional activity (Anderson and Smith 1977, 1978). It should be kept in mind, however, that these oocytes are heterogeneous within one ovary as well as among different females (cf. also Anderson and Smith 1977). Moreover, the comparison of, e.g., maximal activity in stage 3 and 4 in fully active oocytes of hormone-treated frogs with the activity of stage 6 oocytes has indicated a reduction by 40–50% in the later stage (Anderson and Smith 1978). It is to be emphasized, however, that some nucleoli of late stages of *Xenopus* oogenesis do clearly show absence of matrix units in some regions, as opposed to residual transcriptional activity in other regions of the same strand (cf. Scheer 1978).

This phenomenon has been interpreted to reflect the conversion of chromatin from the nucleosomal to the extended state, prior to transcription (Foe 1978), and vice versa, shortly after cessation of transcription (Franke et al. 1976a; Scheer et al. 1976a; Scheer 1978).

5. Transcriptionally inactive nucleolar chromatin can be identified in early stages of embryogenesis of some insects (Foe et al. 1976; McKnight and Miller 1976; Foe 1978), in early, i.e., previtellogenic, and very late, i.e., during or after the lampbrush chromosome loop retraction phase, stages of amphibian or insect oogenesis (Scheer et al. 1976a; Scheer 1978; Scheer and Zentgraf 1978) and after drug-induced inhibition of transcription (Scheer et al. 1975; Franke et al. 1978a,b). Here chromatin regions that contain pre-rRNA genes seem to be packed in nucleosomes indistinguishable from those present in other kinds of inactive chromatin (Franke et al. 1976a, 1978a,b; Foe 1978; Scheer 1978; Scheer and Zentgraf 1978). This demonstrates that inactive nucleolar chromatin assumes the nucleosomal form, resulting in a contraction of the rDNA similar to that described in various other kinds of DNA.

B. Non-Nucleolar Chromatin

1. The organization of intensely transcribed, non-nucleolar genes has been studied in lampbrush chromosome loops (Fig. 3; for refs. see Scheer et al. 1976b, 1978) and in the putatively identified transcriptional unit of the silk fibroin gene of *Bombyx mori* (cf. McKnight et al. 1978). Both situations are characterized by dense packing of transcriptional complexes, absence of additional nucleosome-sized particles, and an apparently extended DNA (Franke et al. 1976a, 1978a,b; Scheer et al. 1976b, 1978; Franke and Scheer 1978; McKnight et al. 1978).

2. Transcriptional units of non-nucleolar genes that show a markedly reduced transcriptional activity, as seen from the relatively large distances between transcriptional complexes ("gaps"), often but not always (Franke et al. 1976a) reveal nucleosome-like particles located in the chromatin axis intercepts between the transcriptional complexes (cf. Foe et al. 1976; Laird et al. 1976; McKnight and Miller 1976; Foe 1978; Franke et al. 1978b; McKnight et al. 1978; Scheer 1978). The nucleosomal nature of these beads located between transcriptional complexes has also been demonstrated by their disappearance during de-histonization using Sarkosyl (Scheer 1978). This situation seems to be also present in transcriptionally active SV40 chromatin in which only few transcriptional complexes seem to be present per molecule (for discussion see Keller et al. 1978; Müller et al. 1978).

From these observations we conclude that fully active non-nucleolar chromatin is also present in a non-nucleosomal form, i.e., with its DNA extended. However, in moderately or weakly active chromatin, which in most nuclei seems to represent the bulk of the total transcribed chromatin, chromatin intercepts that are located between two transcriptional complexes assume, probably as a transient state, the nucleosomal form (for detailed discussion see Franke et al. 1978a,b; Scheer 1978). In other words, transcriptionally inactive regions within transcribed genes often seem to assume the same organization as that present in other inactive chromatin. This suggests that unfolding and reformation of nucleosomes can take place in relatively short periods of time. The rates of such conformational changes of

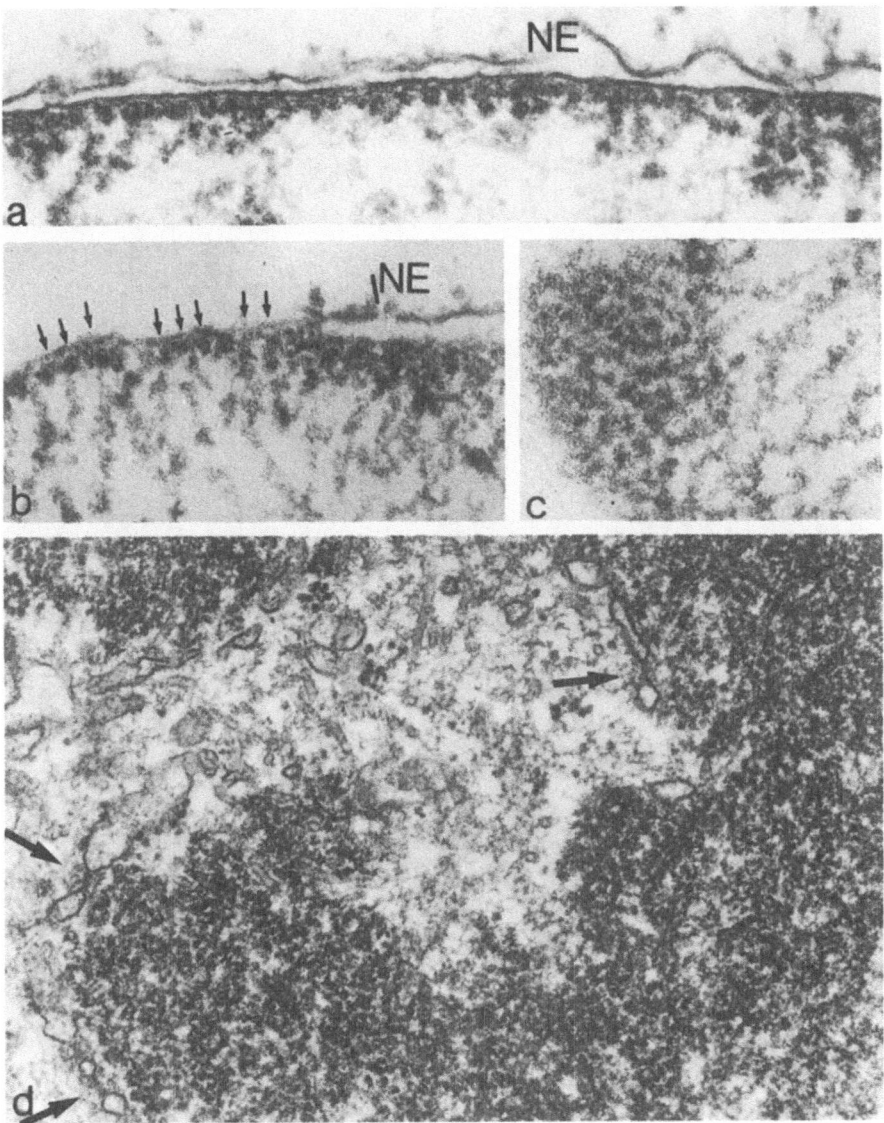

Fig. 6a–d. Electron micrographs of ultrathin sections showing large ("supranucleosomal") granules of chromatin to be especially prominent in the peripheral condensed chromatin (**a–c**) and in mitotic chromosomes (**d**). **a** shows large chromatin granules in the most peripheral layer of chromatin associated with the inner membrane of the nuclear envelope (NE) of a cultured murine 3T3 cell (cf. Franke et al. 1978a) **b** and **c** show, in cross section (**b**) and in grazing section (**c**), the relatively high resistance of the large chromatin granules (some are denoted by *arrows* in **b**) attached to the inner nuclear membrane to treatment with low salt media that promote dispersion and swelling of the chromatin (examples are in isolated calf thymocytes; cf. Franke and Scheer 1974). The late anaphase chromosomes of cultured thymic epithelial cells of the rat (**d**), attached with small pieces of reforming nuclear envelope (*arrows* denote pore complexes; note also ribosomes associated with both sides of some of these pore complex-containing units, indicative of their origin from endoplasmic reticulum), show most of their chromatin arranged in 18–32 nm large granules. **a:** ×60,000; **b, c:** ×100,000 **d:** ×63,000

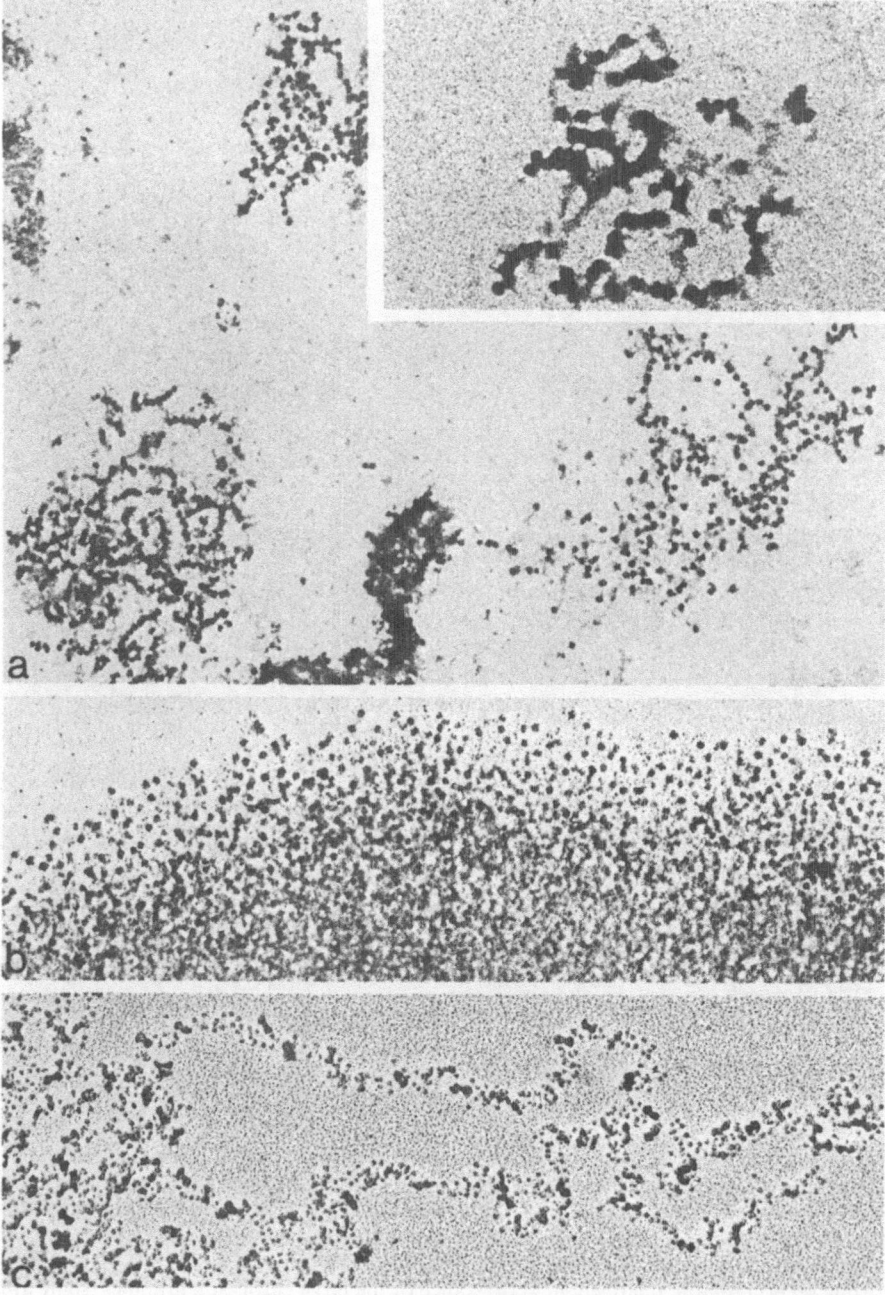

Fig. 7a-c. Occurrence of large (18–32 nm) chromatin granules representing supranucleosomal orders of chromatin packing as revealed after dispersion of chromatin in low salt buffers and spreading preparation according to Miller and Beatty (1969; for specific modification see Franke et al. 1976a). **a** shows typical arrays of "superbeads" seen in a cultured cell (monkey kidney CV 1; for similar observations made in other cultured cells and for details of preparation see Franke et al. 1976a, 1977). The *insert* in **a** shows the chain-like arrays of chromatin organization during transcription may vary in different situations and different genes. A certain retention phase of the conformational changes accompanying a transcriptional event is also suggested by some biochemical results indicating that the selectively higher DNAse I sensitivity characteristic of transcribed chromatin remains for some time after cessation of transcription (e.g., Weintraub and Groudine 1976; Palmiter et al. 1978; see also Chambon 1978).

The concept of a non-nucleosomal organization of transcribed chromatin does not imply the absence of histones in such regions. Evidence available, though still rather scarce, suggests that histones are present in transcriptionally active chromatin (e.g., McKnight et al. 1978), including nucleolar chromatin (Higashinakagawa et al. 1977; Reeder et al. 1978b) and lampbrush chromosome loops (U. Scheer and R. Sperling, unpublished observations by immunofluorescence microscopy) of amphibian oocytes. The presence of histones in transcribed chromatin regions is also suggested from the results obtained by digestion of DNA containing genes assumed to be transcribed using micrococcal nuclease (for refs. see above and Franke et al. 1978a,b; Mathis and Gorovsky 1978; Reeves 1978). The specific molecular architecture of the histone arrangement in transcribed chromatin will have to be elucidated in future experiments.

V. Supranucleosomal Organization Observed in Transcriptionally Inactive Chromatin

Chromatin, which is transcriptionally inactive for extended periods of time or is constitutively nontranscribable (see also definition of constitutive heterochromatin, the refs. given in Introduction), frequently shows an especially prominent condensed appearance. Such chromatin appears to be not only completely compacted in nucleosomes but also presents a distinct form of higher order packing of the "nucleofilament" (for terminology see Klug 1978; Worcel 1978). Certain highly condensed chromatin structures such as pericentromeric and telomeric heterochromatin, the nuclear envelope-attached condensed chromatin, chromatin predominant in spermatids and spermatozoa of various species, chromatin of pyknotic nuclei and maturing nucleated erythrocytes show, in ultrathin sections, arrays of chain-like large closely juxtaposed chromatin granules (diameters 18–32 nm; cf. Fig. 6; see Franke and Scheer 1974, 1978; Zentgraf et al. 1975; Philipp et al. 1976; Franke et al. 1978a,b; see these refs. for further literature).

◄

supranucleosomal granules at higher magnification. **b** and **c** show the appearance of chromatin of sea urchin sperm *(Psammechinus miliaris)* in either predominantly large granular (**b**) or more solenoid-like (**c**) arrangements. The chromatin shown in Fig. b has been prepared by incubation in—and extensive dialysis (12 h, 4 °C) against—buffer without chelate-forming components (200 mM KCl, 10 mM pipes, 2 mM $CaCl_2$, pH 6.8), followed by direct transfer onto a "cushion" of 0.1 M sucrose in low salt borate buffer (pH 8.5) containing 1% formaldehyde and centrifugation (cf. Franke et al. 1976a); the preparation has been positively stained and shadow cast (see Franke et al. 1976a). In the preparation shown in **c** sperm has been incubated for 15 min in slightly alkaline low salt (ca. 0.1 mM) borate buffer (pH 9) and prepared as described (cf. Miller and Beatty 1969; Franke et al. 1976a).
a: × 32,000; insert in **a:** × 76,000; **b:** × 23,000; **c:** × 48,000

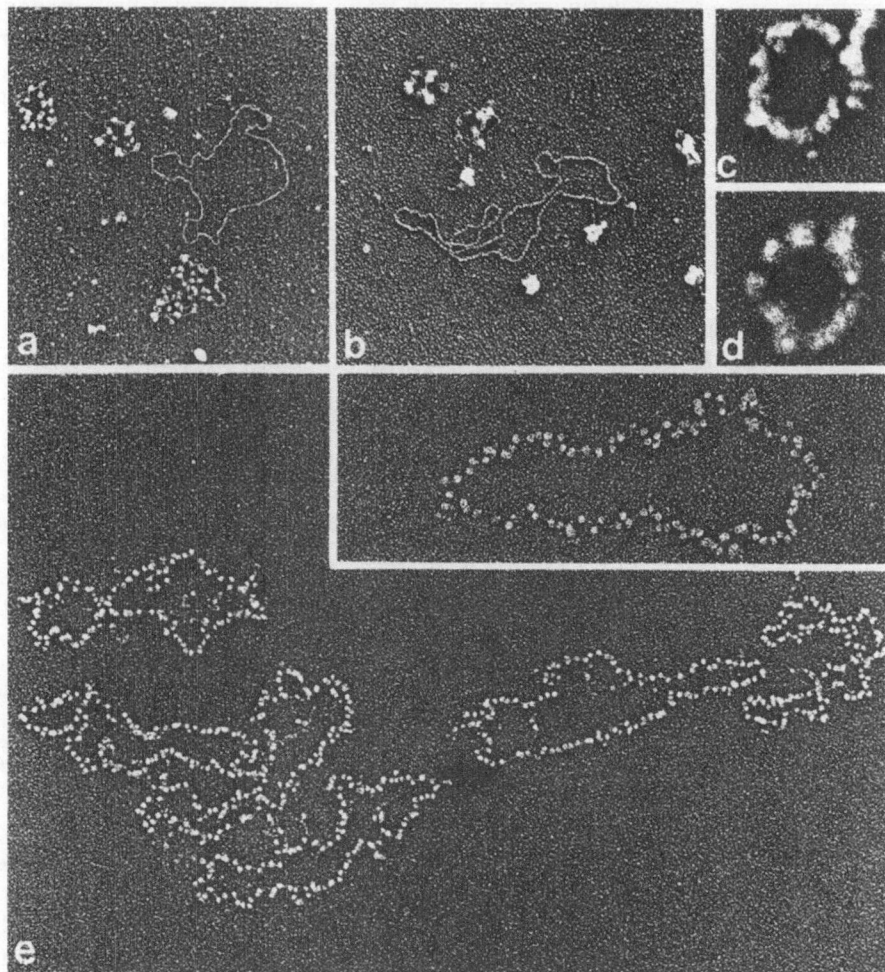

Fig. 8a-e. Nucleosomal and supranucleosomal packing of circular DNA molecules as seen in electron micrographs of spread preparations of chromatin. At the nucleosomal level, the SV40 chromatin prepared from infected cells (for details see Keller et al. 1978; Müller et al. 1978; Zentgraf et al. 1978) shows the typcical beaded appearance and the foreshortening expected for nucleosomal packing (**a**), whereas the supranucleosomal organization of the SV40 chromatin is identified as a compact large globular structure (**b**); de-proteinized SV40 DNA has been added in **a** and **b** to demonstrate the packing of the DNA. Supranucleosomal packing is also observed in transcriptionally inactive rDNA circles isolated from nucleolar chromatin of previtellogenic *Dytiscus marginalis* oocytes (**c,d**). The globular higher order structures are significantly larger than nucleosomes (compare Fig. 4), and the foreshortening is much greater (20-fold and more) than in the nucleofilament form. **e** shows the nucleosomal packing of purified mitochondrial DNA from *Xenopus laevis* oocytes injected into nuclei ("germinal vesicles") of the same cell type. *Xenopus laevis* oocytes have been injected with 2–4 ng of purified mitochondrial DNA, aimed for the oocyte nuclei, and incubated for 1 day. The oocyte nuclei have then been manually isolated and their contents spread for electron microscopy as described (Trendelenburg et al. 1978). Almost all injected circular molecules are visualized as newly assembled chromatin rings with a typical "beaded" ultrastructure (**e**); maximally 80–90 beads per chromatin circle; the *insert* in **e** shows a clearly traceable nucleosome-packed circle of mitochondrial DNA at higher magnification. The molecules

Such large chromatin granules, especially those attached to the nuclear envelope, have also been shown to have a relatively high resistance to swelling and dispersion in solutions of low ionic strength (Fig. 6b,c; cf. Franke and Scheer 1974), compared to the bulk of the chromatin. Apparently it is this property that often allows their demonstration in spread preparations (see above) after relatively short periods of time of incubation in the low salt buffer (Fig. 7; Franke et al. 1976a, 1978a,b; Kiryanov et al. 1976; Olins 1978). During progressive incubation in buffer solutions of low ionic strength one often recognizes transitions from the large granular form to more continuous coil-like fibers, suggestive of a solenoidal arrangement of the nucleosomal chains, and, finally, the extended nucleofilaments (e.g., Fig. 7b,c; cf. Franke et al. 1976a, 1978a,b). Distinct granules ("superbeads", "nucleogranules") 18–32 nm large and containing 6–10 nucleosomes (a maximum of eight has been reported for the granules prepared from rat liver; cf. Strätling et al. 1978a,b) have been isolated by controlled, limited digestion with micrococcal nuclease (Hozier et al. 1977; Renz et al. 1977, 1978; Strätling et al. 1978a,b), and it has been demonstrated that histone H1 is required for this granular form of regular supranucleosomal packing (same refs.). Related observations of a granular higher order packing of a nucleosomal chain have also been reported from studies on the SV40 chromatin ("minchromosomes") which contains a circular chain of 24–26 nucleosomes that are packed, at nearly physiological ionic strength, in large (about 30 nm) aggregates, sometimes allowing the distinction of three or four 17–19 nm large granular constituents, which in low salt buffers extend into the unfolded, nucleofilament-like state and reversibly fold back to the aggregate state at increasing salt, as long as H1 is present (Fig. 8a,b; cf. Griffith 1975; Griffith and Christiansen 1978; Keller et al. 1978; Müller et al. 1978; Varshavsky et al. 1978; Zentgraf et al. 1978). Higher orders of granular packing of nucleosomes have also been described in another circular DNA molecule, the amplified rDNA of *Dytiscus marginalis* (Fig. 8c,d; cf. Scheer and Zentgraf 1978). Whether or not the nucleosomal arrays that are newly formed upon exposure of various kinds of DNA to egg cytoplasm or oocyte nuclei of *Xenopus laevis* (Fig. 8e; cf. Laskey et al. 1978; Trendelenburg and Gurdon 1978; Trendelenburg et al. 1978) are also packed into such granular forms of higher order packing, probably depending on the presence of H1, remains to be clarified.

VI. Nonhistone Proteins and the Organization of Transcriptionally Active and Inactive Chromatin

It is obvious that nonhistone proteins play an important role in the morphology of transcriptionally active chromatin since the appearance of these chromatin

---

shown have mean contour lengths of 2.4–2.5 μm, which has to be compared with the length of the purified circular mitochondrial DNA prior to injection (5.0–5.4 μm; for refs. see Borst and Kroon 1969; Pinon et al. 1978). This shows that, under the specific preparation and spreading conditions used, the mitochondrial DNA is compacted and foreshortened approx. 2.2-fold **a:** × 50,000; **b:** × 78,000; **c, d:** × 110,000; **e:** × 50,000; insert in **e:** × 70,000

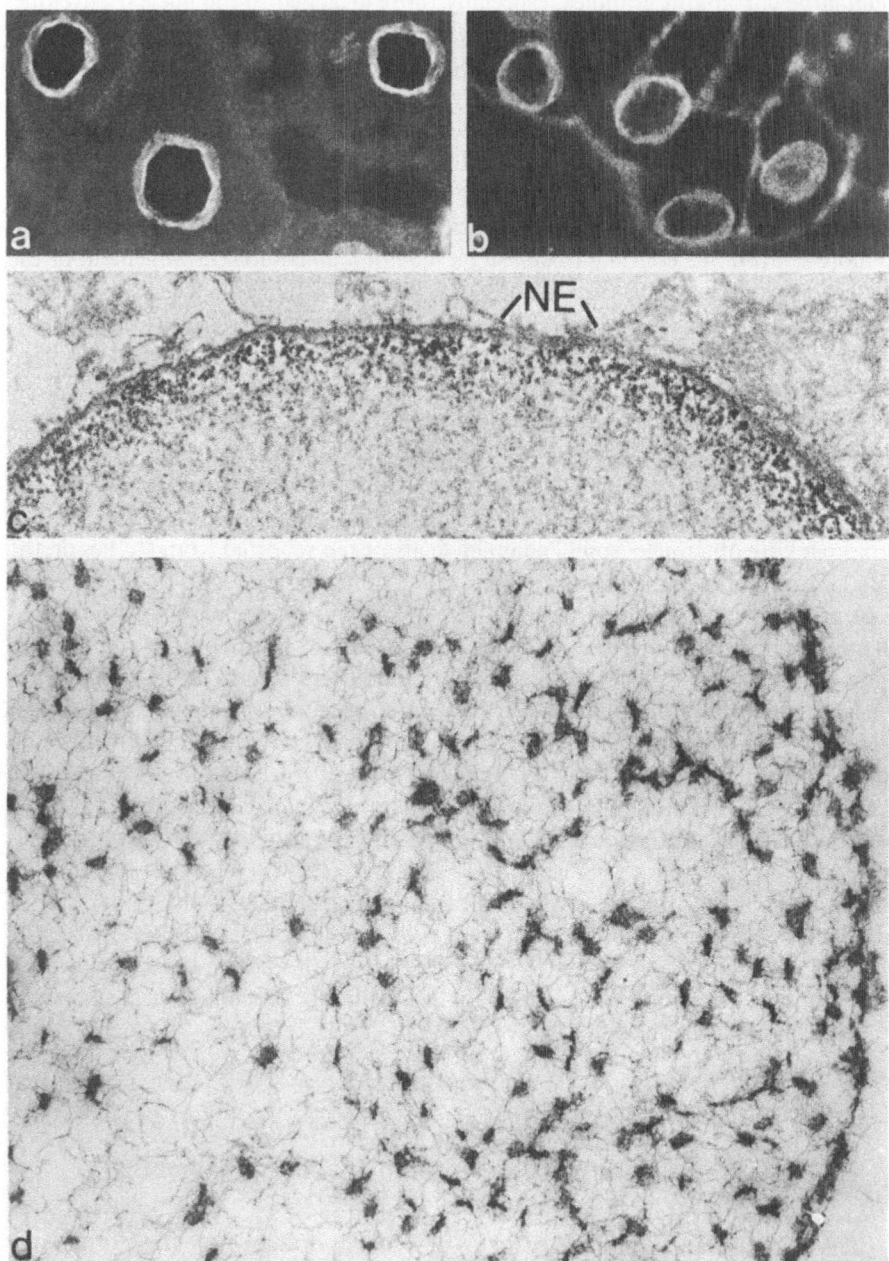

Fig. 9a-d. Demonstration of the involvement of structures containing nonhistone proteins in chromatin organization and topology. **a-c** present the localization of an "insoluble" major nonhistone protein specifically associated with peripheral chromatin (NE, nuclear envelope) in frozen tissue sections of rat (**a, c**) and toad (**b**) *Xenopus laevis* liver by indirect immunofluorescence microscope (**a, b**) and electron microscopy using peroxidase labelling of antibodies (**c**) for details see Krohne et al. 1978). One of the three major polypeptides of the "nuclear pore complex-lamina" fraction from rat liver, prepared essentially as described by regions is characterized and often dominated by the associated RNA polymerases and the proteins associated with the nascent RNA's (see above). However, in view of the appearance of the highly condensed chromatin as aggregates of nucleosomes and supranucleosomal granules, respectively, it is questionable whether the condensed chromatin represents an aggregate of nucleohistone material or whether it contains additional nonhistone proteins. Various authors have described the location of nonhistone proteins in chromatinous and interchromatinous nuclear structures, especially those characterized by a pronounced resistance to treatment with solutions of high and low ionic strength as well as nondenaturing detergents (cf. Georgiev and Chentsov 1960; Berezney and Coffey 1974; Comings and Okada 1976; Dwyer and Blobel 1976; Cobbs and Shelton 1978; Herman et al. 1978; Krohne et al. 1978; Laemmli et al. 1978; Miller et al. 1978; see these for earlier references). In the context of the present discussion we want to present two examples illustrating that some of the "residual" (i.e., those insoluble in low and high salt buffers and nondenaturing detergents) nonhistone proteins are enriched in—and specifically associated with—condensed portions of chromatin. (1) A group of major nonhistone polypeptides appears in residual fractions of whole chromatin of nuclear envelope, and in "matrix" preparations in the form of a characteristic triplet band recognized by gel electrophoresis (approx. mol. wt. ranging from 63,000 to 80,000; for refs. see above). Using antibodies raised against the individual polypeptides of this protein fraction, it has been shown that some of them are specifically located, in somatic interphase nuclei, in close association with the peripheral condensed chromatin (Fig. 9a-c; cf. Krohne et al. 1978). The broad range of immunological cross reaction of these proteins in different species (cf. Fig. 9a, b) suggests that they have been strongly conserved during evolution. (2) Another proteinaceous component located in condensed (not histone containing) chromatin of mammalian spermatids and spermatozoa is the "residual" protein (one major polypeptide of ca. 65,000 mol. wt.) that forms a characteristic three-dimensional meshwork with densely aggregated, somewhat regularly distributed "nodules" (Fig. 9d; H. Zentgraf, G. Heil, and W.W. Franke, in preparation; cf. Evenson et al. 1978). The biochemical nature and the functional significance of such major nonhistone proteins that seem to be interspersed as some sort of a fibrillar "cement" between the 18–32 nm large chromatin granules will be subject to future experimental work.

Dwyer and Blobel (1976), was used as antigen (polypeptide "tri-2"; apparent mol. wt. ca. 73,000). Antisera have been produced in chicken (for details see Ely et al. 1978). It cannot be decided whether the decorated protein material is located in large granules of inactive condensed peripheral chromatin or is interspersed between these granules. Essentially similar results have been obtained in frozen sections stained with antibodies against polypeptides "tri-1" and "tri-3". d shows an ultrathin section through the achromatinous residual skeleton of a human spermatozoan nucleus obtained after prolonged treatment (12 h, 4 °C) with buffer containing high salt concentrations and a nonionic detergent (1.5 M KCl, 50 mM Tris-HCl, pH 7.2, 1% Triton X-100). After centrifugation at 4,000 g the resulting pellet has been resuspended (3 h at room temperature) in low salt buffer (10 mM Tris-HCl, pH 7.2; 1 mM EDTA) made 1% with respect to Sarkosyl NL-97. After washing in low salt buffer without detergent the material has been fixed with buffered 2.5% glutaraldehyde, postfixed with osmium tetroxide and processed for electron microscopy as described (Franke et al. 1976c). **a:** × 1,000; **b:** × 1,200; **c:** × 23,000; **d:** × 41,500

*Acknowledgments.*We are indebted to C. Grund and K. Mähler for valuable technical assistance. We are also thankful to Drs. J. Gurdon (MRC Cambridge, England) for help with the DNA injection experiments, W. Keller (University of Heidelberg) for discussion about chromatin organization, and C. Petzelt (German Cancer Research Center) for supply will sea urchins. The work has been supported by the Deutsche Forschungsgemeinschaft.

References

Allfrey, V.G., Johnson, E.M., Sun, I.Y.-C., Littau, V.C., Matthews, H.R., Bradbury, E.M.: Structural organization and control of the ribosomal genes in *Physarum.* Cold Spring Harbor Symp. Quant. Biol. 42, 505-514 (1978)

Anderson, D.M., Smith, L.D.: Synthesis of heterogeneous nuclear RNA in fullgrown oocytes of *Xenopus laevis* (Daudin). Cell 11, 663-671 (1977)

Anderson, D.M., Smith, L.D.: Patterns of synthesis and accumulation of heterogeneous RNA in lampbrush stage oocytes of *Xenopus laevis* (Daudin). Dev. Biol. 67, 274-285 (1978)

Beermann, W. (ed.): Developmental Studies on Giant Chromosomes. In: Results and Problems in Cell Differentiation, Vol. 4. Berlin-Heidelberg-New York: Springer 1972

Bellard, M., Gannon, F., Chambon, P.: The structure and transcriptional activity of the chromatin containing the ovalbumin and globin genes in chick oviduct nuclei. Cold Spring Harbor Symp. Quant. Biol. 42, 779-791 (1978)

Berezney, R., Coffey, D.S.: Identification of a nuclear protein matrix. Biochem. Biophys. Res. Commun. 60, 1410-1417 (1974)

Bloom, K.S., Anderson, J.N.: Fractionation of hen oviduct chromatin into transcriptionally active and inactive regions after selective micrococcal nuclease digestion. Cell 15, 141-150 (1978)

Bonner, J., Wallace, R.B., Sargent, T.D., Murphy, R.F., Dube, S.K.: The expressed portion of eukaryotic chromatin. Cold Spring Harbor Symp. Quant. Biol. 42, 851-857 (1978)

Borst, P., Kroon, A.M.: Mitochondrial DNA: Physicochemical properties, replication and genetic function. Int. Rev. Cytol. 26, 107-190 (1969)

Brown, S.W.: Heterochromatin. Science 151, 417-425 (1966)

Chambon, P.: Summary: The molecular biology of the eukaryotic genome is coming of age. Cold Spring Harbor Symp. Quant. Biol. 42, 1209-1234 (1978)

Cobbs, C.S., Shelton, K.R.: Major oligomeric structural proteins of the HeLa nucleus. Arch. Biochem. Biophys. 189, 323-335 (1978)

Comings, D.E., Okada, T.A.: Nuclear proteins. The fibrillar nature of the nuclear matrix. Exp. Cell Res. 103, 341-360 (1976)

Dwyer, N., Blobel, G.: A modified procedure for the isolation of a pore complex-lamina fraction from rat liver nuclei. J. Cell Biol. 70, 581-591 (1976)

Ely, S., D'Arcy, A., Jost, E.: Interaction of antibodies against nuclear envelope-associated proteins from rat liver nuclei with rodent and human cells. Exp. Cell Res. 116, 325-331 (1978)

Evenson, D.P., Witkin, S.S., De Harven, E., Bendich, A.: Ultrastructure of partially decondensed human spermatozoan chromatin. J. Ultrastruct. Res. 63, 178-187 (1978)

Felsenfeld, G.: Chromatin. Nature (London) 271, 115-122 (1978)

Flint, S.J., Weintraub, H.M.: An altered subunit configuration associated with the actively transcribed DNA of integrated adenovirus genes. Cell 12, 783-794 (1977)

Foe, V.E.: Modulation of ribosomal RNA synthesis in *Oncopeltus fasciatus:* An electron microscopic study of the relationship between changes in chromatin structure and transcriptional activity. Cold Spring Harbor Symp. Quant. Biol. 42, 723-740 (1978)

Foe, V.E., Wilkinson, L.E., Laird, C.D.: Comparative organization of active transcription units in *Ocopeltus fasciatus.* Cell 9, 131-146 (1976)

Franke, W.W., Scheer, U.: Structures and functions of the nuclear envelope. In: The Cell Nucleus (ed. H. Busch), Vol. I, pp. 219-347. New York-London: Academic Press 1974

Franke, W.W., Scheer, U.: Morphology of transcriptional units at different states of activity. Philos. Trans. R. Soc. London Ser. B 283, 333-342 (1978)

Franke, W.W., Scheer, U., Trendelenburg, M.F., Spring, H., Zentgraf, H.: Absence of nucleosomes in transcriptionally active chromatin. Cytobiology 13, 401-434 (1976a)

Franke, W.W., Scheer, U., Spring, H., Trendelenburg, M.F., Krohne, G.: Morphology of transcriptional units of rDNA. Exp. Cell Res. 100, 233-244 (1976b)

Franke, W.W., Lüder, M.R., Kartenbeck, J., Zerban, H., Keenan, T.W.: Involvement of vesicle coat material in casein secretion and surface regeneration. J. Cell Biol. 69, 173-195 (1976c)

Franke, W.W., Scheer, U., Trendelenburg, M.F., Zentgraf, H., Spring, H.: Morphology of transcriptionally active chromatin. Cold Spring Harbor Symp. Quant. Biol. 42, 755-772 (1978a)

Franke, W.W., Zentgraf, H., Scheer, U.: Supranucleosomal and non-nucleosomal chromatin configurations. 9th Int. Congr. Electron Microsc. Toronto, Vol. III, pp. 573-586. Toronto: Microsc. Soc. of Canada 1978b

Gall, J.G., Callan, H.G.: ^{3}H-uridine incorporation in lampbrush chromosomes. Proc. Natl. Acad. Sci. USA 48, 562-570 (1962)

Garel, A., Axel, R.: Selective digestion of transcriptionally active ovalbumin genes from oviduct nuclei. Proc. Natl. Acad. Sci. USA 73, 3966-3970 (1976)

Garel, A., Axel, R.: The structure of the transcriptionally active ovalbumin genes in chromatin. Cold Spring Harbor Symp. Quant. Biol. 42, 701-708 (1978)

Georgiev, G.P., Chentsov, Y.S.: On the structure of the cell nucleus. Proc. Acad. Sci. USSR 132, 199-202 (1960)

Gottesfeld, J.M.: Organization of transcribed regions of chromatin. Philos. Trans. R. Soc. London Ser. B 283, 343-357 (1978)

Gottesfeld, J.M., Butler, P.J.G.: Structure of transcriptionally active chromatin subunits. Nucleic Acids Res. 4, 3155-3173 (1977)

Gottesfeld, J.M., Melton, D.A.: The length of nucleosome-associated DNA is the same in both transcribed and nontranscribed regions of chromatin. Nature (London) 273, 317-319 (1978)

Griffith, J.D.: Chromatin structures: Deduced from a minichromosome. Science 187, 1202-1203 (1975)

Griffith, J.D., Christiansen, G.: The multifunctional role of histone H1, probed with the SV40 minichromosome. Cold Spring Harbor Symp. Quant. Biol. 42, 215-226 (1978)

Heitz, E.: Heterochromatin, Chromocentren, Chromomeren. Ber. Dtsch. Bot. Ges. 47, 274-284 (1929)

Heitz, E.: Die Herkunft der Chromocentren. Planta 18, 571-637 (1932)

Heitz, E.: Die Chromosomenstruktur im Kern während der Kernteilung und der Entwicklung des Organismus. In: Chromosomes. Lectures held at the Conference on Chromosomes. Wageningen, pp. 5-26. Zwolle: Tjeenk Willink 1956

Herman, R., Weymouth, L., Penman, S.: Heterogeneous nuclear RNA-protein fibers in chromatin depleted nuclei. J. Cell Biol. 78, 663-674 (1978)

Higashinakagawa, T., Wahn, H.L., Reeder, R.H.: Isolation of ribosomal gene chromatin. Dev. Biol. 55, 375-386 (1977)

Hozier, J., Nehls, P., Renz, M.: The chromosome fiber: evidence for an ordered superstructure of nucleosomes. Chromosoma 62, 301-317 (1977)

Jackson, V., Chalkley, R.: The effect of urea on staphylococcal digestion on chromatin. Biochem. Biophys. Res. Commun. 67, 1391-1400 (1975)

Keller, W., Müller, U., Eicken, I., Wendel, I., Zentgraf, H.: Biochemical and ultrastructural analysis of SV40 chromatin. Cold Spring Harbor Symp. Quant. Biol. 42, 227-244 (1978)

Kiryanov, G.I., Manamshjan, T.A., Polyakov, V.V., Fais, D., Chentsov, J.S.: Levels of granular organization of chromatin fibers. FEBS Lett. 67, 323-327 (1976)

Klug, A.: Structure of chromatin. Philos. Trans. R. Soc. London Ser. B 283, 233-239 (1978)

Kornberg, R.D.: Chromatin structure: a repeating unit of histones and DNA. Science 184, 868-871 (1974)

Kornberg, R.D.: Structure of chromatin. Annu. Rev. Biochem. 46, 931-954 (1977)

Krohne, G., Franke, W.W., Ely, S., D'Arcy, A., Jost, E.: Localization of a nuclear envelope-associated protein by indirect immunofluorescence microscopy using antibodies against a major polypeptide from rat liver fractions enriched in nuclear envelope-associated material. Cytobiol. 18, 22-38 (1978)

Laemmli, U.K., Cheng, S.M., Adolph, K.W., Paulson, J.R., Brown, J.A., Baumbach, W.R.: Metaphase chromosome structure: The role of nonhistone proteins. Cold Spring Harbor Symp. Quant. Biol. 42, 351-360 (1978)

Laird, C.D., Wilkinson, L.E., Foe, V.E., Chooi, W.Y.: Analysis of chromatin-associated fiber arrays. Chromosoma 58, 169-190 (1976)

Laskey, R.A., Honda, B.M., Mills, A.D., Morris, N.R., Wyllie, A.H., Mertz, J.E., De Robertis, E.M., Gurdon, J.B.: Chromatin assembly and transcription in eggs and oocytes of Xenopus laevis. Cold Spring Harbor Symp. Quant. Biol. 42, 171-178 (1978)

Levy, B.W., Wong, N.C.W., Watson, D.C., Peters, E.H., Dixon, G.H.: Structure and function of the low-salt extractable chromosomal proteins. Preferential association of trout testis proteins H6 and HMG-T with chromatin regions selectively sensitive to nucleases. Cold Spring Harbor Symp. Quant. Biol. 42, 793-801 (1978)

Levy-Wilson, B., Gjerset, R.A., McCarthy, B.J.: Acetylation and phosphorylation of Drosophila histones. Distribution of acetate and phosphate groups in fractionated chromatin. Biochim. Biophys. Acta 475, 168-175 (1977)

Mathis, D.J., Gorovsky, M.A.: Structure of rDNA-containing chromatin of Tetrahymena pyriformis analyzed by nuclease digestion. Cold Spring Harbor Symp. Quant. Biol. 42, 773-778 (1978)

McKnight, S.L., Miller, O.L.: Ultrastructural patterns of RNA synthesis during early embryogenesis of Drosophila melanogaster. Cell 8, 305-319 (1976)

McKnight, S.L., Sullivan, N.L., Miller, O.L.: Visualization of the silk fibroin transcription unit and nascent silk fibroin molecules on polyribosomes of Bombyx mori. Progr. Nucleic Acids Res. 19, 313-318 (1976)

McKnight, S.L., Bustin, M., Miller, O.L.: Electron microscopic analysis of chromosome metabolism in the Drosophila melanogaster embryo. Cold Spring Harbor Symp. Quant. Biol. 42, 741-754 (1978)

Miller, O.L., Bakken, A.H.: Morphological studies of transcription. Acta Endocrinol. Suppl. 168, 155-177 (1972)

Miller, O.L., Beatty, B.R.: Visualization of nucleolar genes. Science 164, 955-957 (1969)

Miller, T.E., Huang, C.Y., Pogo, A.O.: Rat liver nuclear skeleton and ribonucleo-protein complexes containing HnRNA. J. Cell Biol. 76, 675-691 (1978)

Moudrianakis, E.N., Anderson, P.L., Eickbush, H.T., Longfellow, D.E., Pantazis, P., Rubin, R.L. In: The Molecular Biology of the Mammalian Genetic Apparatus (ed. P. T'so), p. 301. Amsterdam: Elsevier/North Holland 1977

Müller, U., Zentgraf, H., Eicken, I., Keller, W.: Higher order structures of SV40 chromatin. Science 201, 406-419 (1978)

Olins, A.: ν-Bodies are close-packed in chromatin fibers. Cold Spring Harbor Symp. Quant. Biol. 42, 325-329 (1978)

Olins, A.L., Olins, D.E.: Spheroid chromatin units (ν-bodies). J. Cell Biol. 59, 252a (1973)

Olins, A.L., Olins, D.E.: Spheroid chromatin units (ν-bodies). Science 183, 330-332 (1974)

Oudet, P., Gross-Bellard, M., Chambon, P.: Electron microscopic and biochemical evidence that chromatin structure is a repeating unit. Cell 4, 281-300 (1975)

Oudet, P., Germond, J.E., Bellard, M., Spadafora, C., Chambon, P.: Nucleosome structure. Philos. Trans. R. Soc. London Ser. B 283, 241-258 (1978a)

Oudet, P., Spadafora, C., Chambon, P.: Structure of the SV40 minichromosome and electron microscopic evidence for reversible transitions of the nucleosome structure. Cold Spring Harbor Symp. Quant. Biol. 42, 301-312 (1978b)

Palmiter, R.D., Mulvihill, E.R., McKnight, G.S., Senear, A.W.: Regulation of gene expression in the chick oviduct by steroid hormones. Cold Spring Harbor Symp. Quant. Biol. 42, 639-647 (1978)

Paul, J., Zollner, E.J., Gilmour, R.S., Birnie, G.D.: Properties of transcriptionally active chromatin. Cold Spring Harbor Symp. Quant. Biol. 42, 597-603 (1978)

Pelling, C.: Ribonukleinsäure-Synthese der Riesenchromosomen. Autoradiographische Untersuchungen an Chironomus tentans. Chromosoma 15, 71-122 (1964)

Philipp, E., Franke, W.W., Keenan, T.W., Stadler, J., Jarasch, E.: Characterization of nuclear membranes and endoplasmic reticulum isolated from plant tissue. J. Cell Biol. 68, 11-29 (1976)

Pinon, H., Buvat, M., Tourte, M., Dufresne, C., Mounolou, J.C.: Evidence for a mitochondrial chromosome in Xenopus laevis oocytes. Chromosoma 65, 383-389 (1978)

Puvion-Dutilleul, F., May, E.: Visualization of mouse DNA transcriptional complexes in mouse kidney cells infected with SV40 virus. Cytobiol. 18, 294-308 (1978)

Reeder, R.H., McKnight, S.L., Miller, O.: Contraction ratio of the nontranscribed spacer of Xenopus rDNA chromatin. Cold Spring Harbor Symp. Quant. Biol. 42, 1174-1177 (1978a)

Reeder, R.H., Wahn, H.L., Botchan, P., Hipskind, R., Sollner-Webb, B.: Ribosomal genes and their proteins from Xenopus. Cold Spring Harbor Symp. Quant. Biol. 42, 1167-1177 (1978b)

Reeves. R.: Structure of Xenopus ribosomal gene chromatin during changes in genomic transcription rates. Cold Spring Harbor Symp. Quant. Biol. 42, 709-722 (1978)

Reeves, R., Jones, A.: Genomic transcriptional activity and the structure of chromatin. Nature (London) 260, 495-500 (1976)

Renz, M., Nehls, P., Hozier, J.: Involvement of histone H1 in the organization of the chromosome fiber. Proc. Natl. Acad. Sci. USA 74, 1879-1883 (1977)

Renz, M., Nehls, F., Hozier, J.: Histone H1 involvement in the structure of the chromosome fiber. Cold Spring Harbor Symp. Quant. Biol. 42, 245-252 (1978)

Rungger, D., Crippa, M., Trendelenburg, M.F., Scheer, U., Franke, W.W.: Visualization of rDNA spacer transcription in Xenopus oocytes treated with fluorouridine. Exp. Cell Res. 116, 481-486 (1978)

Scheer, U.: Nuclear pore flow rate of ribosomal RNA and chain growth rate of its precursor during oogenesis of Xenopus laevis. Dev. Biol. 30, 13-28 (1973)

Scheer, U.: Changes of nucleosome frequency in nucleolar and non-nucleolar chromatin as a function of a transcription: an electron microscopic study. Cell 13, 535-549 (1978)

Scheer, U., Zentgraf, H.: Nucleosomal and supranucleosomal organization of transcriptionally inactive rDNA circles in Dytiscus oocytes. Chromosoma 69, 243-254 (1978)

Scheer, U., Trendelenburg, M.F., Franke, W.W.: Transcription of ribosomal RNA cistrons. Exp. Cell Res. 80, 175-190 (1973)

Scheer, U., Trendelenburg, M.F., Franke, W.W.: Effects of actinomycin D on the association of newly formed ribonucleoproteins with the cistrons of ribosomal RNA in Triturus oocytes. J. Cell Biol. 65, 163-179 (1975)

Scheer, U., Trendelenburg, M.F., Franke, W.W.: Regulation of transcription of genes of ribosomal RNA during amphibian oogenesis. J. Cell Biol. 69, 465-489 (1976a)

Scheer, U., Franke, W.W., Trendelenburg, M.F., Spring, H.: Classification of loops of lampbrush chromosomes according to the arrangement of transcriptional complexes. J. Cell Sci. 22, 503-519 (1976b)

Scheer, U., Trendelenburg, M.F., Krohne, G., Franke, W.W.: Lengths and patterns of transcriptional units in the amplified nucleoli of oocytes of Xenopus laevis. Chromosoma 60, 147-167 (1977)

Spring, H., Trendelenburg, M.F., Scheer, U., Franke, W.W., Herth, W.: Structural and biochemical studies of the primary nucleus of two green algal species, Acetabularia mediterranea and Acetabularia major. Cytobiology 10, 1-65 (1974)

Spring, H., Krohne, G., Franke, W.W., Scheer, U., Trendelenburg, M.F.: Homogeneity and heterogeneity of sizes of transcriptional units and spacer regions in nuclear genes of Acetabularia. J. Microsc. Biol. Cell 25, 107-116 (1976)

Spring, H., Scheer, U., Franke, W.W., Trendelenburg, M.F.: Lampbrush-type chromosomes in the primary nucleus of the green alga Acetabularia mediterranea. Chromosoma 50, 25-43 (1975)

Strätling, W.H., Müller, U., Zentgraf, H.: Supranucleosomal structure of chromatin. Cell Biol. Int. Rep. 2, 495-499 (1978a)

Strätling, W.H., Müller, U., Zentgraf, H.: The higher order repeat structure of chromatin is built up of globular particles containing eight nucleosomes. Exp. Cell Res. 117, 301-311 (1978b)

Stüber, D., Bujard, H.: Electron microscopy of DNA: Determination of absolute molecular weights and linear density. Mol. Gen. Genet. 154, 299-303 (1977)

Suzuki, Y., Ohshima, Y.: Isolation and characterization of the silk fibroin gene with its flanking sequences. Cold Spring Harbor Symp. Quant. Biol. 42, 947-957 (1978)

Trendelenburg, M.F., Gurdon, J.B.: Transcription of cloned *Xenopus* ribosomal genes visualized after injection into oocyte nuclei. Nature (London) 276, 292-294 (1978)

Trendelenburg, M.F., Scheer, U., Franke, W.W.: Structural organization of the transcription of ribosomal DNA in oocytes of the house cricket. Nature New Biol. 245, 167 (1973)

Trendelenburg, M.F., Scheer, U., Zentgraf, H., Franke, W.W.: Heterogeneity of spacer lengths in circles of amplified ribosomal DNA of two insect species, *Dytiscus marginalis* and *Acheta domesticus*. J. Mol. Biol. 108, 453 (1976)

Trendelenburg, M.F., Zentgraf, H., Franke, W.W., Gurdon, J.B.: Transcription patterns of amplified *Dytiscus* genes coding for ribosomal RNA after injection into *Xenopus* oocyte nuclei. Proc. Natl. Acad. Sci. USA 75, 3791-3795 (1978)

Varshavsky, A.J., Bakayev, V.V., Nedospasov, S.A., Georgiev, G.P.: On the structure of eukaryotic, prokaryotic, and viral chromatin. Cold Spring Harbor Symp. Quant. Biol. 42, 457-473 (1978)

Villard, D., Fakan, S.: Visualization des complexes de transcription dans la chromatine étalée de cellules de Mammifères. C. R. Acad. Sci. Ser. D 286, 777-780 (1978)

Weintraub, H., Groudine, M.: Chromosomal subunits in active genes have an altered conformation. Science 193, 848 (1976)

Woodcock, C.L.F.: Ultrastructure of inactive chromatin. J. Cell Biol. 59, 368a (1973)

Woodcock, C.L.F., Frado, L.-L.Y.: Ultrastructure of chromatin subunits during unfolding, histone depletion, and reconstruction. Cold Spring Harbor Symp. Quant. Biol. 42, 43-55 (1978)

Woodcock, C.L.F., Sweetman, H.E., Frado, L.L.: Structural repeating units in chromatin II. Their isolation and partial characterization. Exp. Cell Res. 97, 111-119 (1976)

Worcel, A.: Molecular architecture of the chromatin fiber. Cold Spring Harbor Symp. Quant. Biol. 42, 313-324 (1978)

Yaneva, M., Dessev, G.: Persistence of the ten-nucleotide repeat in chromatin unfolded in urea, as revealed by digestion with deoxyribonuclease I. Nucleic Acids Res. 3, 1761-1767 (1976)

Zentgraf, H., Falk, H., Franke, W.W.: Nuclear membranes and plasma membranes from hen erythrocytes. IV. Characterization of nuclear membrane attached DNA. Cytobiology 11, 10-29 (1975)

Zentgraf, H., Keller, H., Müller, U.: The structure of SV40 chromatin. Philos. Trans. R. Soc. London Ser. B 283, 299-303 (1978)

Visualization of Transcriptional Activity During Xenopus laevis Oogenesis

K. Martin, Y.N. Osheim, A.L. Beyer
and O.L. Miller, Jr.

Department of Biology, University of Virginia, Charlottesville, VA 22901, USA

I. Introduction

Amphibian oogenesis has been divided into six discrete stages on the basis of morphological changes which occur during oocyte development (Dumont 1972). Stage 3, the maximum "lampbrush" stage, is characterized by highly extended chromosomes and by lateral loops along the chromosome axes that are known to be active sites of RNA synthesis. Dumont stage 4 *Xenopus laevis* oocytes undergo chromosome condensation and increase in size to become mature stage 6 oocytes which are 1200–1300 μm in diameter and marked by an equatorial white band separating the animal and vegetal hemispheres. Previous cytological studies have indicated that the lampbrush loops of mature oocytes are contracted, suggesting that a decrease in synthetic activity occurs at late stages of oogenesis (Dumont 1972; Müller 1974). However, recent biochemical evidence (Anderson and Smith 1978) indicates that stage 6 *X. laevis* oocytes exhibit rates of nonribosomal RNA synthesis that are as high or higher than those of stage 3. We have reexamined transcriptional activity in stage 3 and stage 6 *X. laevis* oocytes using phase contrast and electron microscopy.

II. Transcriptional Activity in Stage 3 and Stage 6 X. laevis Oocytes

Observation of *X. laevis* chromosomes by phase contrast microscopy shows lateral loops extending from the chromosome axis in both stage 3 and stage 6 oocytes (Fig. 1). There is a diminution in the average length of the loop as the oocyte grows, but many loops persist in the mature, white-banded oocyte.

When isolated germinal vesicle contents are allowed to rapidly disperse in H_2O and then centrifuged onto a grid, nascent transcripts can be readily observed with the electron microscope (Hamkalo and Miller 1973). Nonribosomal transcription units of both stage 3 and stage 6 oocytes have been analyzed using this chromatin spreading method. As shown in Fig. 2, we find active transcription units which are densely packed with RNA polymerase (25 polymerases/1 μm DNA) and generally longer than transcription units of other species observed using the same method (Foe et al. 1976; McKnight and Miller 1976). In the vast majority of nucleoli dispersed from both stage 3 and stage 6 oocytes, the ribosomal genes are active and

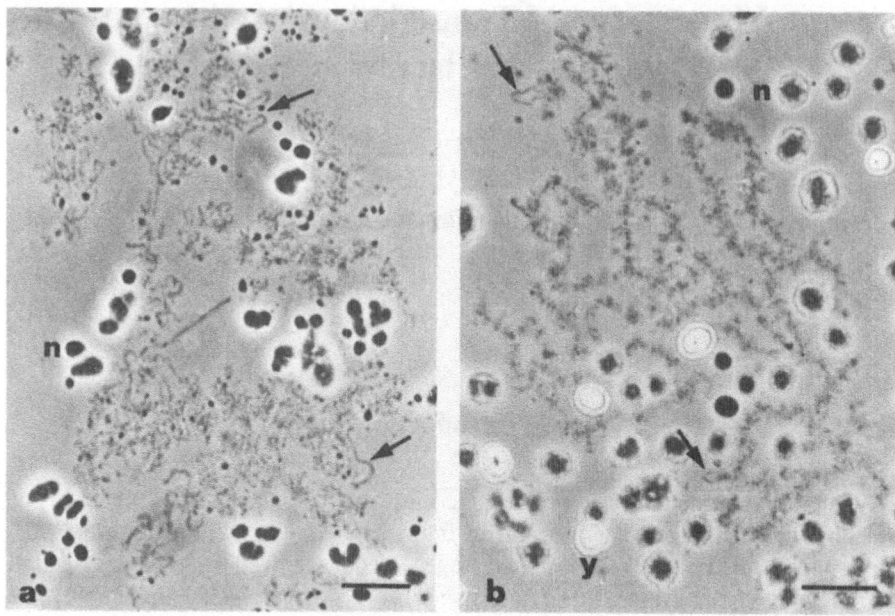

Fig. 1a,b. Phase contrast micrographs of chromosomes from stage 3 and stage 6 *X. laevis* oocytes. In **a** chromosomes from a stage 3 oocyte are displayed. Note the many lateral loops along the chromosome axis. Loops are also observed on chromosomes prepared from a stage 6 oocyte **(b)**. The *small arrows* denote some of the longer loops in the two preparations. Nucleoli *(n)* and yolk platelets *(y)* are observed in the background. The specimens are prepared by dispersing germinal vesicle contents for 1 to 4 h in a solution of 0.0388 M KCl, 0.0078 M NaCl, 0.0073 M Na_2HPO_4, 0.006 M KH_2PO_4 (pH 7) (Müller 1974) and 1% formalin. *Scale bars* represent 20 μm

fully loaded with RNA polymerase (Fig. 3). We are not able to quantify the percentage of the genome that is active in stage 3 or stage 6 oocytes, but we observe no qualitative difference in transcriptional activity in our preparations for electron microscopy.

These combined results for ribosomal and nonribosomal transcription indicate that the high level of transcriptional activity characteristic of mid-sized oocytes continues throughout oocyte growth and is readily observed in full-sized, white-banded oocytes. This high level of transcriptional activity is accompanied by the presence of loops on the extended chromosomes of both stage 3 and stage 6 oocytes. Oogenesis in *X. laevis* extends approximately 6 months, with pre-lampbrush stages taking about 1 month. Therefore, the oocyte spends 5 months engaged in high transcriptional activity. Transcription units which are 8–10 μm or longer and fully loaded with RNA polymerase are routinely observed. These units code for molecules much larger than average mRNA molecules and the high number being transcribed per unit indicates that much post-transcriptional regulation and processing must be occurring in both stages. Anderson and Smith (1978) have shown that ~85% of the nonribosomal RNA transcribed is degraded in the nucleus of stage 3 oocytes and an even higher percentage in stage 6 (~95%). These results

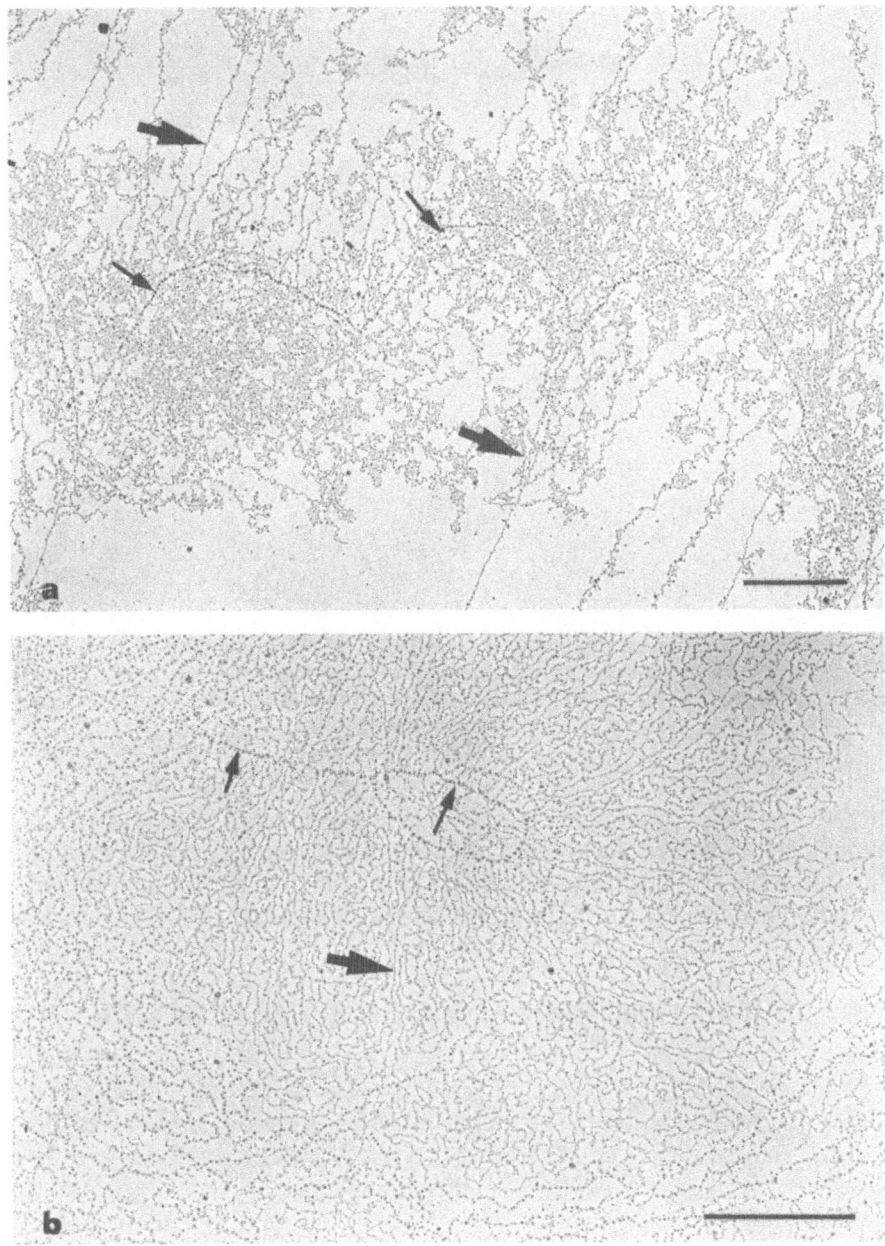

Fig. 2a,b. Electron micrographs of nonribosomal transcription units from stage 3 and stage 6 *X. laevis* oocytes. **a** shows a portion of a stage 3 transcription unit with ∼22 RNA polymerases/μm. **b** shows a portion of a stage 6 transcription unit with approximately the same polymerase packing ratio as the stage 3 unit. In both **a** and **b** *small arrows* denote the polymerase backbone and *large arrows* RNA fibers. *Scale bars* represent 1 μm

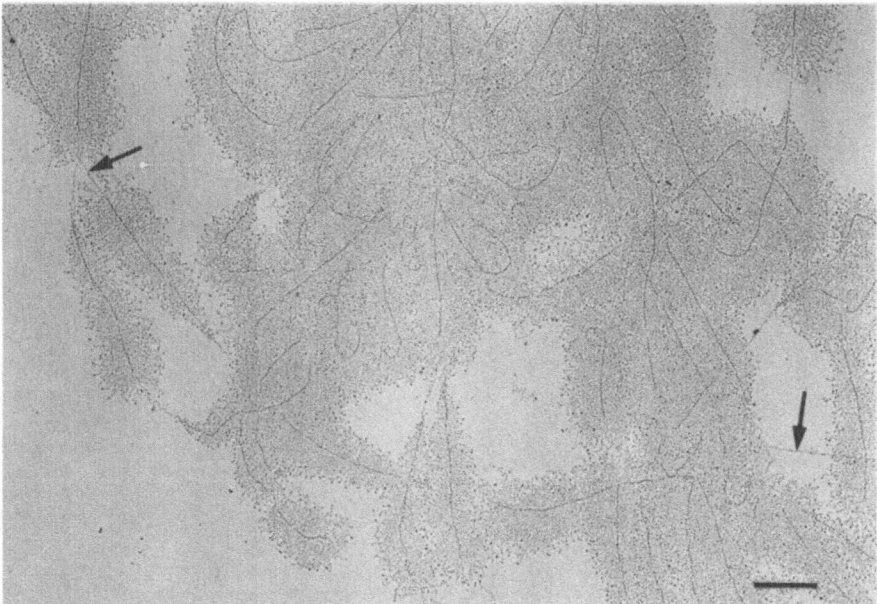

Fig. 3. Electron micrograph of an unwound nucleolar core from a stage 6 *X. laevis* oocyte. The majority of ribosomal genes are active and fully loaded with RNA polymerase. Note the beaded appearance of the spacer region *(arrows)* between the tandemly repeated genes. *Scale bar* represents 1 μm

suggest that the purpose of the high level of transcriptional activity maintained throughout oogenesis is not to accumulate cytoplasmic messenger RNA molecules. The functional significance to the oocyte of maintaining such a high degree of transcriptional activity with little cytoplasmic accumulation of RNA is unclear.

III. Chromatin Structure in Active Transcription Units

The visualization of dispersed nuclear contents in the electron microscope also allows observation of chromatin structure in the transcriptionally active regions of *X. laevis* oocytes. Scheer (1978) reports that nucleosomes are not observed in or adjacent to highly active transcription units of amphibian oocytes. He suggests that the removal of nucleosomes precedes the transcription of highly active units and that a progressive increase in nucleosomes occurs as activity is reduced. Figure 4a shows a group of active ribosomal genes. The spacers between the genes show the typical, beaded appearance of inactive chromatin. One of the genes has a segment lacking polymerases, presumably due to a temporary cutoff of gene activity. This region is shown magnified in Fig. 4b. Nucleosomes, which are distinguishable from RNA polymerase molecules by their smaller size and differential staining properties, can be seen in the gap between the polymerases. Ribosomal genes that are not fully loaded with RNA polymerase are rare in *Xenopus* oocytes, but when

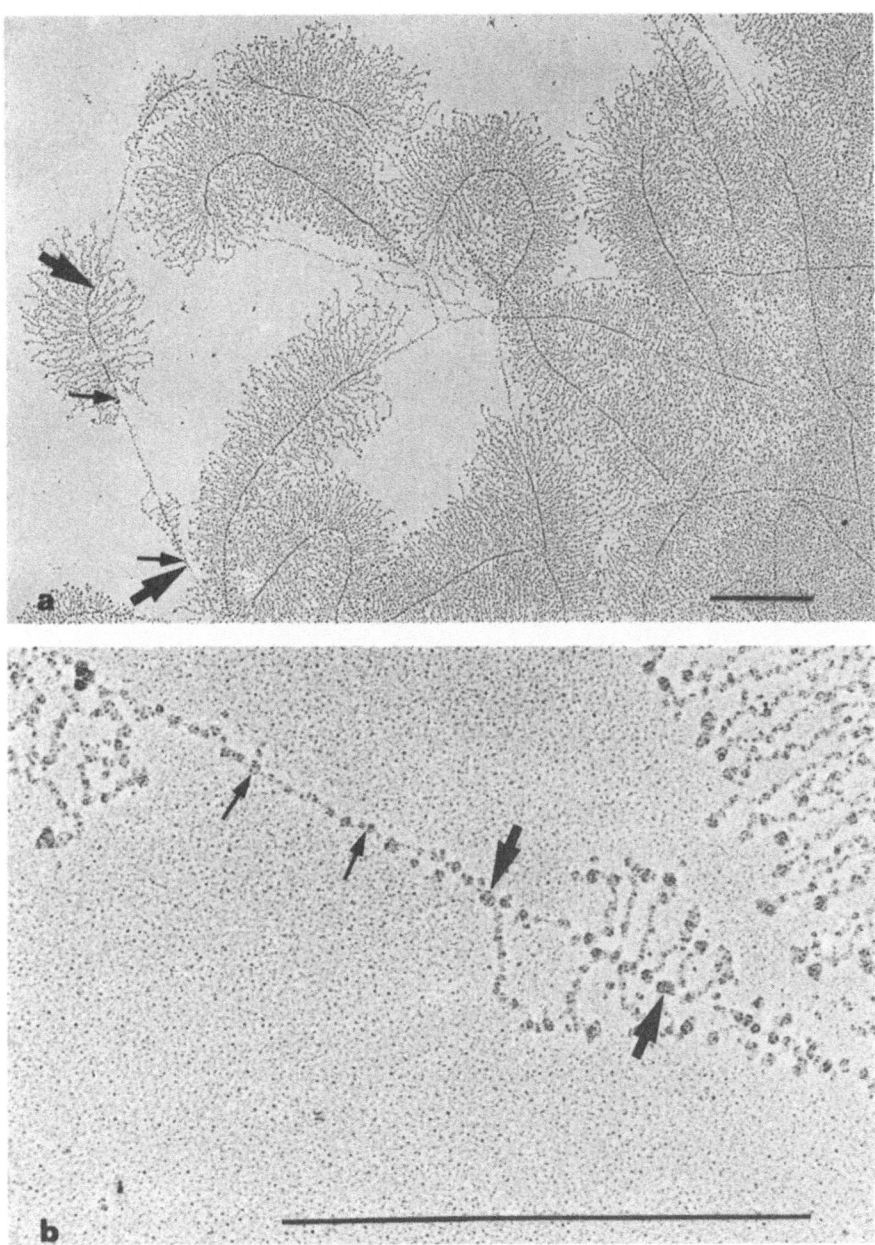

Fig. 4a,b. Chromatin structure in and adjacent to active ribosomal transcription units. In **a** a field of active ribosomal genes is displayed. The *large arrows* define one complete ribosomal gene which contains a gap between polymerases. The region denoted by *small arrows* is magnified in **b** to allow observation of the chromatin structure. Note the beaded appearance of the polymerase-free chromatin within the gene. The *small arrows* point to nucleosomes, the *large* to polymerases. *Scale bars* represent 1 μm

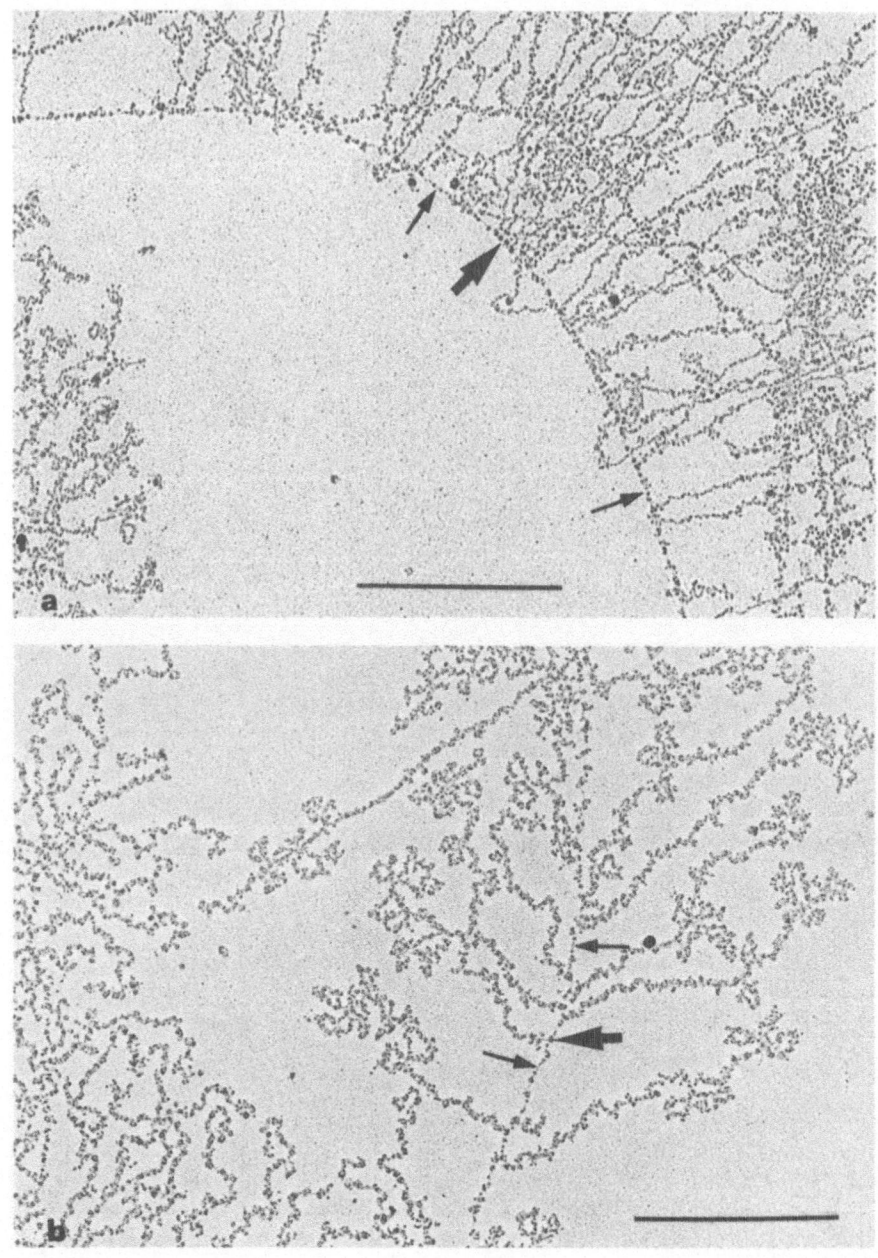

Fig. 5a,b. Electron micrographs of nonribosomal transcription units. **a** is a transcription unit from a stage 3 *X. laevis* oocyte and **b** from a stage 6. The chromatin structure within active regions can be observed. Note the beaded appearance of the chromatin between polymerases. The *small arrows* point to nucleosomes, the *large* to polymerases. *Scale bars* represent 1 μm observed they routinely exhibit nucleosomes between polymerases. Alternatively, other investigators have reported that nucleosomes are absent from ribosomal genes in electron microscope spreads (Foe et al. 1976; Franke et al. 1976; Scheer 1978). It is possible that the nucleosomes within ribosomal genes may have an altered morphology or less stringent association with the rDNA, making them more susceptible to removal by the low ionic conditions under which the preparations are made. Alternatively, other cell types may have a different ribosomal chromatin morphology than *Xenopus* oocytes.

The chromatin of active nonribosomal transcription units of *Drosophila* has been shown to contain histones and to exist in a beaded conformation (McKnight et al. 1977). Figure 5 shows portions of active nonribosomal transcription units from stage 3 and stage 6 *X. laevis* oocytes. Nucleosomes can be seen between polymerases in units that are not maximally active and thus contain regions where the chromatin structure is not obscured by polymerase. Nucleosomes are also observed in the regions adjacent to transcription units. Our observations indicate that all stretches of chromatin not obscured by polymerase are in a nucleosomal configuration, suggesting that high transcriptional activity does not preclude the presence of nucleosomes. Subtle changes in the nucleosome structure in regions of high transcriptional activity can not be detected using this method, but it appears clear that the beaded morphology used to define the presence of nucleosomes is as characteristic of active as it is of inactive chromatin. Furthermore, if nucleosomes are disrupted by RNA polymerase movement, our observations suggest that this disruption is short-lived and the nucleosome arrangement rapidly resumed.

IV. Summary

Phase contrast and electron micrographs of chromosome preparations from stage 3 and stage 6 *X. laevis* oocytes indicate the presence of highly active nonribosomal transcription units in both stages. Furthermore, ribosomal genes from both stages are generally fully loaded with RNA polymerase when observed in electron microscope preparations. These results show that the lampbrush stage of oogenesis presists through oocyte maturation and that the high level of nonribosomal RNA synthesis characteristic of lampbrush chromosomes also remains. Our electron microscope preparations reveal no qualitative difference in amount of transcription nor any alteration in size of transcription units from either stage, although the phase contrast results indicate the longer loops have retracted in stage 6. The chromatin segments adjacent to highly active ribosomal or nonribosomal transcription units exhibit a beaded appearance characteristic of nucleosomes. Nucleosomes are also observed within active transcription units in regions where polymerase molecules do not obscure the chromatin. These results suggest that no persistent gross alteration in chromatin morphology occurs because of transcriptional activity.

Acknowledgments. We thank Lorraine Blanks for excellent technical assistance. This work was supported by NIH Grant GM21020 to O.L.M., Jr.; K.M. is supported by USPHS Training Grant GM01450 and A.L.B. by the Jane Coffin Childs Memorial Fund for Medical Research.

References

Anderson, D.M., Smith, L.D.: Synthesis and accumulation of heterogeneous RNA in lampbrush stage oocytes of *Xenopus laevis* (Daudin). Dev. Biol. 69, 274-285 (1978)

Dumont, J.N.: Oogenesis in *Xenopus laevis*. J. Morphol. 136, 153-164 (1972)

Foe, V.E., Wilkinson, L.E., Laird, C.D.: Comparative organization of active transcription units in *Oncopeltus fasciatus*. Cell 9, 131-146 (1976)

Franke, W.W., Scheer, U., Trendelenburg, M.F., Spring, H., Zentgraf, H.: Absence of nucleosomes in transcriptionally active chromatin. Cytobiologie 13, 401-434 (1976)

Hamkalo, B.A., Miller, O.L., Jr.: Electron microscopy of genetic activity. Annu. Rev. Biochem. 42, 379-396 (1973)

McKnight, S.L., Miller, O.L., Jr.: Ultrastructural patterns of RNA synthesis during early embryogenesis of *Drosophila melanogaster*. Cell 8, 305-319 (1976)

McKnight, S.L., Bustin, M., Miller, O.L., Jr.: Electron microscopic analysis of chromosome metabolism in the *Drosophila melanogaster* embryo. Cold Spring Harbor Symp. Quant. Biol. 42, 741-754 (1977)

Müller, W.P.: The lampbrush chromosomes of *Xenopus laevis*. Chromosoma 47, 283-296 (1974)

Scheer, U.: Changes of nucleosome frequency in nucleolar and non-nucleolar chromatin as a function of transcription: an electron microscopic study. Cell 13, 535-549 (1978)

Differential Histone Phosphorylation During Drosophila Development

M. BLUMENFELD, P.C. BILLINGS, J.W. ORF, C.G. PAN
D.K. PALMER, and L.A. SNYDER

*Department of Genetics and Cell Biology, The University of Minnesota
St. Paul, MN 55108, USA*

I. Differential Replication of Satellite DNA

During the ontogeny of *Drosophila* larval salivary gland cells, nuclear DNA is endoreduplicated and organized into polytene chromosomes. All DNA sequences are not reduplicated to the same extent. Satellite DNA's, which are concentrated in the centromeric heterochromatin of diploid cells, are coalesced in the chromocenter of polytene cells (Gall et al. 1971) and selectively underreplicated (Dickson et al. 1971; Gall et al. 1971). In *D. virilis*, for instance, three satellites comprise about 45% of the diploid DNA (Gall et al. 1971; Blumenfeld and Forrest 1972) but less than 1% of polytene salivary gland cell DNA (Gall et al. 1971). Underreplication presumably involves specific interactions between chromosomal proteins and satellite DNA's, or nucleotide sequences lying near satellites. Proteins that can be correlated with satellites migth be involved with these interactions. With this in mind, consider that the underreplication of satellite DNA is correlated with the decreased phosphorylation of Hl histones (Blumenfeld et al. 1978).

II. Correlation Between Satellite DNA's and Phosphorylated H 1 Histones

Electrophoresis on Triton DF-16/urea/acetic acid gels resolves the Hl histones of *D. virilis* embryos into four major bands—Hlb, Hlc, Hld, and Hle (Fig. 1; Blumenfeld et al. 1978). The Hl histones of embryos and adult heads—two disparate populations of diploid cells—are electrophoretically similar (Billings et al. 1979). Hl histones extracted from four diverse stocks of *D. virilis* each contain the same four bands. Consequently, the bands do not represent an obvious polymorphism (Blumenfeld, unpublished). Alkaline phosphatase digestion reveals that Hle is the phosphorylated form of Hld and Hlc is the phosphorylated form of Hlb (Fig. 1; Blumenfeld et al. 1978). Hlc and Hle represent nearly 50% of the Hl histone in diploid cells. Polytene salivary gland nuclei contain Hlb and Hld, but little, if any, Hlc and Hle (Fig. 1). Thus, the underreplication of satellite DNA in polytene nuclei is paralleled by the decreased phosphorylation of Hl histones. It is not paralleled by the decreased phosphorylation of H2A histones. Approximately 60% of H2A is phosphorylated in nuclei of embryos, adult heads, and salivary glands (data not shown).

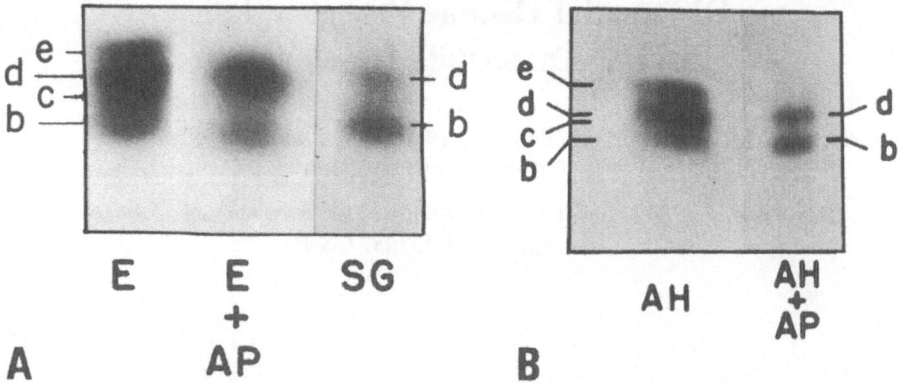

Fig. 1A,B. Electrophoresis of *D. virilis* H1 histones on Triton DF-16, urea, acetic acid gels (Blumenfeld et al. 1978). *E* embryos, *AH* adult heads, *SG* salivary glands, *AP* digested with alkaline phosphatase prior to electrophoresis

In mammals and slime molds, Hl histone phosphorylation is correlated with cell replication (Balhorn et al. 1972; Bradbury et al. 1973; Hohmann et al. 1976). The absence of phosphorylated Hl in nondividing *Drosophila* polytene cells would appear to be yet another such correlation. However, Hl is heavily phosphorylated in cells of the adult head which are diploid and predominantly nondividing (Billings et al. 1979). Therefore, decreased phosphorylation of Hl in polytene cells is correlated with the underreplication of satellite DNA, and not with decreased cell replication.

Hl and H2A are differentially phosphorylated and/or dephosphorylated during *Drosophila* development. Differential phosphorylation might involve site-specific histone kinases and/or histone phosphate phosphatases of the types described by Langan and co-workers (Langan 1969; Meisler and Langan 1969).

III. Phosphorylation and the Conformation of Drosophila H1 Histones

The Hl histones of *D. virilis* embryos or adult heads are resolved into two bands by electrophoresis on SDS gels. Alkaline phosphatase digestion of total Hl leads to the disappearance of the upper band, and the intensification of the lower band (Fig. 2). These results suggested that the alkaline phosphatase-sensitive subfractions Hlc and Hle could be equated with the slower-moving band in SDS gels. This suggestion was supported by two-dimensional electrophoresis and confirmed by ^{32}P-incorporation studies. Embryonic cells or larvae, grown on media containing ^{32}P, incorporate the isotope into the slower-moving band of Hl (Billings et al. 1979). Therefore, the phosphorylated and unphosphorylated Hl histones of *D. virilis* have different mobilities on SDS gels. In contrast, the phosphorylated and unphosphorylated forms of H2A are indistinguishable on SDS gels.

Control experiments rule out ADP-ribosylation and partial protease digestion as explanations for the unusual electrophoretic behavior of the phosphorylated Hl histones (Billings et al. 1979). We are aware of no other reported instance in which

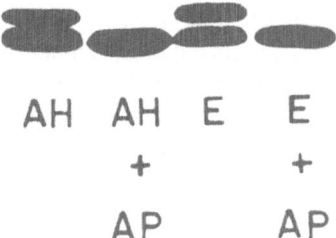

AH AH E   E
+   +
AP   AP

Fig. 2. Electrophoresis of *D. virilis* H1 histones on acrylamide-SDS gels according to Thomas and Kornberg (1975). Abbreviations are the same as in Fig. 1

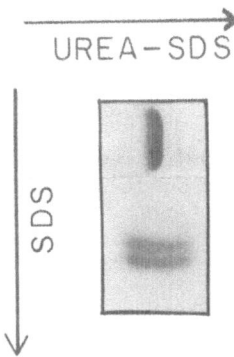

Fig. 3. Electrophoresis of *D. virilis* embryo H1 histones in two dimensions. The first dimension was urea-SDS; the second dimension was SDS alone

the mobility of a phosphoprotein in SDS gels increased after alkaline phosphatase digestion.

H1's migrate as one band when electrophoresed in urea-SDS gels (Fig. 3). Under these extreme denaturing conditions, both forms of H1 apparently have the same conformation. Electrophoresis of this band in a second dimension, in a gel containing only SDS, resolves the phosphorylated and unphosphorylated forms (Fig. 3). Therefore, urea reversibly obliterates an SDS-resistant difference in tertiary structure between the phosphorylated and unphosphorylated H1's. This demonstrates that the H1 subfractions differ in conformation rather than in net charge. As a developmental phenomenon, it is intriguing that the relative amounts of two alternative conformational states of a major class of chromosomal proteins are time- and tissue-specific. The balance between the two could be shifted by enzymatic phosphorylation or dephosphorylation.

IV. A Possible Function for Phosphorylated H1 Histomes in Drosophila

Histones H2A, H2B, H3 and H4 package eukaryotic DNA into a repeating subunit structure termed the nucleosome (reviewed by Kornberg 1977; Felsenfeld

1978). Histone Hl is not an integral part of the nucleosome. Rather, it is thought to lie between nucleosomes and be involved in the higher order folding of chromatin. The correlation between Hl phosphorylation and cell replication suggests that phosphorylation might be involved in mitotic chromatin condensation (Balhorn et al. 1972; Bradbury et al. 1973; Hohmann et al. 1976). The absence of Hl histone in mitotic micronuclei of Tetrahymena (Gorovsky and Keevert 1975) demonstrates that Hl phosphorylation does not always precede mitosis. In *Drosophila*, the converse occurs. High levels of Hl phosphorylation occur in the absence of mitotic correlation (Billings et al. 1979). This suggests that the phosphorylated Hl histones of *Drosophila* might have a function apart from mitotic condensation. *D. virilis* Hl histones preferentially bind satellite DNA's in vitro, whereas nucleosomal core histones do not (Blumenfeld et al. 1978). The phosphorylated Hl's are correlated with satellite DNA's and have unusual conformation in SDS. They have been proposed to compact satellite DNA's into heterochromatin (Blumenfeld et al. 1978). The conformational change in Hl, imposed by phosphorylation, could be related to the compaction of heterochromatin.

Acknowledgment. Supported by GM 22138-04 from the National Institutes of Health. J.W. Orf is a postdoctoral trainee supported by GM 01156.

References

Balhorn, R., Chalkley, R., Granner, D.: Lysine-rich histone phosphorylation: A positive correlation with cell replication. Biochemistry 11, 1094-1098 (1972)

Billings, P.C., Orf, J.W., Palmer, D.K., Talmage, D.A., Pan, C.G., Blumenfeld, M.: Anomalous electrophoretic mobility of *Drosophila* phosphorylated H1 histone: Is it related to the compaction of satellite DNA into heterochromatin? Nucleic Acids Research 6, 2151-2164 (1979)

Blumenfeld, M., Forrest, H.S.: Differential underreplication of satellite DNAs during *Drosophila* development. Nature New Biol. 239, 170-172 (1972)

Blumenfeld, M., Orf, J.W., Sina, B.J., Kreber, R.A., Callahan, M.A., Mullins, J.I., Snyder, L.A.: Correlation between phosphorylated H1 histones and satellite DNAs in *Drosophila virilis*. Proc. Natl. Acad. Sci. USA 75, 866–870 (1978)

Bradbury, E.M., Inglis, R.J., Matthews, H.R., Sarner, N.: Phosphorylation of very-lysine-rich histone in Physarum polycephalum. Correlation with chromatin condensation. Eur. J. Biochem. 33, 131-139 (1973)

Dickson, E., Boyd, J., Laird, C.D.: Sequence diversity of polytene chromosome DNA from *Drosophila hydei*. J. Mol. Biol. 61, 615-627 (1971)

Felsenfeld, G.: Chromatin. Nature 271, 115-122 (1978)

Gall, J.G., Cohen, E.H., Polan, M.L.: Repetitive DNA sequences in *Drosophila*. Chromosoma 33, 319-344 (1971)

Gorovsky, M.A., Keevert, J.B.: The absence of histone F1 in a mitotically dividing, genetically inactive nucleus. Proc. Natl. Acad. Sci. USA 72, 2672-2676 (1975)

Hohmann, P., Tobey, R.A., Gurley, L.R.: Phosphorylation of distinct regions of F1 histone. Relationship to the cell cycle. J. Biol. Chem. 251, 3685-3692 (1976)

Kornberg, R.D.: Structure of chromatin. Ann. Rev. Biochem. 46, 931-954 (1977)

Langan, T.A.: Action of adenosine 3',5'-monophosphate-dependent histone kinase in vivo. J. Biol. Chem. 244, 5763-5765 (1969)

Meisler, M.H., Langan, T.A.: Characterization of a phosphatase specific for phosphorylated histones and protamine. J. Biol. Chem. 244, 4961-4968 (1969)

Thomas, J.O., Kornberg, R.D.: An octamer of histones in chromatin and free in solution. Proc. Natl. Acad. Sci. USA 72, 2676-2680 (1975)

Nonhistone Proteins and Chromosome Structure[1]

D.E. COMINGS and T.A. OKADA

Department of Medical Genetics, City of Hope Medical Center
Duarte, CA 91010, USA

One of the major unanswered questions about chromosome structure concerns the manner in which the chromatin fiber is arranged to form the chromosome bands, chromomeres, and chromatid. It is now clear to all but the most recalcitrant holdout that chromosomes are uninemic structures containing a single DNA molecule that somehow makes its way from one telomere to the other. The advent of chromosome banding indicates that straightforward models of multiple folding of the chromatin fiber (DuPraw 1965; Prescott 1970; Comings 1972) are not adequate. The fact that the chromosome bands of mitotic chromosomes correlate precisely with the chromomeres of meiotic pachytene chromosomes (Okada and Comings 1974; Luciani et al. 1975) indicates that these condensations are a fundamental substructure of the chromosome. Examination of banding patterns in partially condensed prometaphase chromosomes (Yunis 1976) and prematurely condensed interphase chromosomes (Unakul et al. 1973) show that each mitotic chromosome band is composed of multiple smaller sub-bands with a total of 3000 per genome. The question is, how far can this subdivision of bands be carried? An intriguing answer is that it can be carried to the same end point that is already well known from the giant polytene chromosomes of *Drosophila* salivary glands. The chromomeres of such polytene chromosomes appear to be approximately equivalent to one gene. In *Drosophila melanogaster* there are approximately 5000 chromomeres.

Electron microscopy of *Drosophila* polytene chromosomes and egg chromatin suggests the chromomeres can be unfolded into loop-like structures with stretches of interloop DNA corresponding to interchromomeric chromatin (Sorsa 1972, 1976; Comings and Okada 1973). We have been intrigued by the possibility that mammalian and *Drosophila* chromosomes might be organized in basically the same manner and a number of years ago examined mammalian interphase nuclei by a number of spreading techniques, looking for such a loop-interloop arrangement. This was to no avail. No clear patterns of this type were seen (Comings and Okada 1973). McCready et al. (1977) recently examined yeast chromosomes by suspending them in 4 M ammonium acetate and spreading them on a surface of distilled water.

---

1 Supported by NIH Grant GM 15886

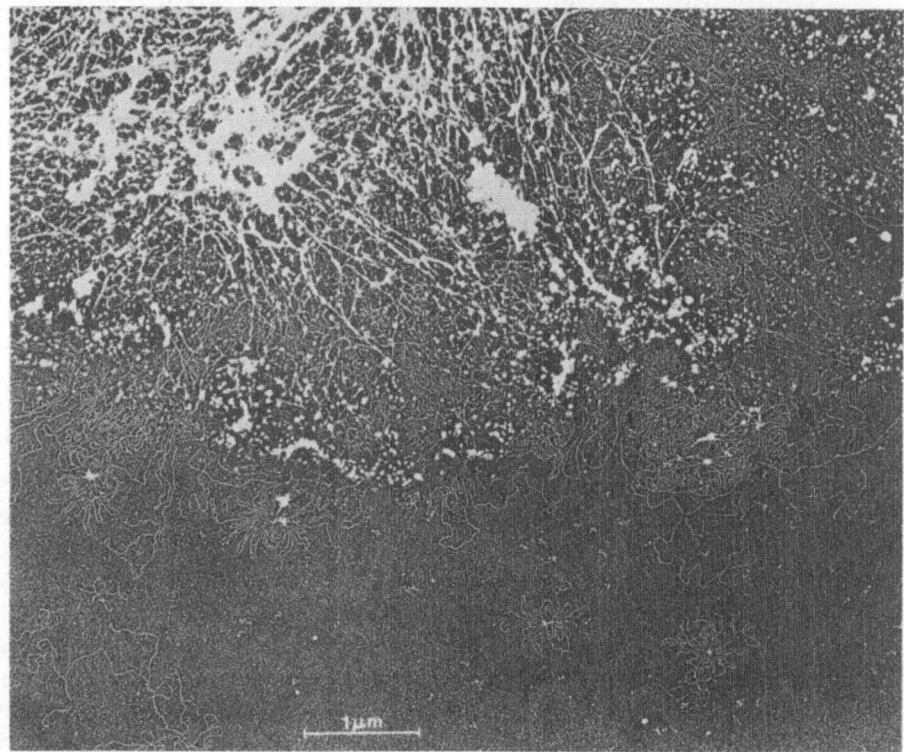

Fig. 1. Periphery of an isolated Chinese hamster metaphase chromosome suspended in 4 M ammonium acetate and then spread on the surface of distilled water. Note the numerous rosette configurations of DNA released from the chromosome

This resulted in an interesting pattern of supercoiled loops of DNA connected with interloop DNA. We felt it would be of interest to examine mammalian chromosomes in a similar manner.

The results of this work are reported in detail elsewhere (Okada and Comings 1979). Figures 1 and 2 serve to illustrate basically what was found. The DNA of chromosomes spread out around a more central mass of less disrupted chromatin similar to that reported by Paulson and Laemmli (1977) using dextran-heparin treatment to dehistonize the chromosomes. However, in contrast to their results we observed a striking tendency for the DNA to spill out into a rosette-interrosette pattern. The mean length of the rosette associated DNA was 13.7 μm $\pm$ 4.8 μm. The mean length of the interrosette DNA was 4.2 μm $\pm$ 1.7 μm. We fully realize that this might simply be an artifact of the spreading technique. The evidence for these being artifacts versus the evidence for this being a real substructure of the chromosome is given in the full report. We nevertheless were intrigued by the fact that the average amount of DNA in the rosettes of 14 μm is very similar to the average of the 10 μm per *Drosophila* chromomere (Edström 1965; Rasch et al. 1971). Assuming approximately three-fourths of the DNA in the rosette, there would be 54,000 such rosettes per haploid genome.

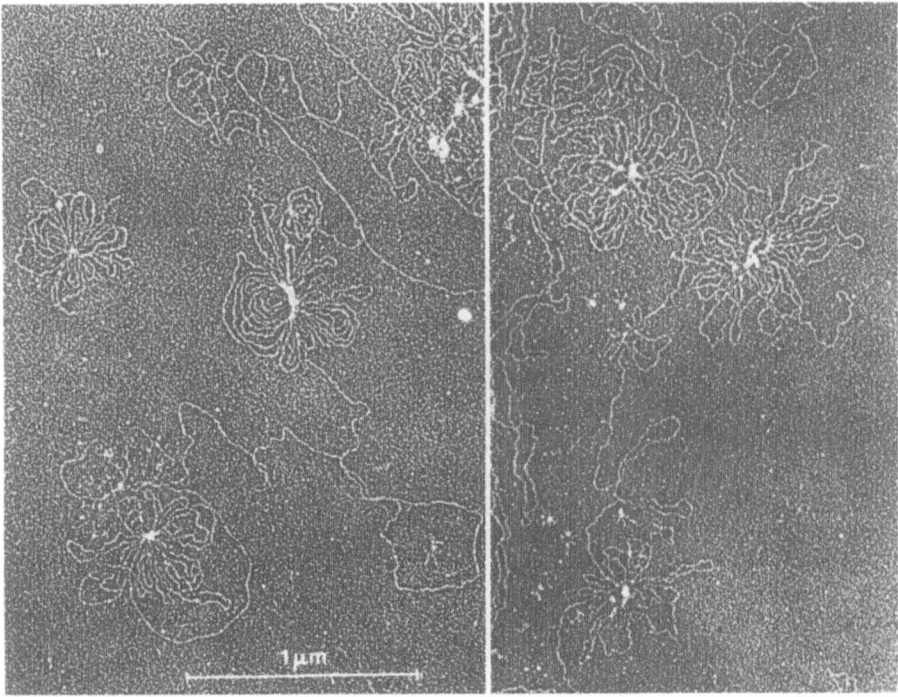

Fig. 2. Higher magnification with two sets of rosette-interrosette DNA released from chromosomes

Studies of chromosome banding indicate the bands are composed of relatively AT-rich DNA, while the interbands are composed of relatively GC-rich DNA (Comings 1978). Denaturation studies are underway to see if the rosette DNA is relatively AT-rich compared to interrosette DNA.

In the interphase nucleus, DNA is attached at many sites to the extensive nuclear matrix (Berezney and Coffey 1976; Comings and Okada 1976). Comparison by SDS acrylamide gel electrophoresis of the nonhistone proteins of isolated Chinese hamster chromosomes pelleted through 4 M ammonium acetate, with isolated Chinese hamster interphase nuclei pelleted through 4 M ammonium acetate, and with isolated Chinese hamster liver nuclear matrix, indicate the nuclear matrix proteins are also present in the chromosomes. We have recently completed studies of the DNA binding properties of nuclear matrix proteins which indicate they preferentially bind to AT-rich DNA (Comings and Wallack 1978). Such nuclear matrix proteins, preferentially binding to AT-rich sequences, may be responsible for the formation of these rosette-interrosette patterns and for chromomeres (Comings 1977).

Additional studies of meiotic and polytene chromosomes, using the same technique, are in progress to see if this pattern is a real substructural feature of chromosomes.

References

Berezney, R., Coffey, D.S.: The nuclear protein matrix: isolation, structure and functions. In: Advances in Enzyme Regulation, Vol. XIV (ed. G. Weber), pp. 63-100. Oxford-New York-Toronto: Pergamon Press 1976

Comings, D.E.: The structure and function of chromatin. In: Advances in Human Genetics, Vol. III (eds. H. Harris, K. Hirschhorn), pp. 237-431. New York: Plenum Press 1972

Comings, D.E.: Mammalian chromosome structure. In: Chromosomes Today, Vol. VI (eds. A. De La Chapelle, M. Sorsa), pp. 19-26. Amsterdam: Elsevier/North Holland 1977

Comings, D.E.: Mechanisms of chromosome banding and implication for chromosome structure. Ann. Rev. Genet. 12, 25-46 (1978)

Comings, D.E., Okada, T.A.: Some aspects of chromosome structure in eukaryotes. Cold Spring Harbor Symp. Quant. Biol. 38, 145-153 (1973)

Comings, D.E., Okada, T.A.: Nuclear proteins, III. The fibrillar nature of the nuclear matrix. Exp. Cell Res. 103, 341-360 (1976)

Comings, D.E., Wallack, A.S.: DNA-binding properties of nuclear matrix proteins. J. Cell Sci. 34, 233-246 (1978)

DuPraw, E.J.: Macromolecular organization of nuclei and chromosomes a folded fibre model based on whole-mount electron microscopy. Nature 206, 338-343 (1965)

Edström, J.E.: Chromosomal RNA and other nuclear fractions. In: The Role of Chromosomes in Development (ed. M. Locke), pp. 137-152. New York: Academic Press 1965

Luciani, J.M., Morazzani, M.R., Stahl, A.: Identification of pachytene bivalents in human male meiosis using G-banding techniques. Chromosoma 52, 275-282 (1975)

McCready, S.J., Cox, B.S., McLaughlin, C.S.: Superhelical DNA in yeast chromosomes. Exp. Cell Res. 108, 473-477 (1977)

Okada, T.A., Comings, D.E.: Mechanisms of chromosome banding III. Similarity between G-bands of mitotic chromosomes and chromomeres of meiotic chromosomes. Chromosoma 48, 65-71 (1974)

Okada, T.A., Comings, D.E.: Higher Order Structure of Chromosomes. Chromosoma 72, 1-14 (1979)

Paulson, J.R., Laemmli, U.K.: The structure of histone-depleted metaphase chromosomes. Cell 12, 817-828 (1977)

Prescott, D.M.: The structure and replication of eukaryotic chromosomes. In: Advances in Cell Biology, Vol. I (eds. D.M. Prescott, L. Goldstein, E. McConkey), pp. 57-117. New York: Appelton-Century Crofts 1970

Rasch, E.M., Barr, H.M., Rasch, R.W.: The DNA content of sperm of Drosophila melanogaster. Chromosoma 33, 1-18 (1971)

Sorsa, V.: Whole mount electron microscopy of core fibrils in salivary-gland chromosomes of Drosophila melanogaster. Hereditas 72, 169-172 (1972a)

Sorsa, V.: Beaded organization of chromatin in the salivary gland chromosome bands of Drosophila melanogaster. Hereditas 84, 213-220 (1972b)

Unakul, W., Johnson, R.T., Rao, P.N., Hsu, T.C.: Giemsa banding in prematurely condensed chromosome obtained by cell fusion. Nature New Biol. 242, 106-107 (1973)

Yunis, J.J.: High resolution of human chromosomes. Science 191, 1268-1270 (1976)

The Current Status of Cloning and Nuclear Reprograming in Amphibian Eggs

M.A. DiBerardino and N.J. Hoffner

Department of Anatomy, The Medical College of Pennsylvania
Philadelphia, PA 19129, USA

I. Introduction

The process of cloning on our planet dates back to the initial reproduction of unicellular plant and animal forms and also to the propagation of multicellular forms by budding. Nuclear transplantation is the name of one of several possible techniques that can give rise to a clone. The first successful transplantation of a living nucleus from one cell to another was achieved in 1934 by Hämmerling in the unicellular alga plant, *Acetabularia*. Five years later, Comandon and de Fonbrune (1939) accomplished this feat in the unicellular animal, *Amoeba*, and in 1952 Briggs and King reported the first successful cloning in a metazoan organism by transplanting embryonic nuclei into enucleated eggs of the frog, *Rana pipiens*.

Nuclear transplantation in the multicellular forms of Amphibia has been applied to a number of biological problems, but especially to the question of how the stability of the differentiated state of multicellular organisms is normally maintained. For almost a century embryologists and geneticists have attacked this problem through various approaches aimed at determining whether the process of cell specialization involves irreversible genetic changes or whether nuclei remain developmentally equivalent to the zygote nucleus. To this day, nuclear transplantation remains the most stringent, functional test of the genetic potentialities of living nuclei.

It is our purpose in this paper, first to examine the evidence in support of nuclear equivalence that has been derived by clonal analyses of amphibian nuclei by means of nuclear transplantation. We shall see that many young embryonic nuclei can be reprogramed to the extent that after transplantation into eggs, they can promote normal and complete development of the egg. However, very few nuclei from advanced cell types have the capacity to undergo complete nuclear reprograming. Second, we shall examine the reasons why they appear to be restricted. Finally, we shall analyze the events that do occur in nuclei from advanced cell types during their limited degree of nuclear reprograming.

II. Developmental Capacity of Nuclei

The nuclear transplantation procedure for amphibia involves activation of the egg, manual removal of the egg nucleus or destruction of the egg nucleus by UV irradiation and finally the transplantation of a test nucleus into the enucleated egg (see reviews by Briggs 1977; McKinnell 1978). The type of development that ensues is then a reflection of the introduced nucleus.

The first successful metazoan nuclear transplantation studies, conducted by Briggs and King in the leopard frog *R. pipiens*, revealed that many of the nuclei from undetermined regions of the blastula and early gastrula promote the test eggs to develop into normal larvae. The larvae when reared could metamorphose into juvenile frogs (for review see Briggs 1977), and develop into sexually mature adult frogs (McKinnell 1962). These results were confirmed in other anuran and also urodelen species (see reviews by Gallien 1966; King 1966; DiBerardino and Hoffner 1970; Gurdon 1974; Briggs 1977; McKinnell 1978), and established unequivocally the totipotency of many nuclei from undetermined regions of the early embryo. Once this answer was obtained, studies were then conducted on nuclei from determined regions of advanced embryos. Nuclear transplanatation tests from various embryonic cell types derived from *Rana, Xenopus, Pleurodeles, Ambystoma,* and *Bufo* by different investigators revealed that there is a progressive decrease in the number of individuals that develop normally, as donor cells are tested from progressively older stages of embryogenesis (see reviews ibid). These results demonstrated that most embryonic nuclei become developmentally restricted during the process of cell type determination.

The main interest in this area today centers around the developmental capacity of nuclei from larvae and adult frogs because their tissues consist of a large proportion of specialized cells, and therefore, provide a critical test for the theory of nuclear equivalence. In what follows we shall examine the cases in support of pluripotent and totipotent nuclei from larval and adult cells and conclude that a few nuclei from specialized cells have been shown to be pluripotent, but, so far, no nucleus from a specialized cell has been demonstrated unequivocally to be totipotent.

A. Larval Cells (Table 1)

Three original nuclei derived from cells of young *Xenopus* larvae have led to the formation of three fertile adult frogs. One female and one male originated from nuclei of cells of the intestine (Gurdon 1962; Gurdon and Uehlinger 1966) and the third frog, a female, derived from the nucleus of a nonciliated epidermal cell (Kobel et al. 1973). These three frogs are the only reported cases of totipotent larval nuclei and represent 0.2–0.3% of the total number of nuclei tested.

Of particular importance is the fact that a small percentage of larval nuclei from different cell types are pluripotent, that is, they contain the genes necessary to promote the injected eggs to undergo embryogenesis and develop into either apparent normal or abnormal larvae. These nuclear transplant larvae contain the various cell types, tissues and organ systems that are present in normal control larvae and they display apparent functional competence in some organ systems such as the cardiovascular, nervous, muscular, and occasionally the digestive system. The percentage of apparent normal larvae ranges from 0.1% to 1.9% of the original somatic nuclei tested, and 0.2–2.3% for abnormal larvae. A few additional cases in support of limited nuclear pluripotency come from nuclear transplants that arrest during post-neurula stages. These individuals have attained early stages of organogenesis.

Table 1. Advanced development of nuclear transplants obtained from larval cells

Source of cells	Total nuclei tested[a]	Post-neurula embryos	Larvae		Adults		Reference
			Abnormal	Normal	Sterile	Fertile	
		(%)	(%)	(%)	(%)	(%)	
Intestine (X)	726	36 (5.0)	17 (2.3)	14 (1.9)	2 (0.3)	2 (0.3)	Gurdon (1962), Gurdon and Uehlinger (1966)
Primordial germ (R)	410	35 (8.5)		31 (7.6)[b]			Smith (1965)
Cell cultures (X)	3686	23 (0.6)[c]	6 (0.2)[c]	3M (0.1)			Gurdon and Laskey (1970)
Ciliated epidermis (X)	174						Kobel et al. (1973)
Non-ciliated epidermis (X)	440	2 (0.4)	1 (0.2)	1 (0.2)		1 (0.2)	Kobel et al. (1973)
Melanophores (X)	200						Kobel et al. (1973)
Cultured melanophores (X)	257	2 (0.8)	2 (0.8)				Kobel et al. (1973)
Intestine (X)							
stgs. 46–48	210	4 (1.9)	3 (1.4)	1 (0.5)[d]			Marshall and Dixon (1977)
stg. 54	110	3 (2.7)	1 (0.9)	2 (1.8)[e]			Marshall and Dixon (1977)
stg. 57	202	1 (0.5)		1 (0.5)[d]			Marshall and Dixon (1977)

a Total number of nuclei tested includes results from serial transplantations
b Later development not reported
c Data have been estimated from graphs
d Died during early larval stages
e One died during early larval stage; the other metamorphosed
X = Xenopus laevis; R = Rana pipiens; M = Metamorphosed

Table 2. Advanced development of nuclear transplants obtained from renal carcinoma of *Rana pipiens*

Donor	Source of cells	Total nuclei tested[a]	Post-neurula embryos	Abnormal larvae	Reference
			(%)	(%)	
Meta-morphosing tadpoles	Kidney carcinoma multiple	324		21 (6.5)	McKinnell et al. (1969)
Adults	Kidney carcinoma multiple	142	1 (0.7)		King and McKinnell (1960)
	single	263	3 (1.1)	3 (1.1)	King and DiBerardino (1965)

[a] Total number of nuclei tested included results from serial transplantations

All of the above cases are derived from tissues that contain cells capable of mitosis in vivo, including stem cells, or from in vitro cell cultures. In contrast to these donor cells, the two cell types that exhibit unequivocal differentiated properties at the time of transplantation are the ciliated epidermal cells and melanophores (Kobel et al. 1973). None of these nuclei promoted development to stages of organogenesis. However, nonciliated epidermal cell nuclei led to the formation of two larvae, one of which developed into a fertile frog, and nuclei of cultured melanophores promoted the development of two abnormal larvae.

Thus far, we have been concerned with the developmental capacity of larval *somatic* nuclei. Tests conducted by Smith (1965) demonstrated that 7.6% of primordial *germ* cell nuclei from young feeding *R. pipiens* larvae promote injected eggs to develop into normal larvae. This yield is 4–76 × greater than that obtained with somatic cell nuclei and demonstrates the greater developmental potential of primordial germ cell nuclei.

B. Renal Carcinoma (Table 2)

When amphibian nuclear transplantation proved to be successful for normal somatic cells, it opened the way to test the developmental capacity of nuclei from malignant neoplasms (King and McKinnell 1960). Fortunately, the renal carcinoma of *R. pipiens* proved to be testable for a number of reasons, including the fact that primary, transplanted tumors and short term cultures of tumors were mainly diploid with very few deviations from the normal karyotype (DiBerardino et al. 1963; DiBerardino and King 1965; DiBerardino and Hoffner 1969).

Both single and multiple nuclei of the tumors have been injected into enucleated eggs. In the case of *single* nuclear transfers, three nuclear transplants (1.1%) attained the feeding larval stage; however, they failed to feed and were morphologically abnormal in all organ systems (DiBerardino and King 1965; King and DiBerardino 1965). Despite the fact that the best cases arrested during early larval stages, these studies demonstrated that *single* nuclei of spontaneous renal carcinoma contain sufficient genetic information to direct the formation and function of cell types, tissues and organ systems found in normal larvae. Their inability to proceed further was due to chromosomal aberrations arising after transplantation into enucleated eggs (DiBerardino and King 1965; see Sect.III). Assurance that the tumor larval nuclear transplants were derived from the injected nucleus and not the egg nucleus was obtained by removal and preservation of the vitelline membranes from nuclear transplants during early embryonic stages. Fixed and stained preparations of vitelline membranes invariably contained the encucleated egg nucleus (King and DiBerardino 1965).

Additional larval nuclear transplants were obtained by injecting *multiple* nuclei from induced renal carcinoma of metamorphosing tadpoles (McKinnell et al. 1969). In these studies 21 (6.5%) of the injected eggs attained the abnormal feeding larval stage but did not feed. Seven of the injected nuclei contained a ploidy nuclear marker, i.e., they were triploids. In addition, 98.5% of cells of the tumor dissociated in calcium free salt solution with or without EDTA have been shown by fluorescent microscopy to be epithelial cells of the tumor and not stromal cells of the connective tissue (McKinnell et al. 1976). Cloning studies of renal carcinoma have shown that the nuclei of this malinant cancer can be partially reversed in the environment of the egg cytoplasm and contain sufficient genetic information to program an enucleated egg to early tadpole development.

C. Normal Adult Cells (Table 3)

Advanced development of nuclear transplants derived from adult nuclei is summarized in Table 3. The most advanced nuclear transplant reported from adult nuclei has been obtained from spermatogonial cells (DiBerardino and Hoffner 1971). Its developmental history was normal until early larval stages. At this time its rate of development was slightly retarded in comparison to the fertilized controls. It fed, produced feces, but subsequently died on the 20th day. That this larva developed from the transplanted nucleus and not the egg nucleus of the host eggs was confirmed microscopically. The vitelline membrane was removed from the nuclear transplant during early embryogenesis. Adhering to the vitelline membrane was the exovate formed at the time when the egg was enucleated. Microscopic sections of the exovate stained with the Feulgen procedure revealed the presence of the egg nucleus.

Among somatic adult nuclei listed in Table 3 there has been one claim that a normal larva resulted and this one originated from crest cell nuclei of the intestine, but this apparent normal larva died during an early larval stage (McAvoy et al. 1975). There are, however, a number of cases in support of pluripotency of adult nuclei in which abnormal nuclear transplant larvae resulted. These cases have been obtained from in vitro cell cultures and also directly from cells in vivo and the percentage of abnormal larvae obtained ranges from 0.3–6%. Some additional cases of limited pluripotency were obtained from nuclear transplants that arrested during post-neurula stages.

Table 3. Advanced development of nuclear transplants obtained from normal adult cells

Source of cells	Total nuclei tested[a]	Post-neurula embryos	Larvae		Reference
			Abnormal	Normal	
		(%)	(%)	(%)	
Cell cultures (X) skin, lung, kidney	2322	26 (1.1)[b]	7 (0.3)[b]		Laskey and Gurdon (1970)
Male germ (R)	116	4 (3.5)	1 (0.9)		DiBerardino and Hoffner (1971)
A-8 cell line (X)	365	2 (0.6)	2 (0.6)		Kobel et al. (1973)
Intestine (X)					
trough	564	5 (0.9)	3 (0.5)		McAvoy et al. (1975)
crest	548	5 (0.9)	2 (0.4)	1 (0.2)[c]	McAvoy et al. (1975)
Cell culture (X)					
skin	129	6 (4.6)	4 (3.1)		Gurdon et al. (1975)
Spleen lymphocytes (X)	100	6 (6.0)	6 (6.0)		Wabl et al. (1975)
Erythrocytes (X)					
broken	440				Brun (1978)
Erythroblasts (X)					
broken	254		1 (0.4)		Brun (1978)
unbroken	188		7 (3.7)		Brun (1978)

[a] Total number of nuclei tested includes results from serial transplantations
[b] Data have been estimated from graphs
[c] Died during an early larval stage
X = *Xenopus laevis*; R = *Rana pipiens*

There has been one claim that is an exception to the data presented in Table 3. Serial nuclear transfer tests were conducted on nuclei derived from organ cultured adult lens cells, and 6% (1.4 individuals) of the first serial transfer complete blastulae (24) went on to metamorphosis (Muggleton-Harris and Pezzela 1972). However, there is no documentation that the nucleus is derived from a specialized lens cell, nor is there evidence presented that the transplanted nucleus and not the egg nucleus is responsible for the development. Since this cell type could serve as an adequate test for a specialized cell type, it would be desirable to extend these studies to include documentation of the differentiated state of the donor cell, as well as presentation of data describing early cleavage histories and enucleation efficiency, or utilization of a nuclear marker in the donor cell nucleus.

The studies involving nuclei of adult cultured skin cells, lymphocytes, and erythroblasts provide the most critical tests of nuclei from normal somatic specialized cells. In the case of the adult cultured skin cells, over 99.9% of the donor cells displayed immunoreactive keratin. Nuclear transfers from these cells yielded four heartbeat larvae (3.1%) and bore the nucleolar marker (1 nu) of the transplanted nucleus (Gurdon et al. 1975). When adult lymphocytes to be used for donor cells were tested, 96.1–98.7% of these cells were shown to be immunoglobulin-bearing cells and six (6%) abnormal larvae resulted. All of these larvae were reported to contain the 1 nu nucleolar marker and were diploid (Wabl et al. 1975). More recently, Brun (1978) has compared the developmental capacity of nuclei from erythrocytes and erythroblasts taken from adult *Xenopus*. Both donor cell types contain hemoglobin. The erythroblast cells are in the proliferative stage and their nuclei promoted the development of eight tadpoles (4.1%), seven of which where diploid and contained a single nucleolus, the marker of the donor nucleus. Erythrocytes in the stationary phase never promoted development beyond the early gastrula stage.

All of the nuclear transplant larvae reported so far from adult nuclei regardless of cell type do not survive, and die during the early stages of larval development. The main points to be emphasized in these studies are the following. There has now accumulated a number of cases in support of pluripotency of adult nuclei. However, the validity of the accurate number of cases is influenced by three important factors. First, it has not been demonstrated in all studies that 100% of the donor cells are specialized cells. Although the error is probably small, an error does exist. Second, it has not been proven in all cases that development is due to the transplanted nucleus and not to gynogenesis resulting from retention of the egg nucleus. Third, we have no current means of evaluating the contribution of maternal templates to the development of the nuclear transplant; therefore, we do not know the relative contribution made by the transplanted nucleus and the maternal templates to the formation of these larvae. It is clear, however, that some contribution is made by the transplanted nucleus to the formation of these larvae, because amphibian eggs lacking a functional nucleus form only partial blastulae (Briggs et al. 1951). In summary, no nucleus of an adult cell whether a specialized cell or a stem cell has promoted an injected egg to develop into an adult frog.

III. Chromosomal Restrictions

Why do nuclear transplants from advanced cell types display developmental restrictions and develop in an abnormal fashion? Some insight into this problem has come from the chromosomal analyses of nuclear transplants (for a recent review see DiBerardino 1979). These studies have shown: (1) that most abnormal nuclear transplants examined contain chromosome abnormalities; (2) that these chromosome aberrations occur in nuclear transplants derived from various cell types and various amphibian species; (3) that there is a progressive increase in the percentage of nuclear transplants possessing chromosomal aberrations when nuclear donors are derived from progressively older stages of development; and (4) that the extent of development of nuclear transplants is correlated with the extent of chromosomal abnormalities; those that contain extensive aneuploidy and severe structural alterations arrest usually by gastrulation, whereas those with lesser aberrations can proceed through organogenesis; for example, larvae usually are euploid but have structural alterations such as deletions and translocations.

Nuclear transplants examined during the earliest stages of cleavage have revealed that chromosome abnormalities can arise in the majority of cases during the first cell cycle of the egg (DiBerardino and Hoffner 1970). In this study, 59% of late gastrula endodermal nuclear transplants displayed gross chromosome aberrations during these early cleavage stages compared to only 10% of nuclear transplants derived from blastula nuclei. These aberrations are highly variable and extensive and are now known to be the cause of developmental arrest and not the consequence of developmental arrest.

These results can be interpreted in one of two ways: (1) the developmental and chromosomal abnormalities are caused by technical damage inflicted on the nucleus during the transfer procedure, or (2) these changes are a reflection of the state of nuclear differentiation of the transplanted nuclei. The evidence in support of nuclear differentiation is derived from various lines of work, but the most convincing evidence against nuclear damage is the existence of karyotypic mosaicism in individual nuclear transplants. We consistently find individual nuclear transplants with an admixture of apparent normal and abnormal karyotypes and we have found this type of mosaicism as early as the 4-cell stage (DiBerardino and Hoffner 1970). The events leading to karyological mosaicism are still unknown, but conceivably could arise through damage to one DNA strand at the time of transplantation and express itself then or later in a delayed manner. In either case this would lead to the formation of one normal and one defective daughter nucleus. However, it is unlikely that technical damage would consistently result in a one-strand hit. It seems more likely that since the transplanted nucleus is out of phase with the mitotic events of the egg (DiBerardino and Hoffner 1970; DiBerardino 1979), it produces during its first DNA synthesis an abnormal replicate, but its original template remains normal. In subsequent cleavage stages, the normal DNA strands may become attuned to the cell cycle program of some regions of the egg and produce normal replicates. In fact, the explanation for the improvement of retransfers over original transfers (Gurdon et al. 1975) must be concerned with chance selection of nuclei with the normal or near normal karyotype from the mosaic nuclear transplant donor. Thus, chromosome abnormalities appear to arise through a failure of normal replication due to an incompatibility between the cell cycle programs of the transplanted nucleus and the host egg.

IV. Nuclear Reprograming

What are the controls existent in chromosomes of advanced embryonic and adult cells that prevent the complete expression of their genetic constitution? The obvious candidates for these controls are the proteins associated with the chromosomes. It was within this framework of the analysis of the problem that we began to examine the process of nuclear reprograming in the mature egg with respect to the nucleo-cytoplasmic exchange of nonhistone proteins.

Although most nuclei from advanced cell types do not promote normal development of the test eggs, they do in most cases undergo some degree of nuclear reprograming when exposed to the cytoplasm of eggs. For example, Subtelny and Bradt (1963) first showed that the transplanted nucleus swells and its compact chromatin decondenses. Subsequently, Graham et al. (1966) demonstrated that, concomitantly with these changes, the nuclei incorporate ^{3}H-thymidine, and later Merriam (1969) showed that during this time there was an influx of egg cytoplasmic proteins into the nuclei.

Recently, we have studied during the process of nuclear reprograming the behavior of nonhistone proteins, labeled with ^{3}H-tryptophan. Late gastrula endodermal nuclei of *Rana* were labeled with ^{3}H-tryptophan in vivo and then transplanted into enucleated eggs. Nuclear transplants were fixed during the first cell cycle of the egg prior to the dissolution of the nuclear membrane. Quantitative analysis of autoradiograms revealed that most of the label is lost from the transplanted nuclei. If, however, donor nuclei are prelabeled with ^{3}H-lysine, most of the label remains in the nucleus after transplantation into eggs (DiBerardino and Hoffner 1975; DiBerardino et al. 1977). These studies demonstrate a loss of nonhistone proteins by transplanted nuclei. Whether this loss occurs through migration of intact proteins from the nucleus to the cytoplasm or through breakdown of proteins in the nucleus and then subsequent migration of the subunits to the cytoplasm is not known.

To answer the question whether transplanted nuclei acquire nonhistone proteins from the egg cytoplasm during the process of nuclear reprograming, late gastrula endodermal nuclei were first transplanted into eggs. Then ^{3}H-tryptophan was injected into the egg cytoplasm. Autoradiograms revealed that a substantial concentration of labeled nonhistone proteins is present in the nuclei within 2 h after transplantation. A similar acquisition of nonhistone proteins by pronuclei of fertilized eggs also occurs (Hoffner and DiBerardino 1977).

The nature of these proteins at this time is unknown. Although no causal relationship has been established between the bidirectional movement of these nonhistone proteins and nuclear reprograming, one could speculate that those egressing from transplanted nuclei include nonhistone proteins required for endodermal determination, whereas those migrating into the nuclei could include enzymes for unwinding chromatin, DNA polymerases and perhaps initiating protein(s) for DNA synthesis.

Ultimately, the process of complete nuclear reprograming of nuclei must involve the reprograming of the chromatin to the extent that both nonhistone and histone proteins of the donor nuclei are exchanged for early embryonic chromatin proteins. This might be more easily accomplished for nonhistone proteins, since they turn over relatively frequently compared to chromosomal histones which appear to be highly conserved. However, some of the progeny of transplanted nuclei might acquire the proper stage-specific histones in their newly formed chromosomes during DNA replication in early cleavage stages. For example, during normal embryogenesis in the sea urchin, different sub-classes of histones are synthesized at different stages of development and are conserved (Cohen et al. 1975). If these histones become integral parts of the chromosomes, and presumably they do, these histones together with the acquisition of stage-specific nonhistone proteins by replicating chromosomes could lead to the replication in some cells of an apparent normal karyotype and account for the karyotypic mosaicism which we consistently observe in nuclear transplants. Although the egg cytoplasmic proteins which enter transplanted nuclei during the first cell cycle must be involved in a major way in nuclear reprograming (Merriam 1969; Hoffner and DiBerardino 1977), it seems more likely that nuclei from advanced cell types remain restricted not because the proper egg proteins are not available, but because the chromosomal proteins of the donor nuclei have not completely egressed from the DNA.

V. Conclusions

A review of the studies on the developmental capacity of amphibian nuclei by means of nuclear transplantation has led to the following conclusions.

1. Nuclear transplantation tests of embryonic nuclei from cells of undetermined regions of young embryos have shown that many of these nuclei are capable of promoting complete and normal development of the eggs and are therefore totipotent, whereas nuclei from advanced embryonic and larval stages display a progressive decrease in developmental capacity.

2. Tests of adult somatic and germ cell nuclei demonstrate that a small percentage of nuclear transplants can undergo embryogenesis and attain the initial stages of larval development, but do not survive thereafter. Thus, cloning studies in amphibian eggs have provided some cases in support of genetic pluripotency but not totipotency for adult nuclei.

3. A few nuclei of the renal carcinoma and of some normal specialized cells have been shown to be pluripotent. These studies indicate that nuclei engaged in activities of this cancer and these normal specialized cells can be partially reversed to the extent that their genes can program for the cell types, tissues, and organ systems required to form young tadpoles.

4. Most of the nuclear transplants derived from advanced embryonic cells, larval cells, adult somatic and germ cells, as well as the renal carcinoma arrest during early embryonic stages. Chromosomal studies of nuclear transplants indicate that the developmental abnormalities are caused by chromosomal aberrations which in most cases arise during the first cell cycle of the egg and are the cause of developmental arrest. The formation of abnormal karyotypes is considered a reflection of the state of nuclear differentiation of the transplanted nucleus which fails to undergo complete nuclear and chromatin reprograming.

5. At the present time information is not yet available to answer the question whether the stable genetic changes occurring during the normal development of an organism are reversible or irreversible. However, since genetically totipotent nuclei of adult male germ cells fail to promote development beyond the early tadpole stage, it seems most probable that the exact requirements for normal replication of these nuclei have not been satisfied in the egg.

6. Soon after transplantation into egg cytoplasm nuclei exhibit a number of morphological and molecular changes, namely nuclear swelling, chromatin decondensation, DNA synthesis, and a bidirectional movement of nonhistone proteins between the nucleus and the cytoplasm. All of these events occur during the first cell cycle of the egg when transplanted nuclei undergo nuclear reprograming. Although no causal relationship has yet been established between these events and nuclear reprograming, we suggest that an understanding of the process of nuclear reprograming might contribute to the question of whether the stable nuclear changes are reversible or irreversible.

Acknowledgment. This work was supported by research grants from the National Science Foundation.

References

Briggs, R.: Genetics of cell type determination. In: Cell Interactions in Differentiation (eds. L. Saxen, L. Weiss), pp. 23-43. New York-London: Academic Press 1977

Briggs, R., King, T.J.: Transplantation of living nuclei from blastula cells into enucleated frogs' eggs. Proc. Natl. Acad. Sci. USA 38, 455-463 (1952)

Briggs, R., Green, E.U., King, T.J.: An investigation of the capacity for cleavage and differentiation in *Rana pipiens* eggs lacking "functional" chromosomes. J. Exp. Zool. 116, 455-500 (1951)

Brun, R.B.: Developmental capacities of *Xenopus* eggs, provided with erythrocyte or erythroblast nuclei from adults. Dev. Biol. 65, 271-284 (1978)

Cohen, L.H., Newrock, K.M., Zweidler, A.: Stage-specific switches in histone synthesis during embryogenesis of the sea urchin. Science 190, 994-997 (1975)

Comandon, J., de Fonbrune, P.: Greffe nucléaire totale, simple ou multiple, chez une Amibe. C.R. Séances Soc. Biol. 130, 744-748 (1939)

DiBerardino, M.A.: Nuclear and chromosomal behavior in amphibian nuclear transplants. In: International Review of Cytology, Suppl. 9 (eds. J.F. Danielli, M.A. DiBerardino), pp. 129-160. New York-London: Academic Press 1979

DiBerardino, M.A., Hoffner, N.: Chromosome studies of primary renal carcinoma from Vermont *Rana pipiens*. In: Recent Results in Cancer Research, Special Suppl.: Biology of Amphibian Tumors (ed. M. Mizell), pp. 261-278. Berlin-Heidelberg-New York: Springer 1969

DiBerardino, M.A., Hoffner, N.: Origin of chromosomal abnormalities in nuclear transplants.—A reevaluation of nuclear differentiation and nuclear equivalence in amphibians. Dev. Biol. 23, 185-209 (1970)

DiBerardino, M.A., Hoffner, N.: Development and chromosomal constitution of nuclear-transplants derived from male germ cells. J. Exp. Zool. 176, 61-72 (1971)

DiBerardino, M.A., Hoffner, N.J.: Nucleo-cytoplasmic exchange of nonhistone proteins in amphibian embryos. Exp. Cell Res. 94, 235-252 (1975)

DiBerardino, M.A., King, T.J.: Transplantation of nuclei from the frog renal adenocarcinoma. II. Chromosomal and histologic analysis of tumor nuclear-transplant embryos. Dev. Biol. 11, 217-242 (1965)

DiBerardino, M.A., King, T.J., McKinnell, R.G.: Chromosome studies of a frog renal adenocarcinoma line carried by serial intraocular transplantation. J. Natl. Cancer Inst. 31, 769-789 (1963)

DiBerardino, M.A., Hoffner, N.J., Matilsky, M.B.: Methods for studying nucleocytoplasmic exchange of nonhistone proteins in embryos. In: Methods in Cell Biology, Vol. XVI (eds. G. Stein, J. Stein, L.J., Kleinsmith), pp. 141-165. New York-London: Academic Press 1977

Gallien, L.: La greffe nucléaire chez les amphibiens. Ann. Biol. 5-6, 241-269 (1966)

Graham, C.F., Arms, K., Gurdon, J.B.: The induction of DNA synthesis by frog egg cytoplasm. Dev. Biol. 14, 349-381 (1966)

Gurdon, J.B.: The developmental capacity of nuclei taken from intestinal epithelium cells of feeding tadpoles. J. Embryol. Exp. Morphol. 10, 622-640 (1962)

Gurdon, J.B.: The genome in specialized cells, as revealed by nuclear transplantation in amphibia. In: The Cell Nucleus, Vol. I (ed. H. Busch), pp. 471-489. New York-London: Academic Press 1974

Gurdon, J.B., Laskey, R.A.: The transplantation of nuclei from single cultured cells into enucleate frogs' eggs. J. Embryol. Exp. Morphol. 24, 227-248 (1970)

Gurdon, J.B., Uehlinger, V.: "Fertile" intestine nuclei. Nature (London) 210, 1240-1241 (1966)

Gurdon, J.B., Laskey, R.A., Reeves, O.R.: The developmental capacity of nuclei transplanted from keratinized cells of adult frogs. J. Embryol. Exp. Morphol. 34, 93-112 (1975)

Hämmerling, J.: Über Genomwirkungen und Formbildungsfähigkeit bei *Acetabularia*. Wilhelm Roux' Archiv Entwicklungsmech. Org. 132, 424-462 (1934)

Hoffner, N.J., Di Berardino, M.A.: The acquisition of egg cytoplasmic nonhistone proteins by nuclei during nuclear reprograming. Exp. Cell Res. 108, 421-427 (1977)

King, T.J.: Nuclear transplantation in amphibia. In: Methods in Cell Physiology, Vol. II (ed. D.M. Prescott), pp. 1-36. New York-London: Academic Press 1966

King, T.J., DiBerardino, M.A.: Transplantation of nuclei from the frog renal adenocarcinoma. I. Development of tumor nuclear-transplant embryos. Ann. N.Y. Acad. Sci. 126, 115-126 (1965)

King, T.J., McKinnell, R.G.: An attempt to determine the developmental potentialities of the cancer cell nucleus by means of transplantation. In: Cell Physiology of Neoplasia, pp. 591-617. Austin: University of Texas Press 1960

Kobel, H.R., Brun, R.B., Fischberg, M.: Nuclear transplantation with melanophores, ciliated epidermal cells, and the established cell line A-8 in Xenopus laevis. J. Embryol. Exp. Morphol. 29, 539-547 (1973)

Laskey, R.A., Gurdon, J.B.: Genetic content of adult somatic cells tested by nuclear transplantation from cultured cells. Nature (London) 228, 1332-1334 (1970)

Marshall, J.A., Dixon, K.E.: Nuclear transplantation from intestinal epithelial cells of early and late Xenopus laevis tadpoles. J. Embryol. Exp. Morphol. 40, 167-174 (1977)

McAvoy, J.W., Dixon, K.E., Marshall, J.A.: Effects of differences in mitotic activity, stage of cell cycle, and degree of specialization of donor cells on nuclear transplantation in Xenopus laevis. Dev. Biol. 45, 330-339 (1975)

McKinnell, R.G.: Intraspecific nuclear transplantation in frogs. J. Hered. 53, 199-207 (1962)

McKinnell, R.G.: Cloning, Nuclear Transplantation in Amphibia. Minneapolis: University of Minnesota Press 1978

McKinnell, R.G., Deggins, B.A., Labat, D.D.: Transplantation of pluripotential nuclei from triploid frog tumors. Science 165, 394-396 (1969)

McKinnell, R.G., Steven, Jr. L.M., Labat, D.D.: Frog renal tumors are composed of stroma, vascular elements and epithelial cells: What type nucleus programs for tadpoles with the cloning procedure? In: Progress in Differentiation Research (ed. N. Müller-Berat), pp. 319-330. Amsterdam-New York: Elsevier 1976

Merriam, R.W.: Movement of cytoplasmic proteins into nuclei induced to enlarge and initiate DNA or RNA synthesis. J. Cell Sci. 5, 333-349 (1969)

Muggleton-Harris, A.L., Pezella, K.: The ability of the lens cell nucleus to promote complete embryonic development through to metamorphosis and its applications to ophthalmic gerontology. Exp. Gerontol. 7, 427-431 (1972)

Smith, L.D.: Transplantation of the nuclei of primordial germ cells into enucleated eggs of Rana pipiens. Proc. Natl. Acad. Sci. USA 54, 101-107 (1965)

Subtelny, S., Bradt, C.: Cytological observations on the early developmental stages of activated Rana pipiens eggs receiving a transplanted blastula nucleus. J. Morphol. 112, 45-59 (1963)

Wabl, M.R., Brun, R.B., Du Pasquier, L.: Lymphocytes of the toad Xenopus laevis have the gene set for promoting tadpole development. Science 190, 1310-1312 (1975)

Control of Early Embryonic Development: An Analysis of a Cytoplasmic Component and Its Mode of Action

A.J. BROTHERS

Department of Zoology, University of California, Berkeley, CA 94720, USA

I. Introduction: Cytoplasmic Determinants Control Early Development

The classical experimental embryologists established that a close relationship exists between components of the egg cytoplasm and specific types of cell determination (Morgan 1927). Apparently, during oogenesis the genome directs the synthesis of substances which are stored in the oocyte and after fertilization of the egg act to control the development of the embryo (Briggs, 1973; Gurdon 1974; Whittaker 1979). This control must involve an interaction between the genetically identical embryonic nuclei and the morphogenetic substances present in the egg cytoplasm. It is believed that those nucleo-cytoplasmic interactions are involved in the critical event whereby cells become determined to different developmental pathways and as a consequence different genes will be expressed in particular cell types and not in other cells, leading eventually to the expression of differentiated molecular and morphological characteristics (Briggs and King 1959). Recently some progress has been described in the characterization of some of these morphogenetic substances, particularly for the initiation of DNA synthesis (Benbow and Ford 1975; Laskey et al. 1977), pole cell determination (Mahowald et al. 1979), and the anterior determinant of *Smittia* eggs (Kalthoff 1979). Direct evidence has been presented to indicate that components of the egg cytoplasm can act to modify or reprogram the transcriptional activity of injected somatic cell nuclei (De Robertis et al. 1977; Etkin 1976). Powerful new techniques have been developed which will allow the identification of the morphogenetic substances through the injection of purified genes into the oocyte nucleus (germinal vesicle) and the subsequent recovery of those DNA segments along with the complexed components of the oocyte nucleoplasm or ooplasm (Mertz and Gurdon 1977; Gurdon 1977).

Another approach to the problem of how the egg cytoplasm acts to control the development of the zygote into the differentiated multicellular organism is provided by the use of maternal effect mutations. This type of mutation acts to cause an alteration in a single morphogenetic component of the egg cytoplasm resulting in a specific effect upon a particular critical event in development (Briggs 1973).

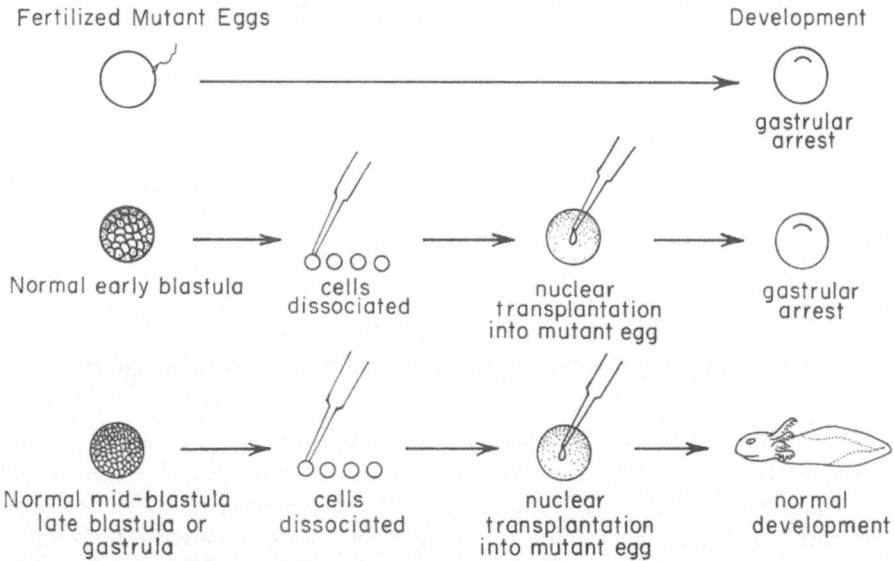

Fig. 1. Nuclear transplantation experiments demonstrated that the $o+$ substance acts to produce a stable activation during blastulation of the nuclear programs required for gastrulation and neurulation

II. The Maternal Effect Mutation o, for Ova Deficient, in Ambystoma mexicanum

Humphrey (1966) described such a maternal effect mutation named o, for ova deficient, in the Mexican axolotl. All eggs spawned by a female homozygous for the mutation *(o/o)* arrest at gastrulation. No neural or axial structures are ever formed in embryos derived from such mutant eggs, regardless of whether or not the normal allele *(o+)* is introduced by the sperm at fertilization (see Brothers 1979). This recessive mutation has 100% expressivity and penetrance. The evidence suggests that it may be located on the fourth chromosome (Carroll 1974). The gastrular arrest can be completely corrected by microinjection of plasm obtained from the nucleus (germinal vesicle) of normal oocytes or the cytoplasm of normal eggs (Briggs and Cassens 1966). Only a very minute portion of the nucleoplasm (1–2%) from a germinal vesicle is required to obtain full correction. It is not species-specific for it can be found in the oocytes of a wide variety of amphibians (Briggs 1972).

The $o+$ substance apparently is synthesized under the direction of the normal allele during oogenesis and accumulates in the germinal vesicle. After hormone stimulation the oocyte matures, germinal vesicle breakdown occurs and its contents are dispersed to the egg cytoplasm. The active component can then be recovered from the cytoplasm of unfertilized eggs (Briggs 1972). Biological activity can be found in fractions from normal fertilized eggs and cleavage-stage embryos,

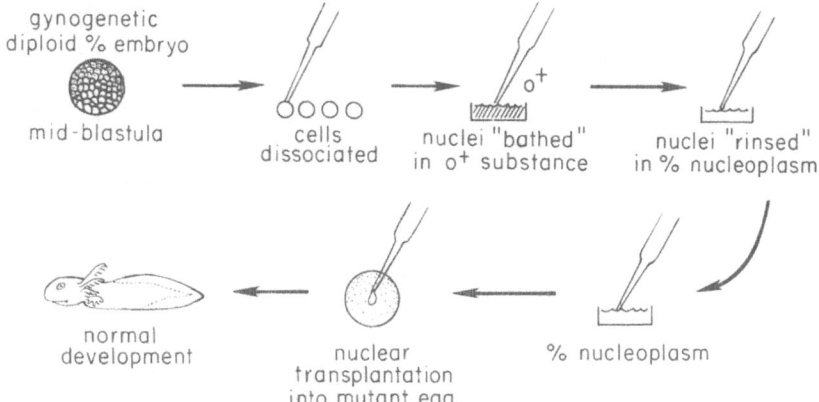

Fig. 2. Nuclei from mutant blastula are given a brief exposure to the $o+$ substance and then transplanted into enucleated recipient mutant eggs, demonstrating that those conditions are sufficient to produce activation

however by late-blastula and gastrula stages activity is no longer detectable, suggesting that perhaps it has become degraded or fixed to some subcellular particle (Briggs and Justus 1968).

The presence in the egg cytoplasm of the $o+$ substance is absolutely required for development beyond gastrulation. Amphibian eggs can develop up through blastulation without a functional genome, however, gastrulation and organo-genesis do require a functional zygote genome. Apparently, new transcripts are produced during blastulation which are required for later differentiation (Davidson 1976). It was therefore suggested that the $o+$ substance might be involved in the activation of those nuclear genes required for development through gastrulation and neurulation.

A. The Interaction Between the $o+$ Substance and the Blastula Nucleus

The question of whether or not the nucleo-cytoplasmic interaction between the blastula nucleus and the $o+$ substance results in a stable activation of those nuclear genes can be tested by transplanting nuclei from normal blastulae (those nuclei have been exposed to the $o+$ substance present in their egg cytoplasm) into enucleated mutant eggs (which lack the $o+$ substance). Transplants of early blastula nuclei show that there is no stable activation of those nuclei—all those nuclear transplants show the mutant phenotype and arrest at gastrulation. However, donor nuclei transplanted from mid-blastulae and later developmental stages can support normal development (Fig. 1). Serial clonal nuclear transplant experiments indicate that the activation is mitotically heritable for at least 30 cell generations (Brothers 1976).

B. Determination of the Mode of Action of the $o+$ Substance

The previously described experiments demonstrated that nuclar activation during blastulation is dependent upon a nucleo-cytoplasmic interaction with the $o+$ substance present in the egg cytoplasm. Those experiments do not show whether a very short exposure to the $o+$ substance is sufficient to produce this activation and whether there is a direct interaction of the $o+$ substance and the blastula nucleus. Preliminary experiments to examine these questions have been performed. The experimental design is given in Fig. 2.

Mutant eggs are fertilized with UV- or X-irradiation sperm and then the eggs are subjected to either cold shock or Cytochalasin B to create gynogenetic diploids of known o/o genotype in mutant egg cytoplasm. Gynogenetic diploids are created by the suppression of the extrusion of the second polar body in the case of cold shock (Fankhuaser and Humphrey 1942). Cytochalasin B can be used to suppress either the extrusion of the second polar body or to suppress cytokinesis at the first cleavage division (Brothers, unpublished). Previous results indicated that the normal allele was not functional during early development (Briggs 1972), however a direct test using donor nuclei of known o/o genotype would eliminate the possibility of any contribution from the activation of the $o+$ allele during these experiments. The mutant gynogenetic diploids are allowed to develop to blastula stages. Some of the blastulae are selected as nuclear donors and the cells are dissociated in the usual manner. The donor cells are then gently picked up with the nuclear transplantation pipet and the pipet then transferred to a small well containing broken germinal vesicles obtained from normal oocytes. This preparation provides a source of concentrated $o+$ substance. A small portion of this $o+$ substance is drawn up into the tip of the pipet, "bathing" the blastula nucleus. Contact time with the $o+$ substance is of about 5 to 7 min duration. Subsequently, the nucleus is "washed" in several successive baths (2 to 4), each of approximately 5 to 10 min duration, in similar wells containing germinal vesicles from oocytes removed from a homozygous mutant (o/o) female. The "bathed" nuclei are then transplanted into enucleated recipient mutant eggs. Some of the resulting blastula are sacrificed at early blastula stage and used as nuclear donors for serial clonal transplants, again using enucleated mutant eggs as recipients. A few cases of development to advanced tail-bud or larval stages have been obtained thus far. The results demonstrate that a brief exposure to the $o+$ substance is sufficient to produce the nuclear activation, the capacity to promote development through gastrulation and neurulation. Those cases do suggest, but do not prove, that there may be a direct interation between the blastula nucleus and the $o+$ substance. Further experiments to examine this question of a direct interaction between the $o+$ substance and the nucleus will be undertaken once homozygous (o/o) mutant females become available.

III. Discussion

Cellular determination is a highly stable and heritable capacity of individual cells (Hadorn 1967) even though the differentiated functions are not expressed (Konigsberg 1963; Cahn and Cahn 1966; Coon 1966). The pattern of gene expression can be modified and controlled by extracellular or cytoplasmic factors (Gurdon 1977), but once it is established, the capacity to express a specific state of determination is mitotically inherited over many cellular generations (Ephrussi 1972). Cytoplasmically controlled stable nuclear determination has been shown for *Paramecium* (Sonneborn 1954) and *Tetrahymena* (Nanney 1956). The differential accumulation of certain classes of proteins in germinal vesicles has been demonstrated (Bonner 1975) and the effects of oocyte cytoplasm have been shown to modify selectively the gene expression of injected somatic cell nuclei (Etkin 1976; DeRobertis and Gurdon 1977; DeRobertis et al. 1977). Exchange of proteins between embryonic nuclei and the egg cytoplasm has been established (Ecker and Smith 1971; DiBerardino and Hoffner 1975), and it is believed that such proteins may be involved in the control of cellular determination.

The $o+$ substance is synthesized during oogenesis and the presence of this substance in the egg cytoplasm is essential for the establishment of a stable nuclear activation. Evidence suggests that it may be a protein(s) or depend upon a protein for its activity (Briggs and Justus 1968; Malacinski, unpublished). The sensitive period for this activation is restricted to one point during embryonic development, mid-blastulation, and a brief exposure of the blastula nucleus to the $o+$ substance is sufficient to evoke this activation. These results are interpreted to suggest that this alteration of the capacity to support development the result of the activation of the nuclear genes is required for gastrulation and neurulation. It is tempting to speculate that the $o+$ substance might be a regulatory substance, such as has been suggested to be involved in eukaryotic gene regulation (Davidson and Britten 1973). Models for eukaryotic regulatory molecules and their interaction with the genome have suggested that those molecules preferentially recognize specific DNA sequences and act to change the conformation of the entire transcriptional unit (Frenster 1976; Yamamoto and Alberts 1976; Davidson et al. 1977). This would be one possible explanation for the stable activation of the nuclear capacity to support development through gastrulation and neurulation. The results of the nuclear transplantation experiments suggest, but do not prove, that there is a direct interaction between the $o+$ substance and the embryonic nucleus. Further experiments are required to demonstrate that there is an interaction of the $o+$ substance and the embryonic genome, such as would be predicted for a regulatory substance.

References

Benbow, R.M., Ford, C.C.: Cytoplasmic control of nuclear DNA synthesis during early development of *Xenopus laevis*: A Cell-free assay. Proc. Natl. Acad. Sci. USA 272, 2437-2441 (1975)

Bonner, W.M.: Protein migration into nuclei. II. Frog oocyte nuclei accumulate a class of microinjected oocyte nuclear proteins and exclude a class of microinjected oocyte cytoplasmic proteins. J. Cell Biol. 64, 431-437 (1975)

Briggs, R.: Further studies on the maternal effect of the *o* gene in the Mexican axolotl. J. Exp. Zool. 181, 271-280 (1972)

Briggs, R.: Developmental genetics of the axolotl. In: Genetic Mechanisms of Development (ed. F.R. Ruddle), pp. 169-199. New York-London: Academic Press 1973

Briggs, R., Cassens, G.: Accumulation in the oocyte nucleus of a gene product essential for embryonic development beyond gastrulation. Proc. Natl. Acad. Sci. USA 55, 1103-1109 (1966)

Briggs, R., Justus, J.T.: Partial characterization of the component from normal eggs which corrects the maternal effect of the gene *o* in the Mexican axolotl (*Ambystoma mexicanum*). J. Exp. Zool. 147, 105-116 (1968)

Briggs, R., King, T.J.: Nucleocytoplasmic interactions in eggs and embryos. In: The Cell, Vol. I. (eds. J. Brachet, A.E. Mirsky), pp. 538-617. New York-London: Academic Press 1959

Brothers, A.J.: Stable nuclear activation dependent on a protein synthesized during oogenesis. Nature (London) 260, 112-115 (1976)

Brothers, A.J.: A specific case of genetic control of early development; the *o* maternal effect mutattion of the Mexican axolotl. In: Determinants of Spatial Organization (eds. S. Subtelny, I.R. Konigsberg), pp. 167-183. New York-London: Academic Press 1979

Cahn, R.D., Cahn, M.B.: Heritability of cellular differentiation: clonal growth and expression of differentiation in retinal pigment cells *in vitro*. Proc. Natl. Acad. Sci. USA 55, 106-114 (1966)

Carroll, C.R.: Comparative study of early embryonic cytology and nucleic acid synthesis of *Ambystoma mexicanum* normal and *o* mutant embryos. J. Exp. Zool. 187, 409-422 (1974)

Coon, H.G.: Clonal stability and phenotypic expression of chick cartilage cells *in vitro*. Proc. Natl. Acad. Sci. USA 55, 66-73 (1966)

Davidson, E.H.: Gene Activity in Early Development, 2nd ed. New York-London: Academic Press 1976

Davidson, E.H., Britten, R.J.: Organization, transcription and regulation in the animal genome. Q. Rev. Biol. 48, 565-613 (1973)

Davidson, E.H., Klein, W.H., Britten, R.J.: Sequence organization in animal DNA and a speculation on hnRNA as a coordinate regulatory transcript. Dev. Biol. 55, 69-84 (1977)

DeRobertis, E.M., Gurdon, J.B.: Gene activation after injection into amphibian oocytes. Proc. Natl. Acad. Sci. USA 74, 2470-2474 (1977)

DeRobertis, E.M., Partington, G.A., Longthorne, R.F., Gurdon, J.B.: Somatic nuclei in amphibian oocytes: evidence for selective gene expression. J. Embryol. Exp. Morphol. 40, 199-214 (1977)

DiBerardino, M.A., Hoffner, N.J.: Nucleo-cytoplasmic exchange of nonhistone proteins in amphibian oocytes. Exp. Cell Res. 94, 235-252 (1975)

Ecker, R.E., Smith, L.D.: The nature and fate of *Rana pipiens* proteins synthesized during maturation and early cleavage. Dev. Biol. 24, 559-576 (1971)

Ephrussi, B.: Hybridization of somatic cells. Princeton, N.J.: Univ. Princeton Press 1972

Etkin, L.D.: Regulation of lactate dehydrogenase (LDH) and alcohol dehydrogenase (ADH) synthesis in liver nuclei, following their transfer into oocytes. Dev. Biol. 52, 201-209 (1976)

Fankhauser, G., Humphrey, R.R.: Induction of triploidy and haploidy in axolotl eggs by cold treatment. Biol. Bull. 83, 367-374 (1942)

Frenster, J.H.: Selective control of DNA Helix openings during gene regulation. Cancer Res. 36, 3394-3398 (1976)

Gurdon, J.B.: The Control of Gene Expression in Animal Development. Cambridge: Harvard University Press 1974

Gurdon, J.B.: The Croonian Lecture, 1976. Egg cytoplasm and gene control in development. Proc. R. Soc. London Ser. B 198, 211-247 (1977)

Hadorn, E.: Dynamics of determination. In: Major Problems in Developmental Biology (ed. M. Locke), pp. 84-104. New York-London: Academic Press 1967

Humphrey, R.R.: A recessive factor (*o*, for ova deficient) determining a complex of abnormalities in the Mexican axolotl *(Ambystoma mexicanum)*. Dev. Biol. 13, 57-76 (1966)

Kalthoff, K.: Analysis of a morphogenetic determinant in an insect egg (*Smittia* spec., *Chironomidae*, Diptera). In: Determinants of Spatial Organization (eds. S. Subtelny, I.R. Konigsberg), pp. 97-126. New York-London: Academic Press 1979

Konigsberg, I.R.: Clonal analysis of myogenesis. Science 140, 1273-1284 (1963)

Laskey, R.A., Mills, A.D., Morris, N.R.: Assembly of SV40 Chromatin in a cell-free system from *Xenopus* eggs. Cell 10, 237-243 (1977)

Mahowald, A.P., Allis, C.D., Karrer, K.M., Underwood, E.M., Waring, G.L.: Germ plasm and pole cells of Drosophila. In: Determinants of Spatial Organization (eds. S. Subtelny, I.R., Konigsberg), pp. 127-146. New York-London: Academic Press 1979

Mertz, J.E., Gurdon, J.B.: Purified DNA's are transcribed after microinjection into *Xenopus* oocytes. Proc. Natl. Acad. Sci. USA 74, 1502-1506 (1977)

Morgan, T.H.: Experimental Embryology. New York: Columbia University Press 1927

Nanney, D.L.: Caryonidal inheritance and nuclear differentiation. Am. Nat. 90, 291-307 (1956)

Sonneborn, T.M.: Patterns of nucleocytoplasmic integration in *Paramecium*. In: Proc. 9th Int. Congr. Genet. Carylogia (Suppl), pp. 307-325 (1954)

Whittaker, J.R.: Cytoplasmic determinants of tissue differentiation in the Ascidian egg. In: Determinants of Spatial Organization (eds. S. Subtelny, I.R. Konigsberg), pp. 29-51. New York-London: Academic Press 1979

Yamamoto, K.R., Alberts, B.M.: Steroid receptors; elements for modulation of eukaryotic transcription. Annu. Rev. Biochem. 45, 722-746 (1976)

Evidence of the First Genetic Activity
Required in Axolotl Development

J. SIGNORET

Laboratoire d'Embryologie, Université de Caen
Caen, France

I. Introduction

Early development of an amphibian egg does not depend on the genetic activity of its nuclei. Rather, early development is guided by a stockpile of gene products accumulated during oogenesis. Later, the zygotic genome participates in development and it is only then that one can detect, in a progressive manner, the intervention of the individual's own genome for any particular developmental event.

This study intends to identify precisely when and how the zygotic genome participation becomes necessary for development to proceed further. This study does not raise the question of when do the embryo's genes begin their function (probably early) or how long and up to what stage is activity of oocyte gene products maintained (certainly rather late). Instead of these considerations, the present study examines the development of an embryo in which gene function is inhibited, or severely disturbed, so as to define when and how this activity becomes critical. Describing the first intrinsic genetic activity required in the course of development enables us to specify in which stage it becomes necessary, which in turn enables us to understand better the mechanisms which account for these observations.

II. Preliminary Approaches

Our first observations with the axolotl concerned the transplantation of somatic nuclei into enucleated eggs (Briggs et al. 1964). Implanting a notochord nucleus into an enucleated egg, for instance, generally elicits normal cleavage that appears identical to that of a fertilized egg control. Nevertheless, the normal-looking blastula so obtained most often is arrested in middle or late cleavage. Cytological study of these arrested blastulae reveals severe chromosomal disorders, with rings, acentric fragments, and lagging chromosomes. The chromosomal aberrations, which account very well for the arrest in development, do not appear to have any consequence on early cleavage events. Similar results are obtained with nuclei of one species implanted into the cytoplasm of a different species and confirm Hennen's (1963) observations with anurans. Lethal hybrids between distant urodelian species are also characterized by normal early cleavage (Chen 1967).

Even the complete absence of functional chromatin does not prevent the characteristic segmentation of the embryo (Briggs et al. 1951). Treatment with an antimetabolite to inhibit gene expression can be considered a chemical enucleation. The antimetabolite, however, in no way affects cleavage but it does impede further development (Brachet and Denis 1963). It is thus classically established that in amphibians, in general, the early events of segmentation do not depend upon the activity of specific genes of the embryo. In the axolotl in particular, accumulated gene products from the oocyte govern development up to the middle or late blastula stage. In order to obtain a better understanding of this early period, a systematic analysis of segmentation has been undertaken.

III. Analysis of Cleavage

When scrutinizing the animal hemisphere of an axolotl embryo during its early development, it is possible to notice a period of active synchronous divisions followed by a desynchronized period of slower cell division (Signoret and Lefresne 1971).

A. The Synchronous Period

Eggs obtained from the first spawning of a 12- to 18-month-old female, when maintained at $20° \pm 0.5$, exhibit cleavage cycles with a mean duration of 80 min. These cycles do not vary more than 5 min from one blastomere to another, and perform a synchronous segmentation up to the end of the ninth cycle. The first sign of the cleavage furrow on the blastomere surface corresponds, in the nucleus, to late telophase. The chromosomes then swell, give rise to chromosomal vesicles, and enter into active DNA replication even before a typical interphase nucleus is reconstituted. The presynthetic phase G_1 is totally absent during this stage of development (Signoret and Lefresne 1976). When the embryo nucleus is severely damaged, or when it is of a genetically inappropriate type for the cytoplasm, the series of synchronous divisions appear perfectly normal. The injection of actinomycin into the egg, or the treatment of an egg fragment with an appropriate antimetabolite solution, are without effect on the early cleavage process.

B. The Asynchronous Period

A dramatic desynchronization occurs beginning with the tenth cycle of blastomere division. Individual cell cycles lengthen and there is a large variability for a given cycle. For example, the 11th cycle lasts from 1.5 h to 3.5 h. The statistical distribution of individual cell cycle lengths is characteristic and does not fit a Gaussian distribution. We have proposed elsewhere a theoretical model that fits reasonably well with the observed distribution of cell cycle lengths (Signoret 1977).

With a series of successive pictures, it is possible to record the origin for a given couple of cells, when the cleavage furrow first appears, and to make autoradiographs at a selected time. Our results show that newly formed nuclei after telophase have not as yet begun DNA replication. Interphase nuclei are observed still in presynthetic phase (G$_1$) 20 or 30 min after telophase. However, in a comparative study of the two nuclei resulting from a division, one may be in phase G$_1$ while the other nucleus may already be actively replicating its DNA. The G$_1$ phase thus appears simultaneously with the lengthening and the variability of the cell cycle, and, moreover, the G$_1$ phase is clearly variable.

Embryos with severe chromosomal aberrations, or those treated with actinomycin, never proceed normally after the onset of the period of asynchronous division. Such embryos are observed arrested during mid to late cleavage with the majority of the embryos blocking about the tenth cycle of segmentation.

C. The Critical Stage

The different experiments discussed above suffer from some methodological problems. Some transcriptional activity may occur from broken chromosomes. The complete inhibition of gene expression by actinomycin, on the other hand, may interfere with metabolic processes other than transcriptional activity. Perhaps the most rigorous specificity is obtained with the action of α-amanitine. This substance binds with RNA polymerase and consequently impedes transcription (Brachet et al. 1972). A technical problem with this substance is the relative impermeability of the egg. To overcome this problem, one can simply inject α-amanitine, or one can dissociate a young blastula in an α-amanitine solution, or one can treat the egg at low temperature (4–6° C) with a slightly hypertonic solution of α-amanitine. Under these conditions of α-amanitine treatment the blastomeres undergo a normal appearing series of synchronous divisions, and then cease to divide before the period which would normally be asynchronous. The blocked blastomeres thus obtained do not seem perceptibly different from the blastomeres of the control embryo at the end of the ninth cell cycle.

IV. Conclusions and Interpretations

It is clearly established that the cell divisions during the period of rapid and synchronous cleavage do not require genetic activity on the part of the embryo. However, cleavage after this initial period, with lengthening and variable cell cycles, is dependent upon the normal functioning of the embryo's DNA.

One can conclude that the first genetic activity directly required in axolotl development occurs at the beginning of the first cycle that is desynchronized. The embryo's first genetic activity coincides with the stage when the cell cycle lengthens and becomes variable, and with the appearance of a variable G$_1$ phase.

We propose that this coincidence should be interpreted as a cause and effect relationship and not as fortuitous.

We know that the egg cytoplasm has the capacity to cause unprotected DNA to undergo replication. In such a milieu, as soon as the anaphase chromatin decompacts, the nucleus enters the S phase with no G$_1$ phase. The factor responsible for these phenomena is abundant in the oocyte, and the factor probably associates with DNA as development proceeds, the cytoplasm being progressively exhausted of it as nuclei multiply. When cell cycles have proceeded to the point that the embryo contains about 1000 nuclei, the factor is probably depleted. Chromatin following telophase becomes decompacted without replicating its DNA. Interphase nuclei enter a presynthetic G_1 phase for the first time beginning with the tenth cycle. Consistent with the hypothesis is the observation that a haploid embryo experiences one more synchronous cycle.

With the onset of the first G_1 phase at the tenth cell cycle, embryonic gene expression proceeds and leads to the de novo production of a factor that commits the cell to complete G_1 and enter S phase. The DNA replication is no longer automatic but is determined by intrinsic gene activity. Our model supposes in addition that the production of the specific factor exhibits characteristic kinetics. We suggest that cells remain at a critical point in G_1, overpassing this step at a characteristic rate. The event that commits the cell to proceed further has a constant probability of occurrence from cell to cell, beginning at the criticial state.

In any case, experimental evidence and analytical clues lead one to conclude that the first genetic activity has a fundamental controlling role in cell division. The consequence of the embryo's genetic activity is to enable the cell in G_1 phase to proceed further and to complete the cycle. Hence, the control of cell proliferation depends upon its conclusion.

The production of a factor responsible for the control of cell multiplication is an important phenomenon worthy of renewed experimental investigation. This is why further studies of the axolotl egg cleavage remain attractive and promising.

References

Brachet, J., Denis, H.: Effects of actinomycin D on morphogenesis. Nature (London) 198, 205-206 (1963)

Brachet, J., Hubert, E., Lievens, A.: The effects of α-amanitin and rifampicin on amphibian egg development. Rev. Suisse Zool. 79, 47-63 (1972)

Briggs, R., Green, E.V., King, T.J.: An investigation of the capacity for cleavage and differentiation in *Rana pipiens* eggs lacking "functional" chromosomes. J. Exp. Zool. 116, 455-500 (1951)

Briggs, R., Signoret, J., Humphrey, R.R.: Transplantation of nuclei of various cell types from neurulae of the Mexican Axolotl. Dev. Biol. 10, 233-246 (1964)

Chen, P.S.: Biochemistry of nucleo-cytoplasmic interactions in morphogenesis. In: The Biochemistry of Animal Development, Vol. II (ed. R. Weber), pp. 115-191. New York-London: Academic Press 1967

Hennen, S.: Chromosomal and embryological analyses of nuclear changes occurring in embryos derived from transfers of nuclei between *Rana pipiens* and *Rana sylvatica*. Dev. Biol. 6, 133-183 (1963)

Signoret, J.: La cinétique cellulaire au cours de la segmentation du germe d'Axolotl: Proposition d'un modèle statistique. J. Embryol. Exp. Morphol. 42, 4-14 (1977)

Signoret, J., Lefresne, J.: Contribution à l'étude de la segmentation de l'oeuf d'Axolotl. I. Définition de la transition blastuléenne. Ann. Embryol. Morphol. 4, 113-123 (1971)

Signoret, J., Lefresne, J.: Le cycle cellulaire au cours de la segmentation de germe d'Axolotl. Bull. Soc. Zool. Fr. 101, 123-127 (1976)

Genetic Manipulation of the Early Mouse Embryo

KARL ILLMENSEE

Department of Biology, University of Geneva, Geneva, Switzerland

I. Introduction

The experimental modification of the genome of a mammalian cell might be envisaged by the following two microsurgical approaches: (1) the removal of genetic information and its phenotypic consequences to differentiation, and (2) the addition of genetic material and its expression during development and neoplasia. During the past few years, both approaches have been realized in the mouse, the most suitable mammalian organism for genetic studies, and I should like briefly to summarize these new results, discuss their relevance and applicability to experimental embryology and cancer research and finally remark upon some promising prospects in genetic manipulation of the mammalian embryo.

II. Uniparental Homozygous-Diploid Mice

Development of the fertilized egg in which only the maternal (gynogenesis) or paternal (androgenesis) genome is retained seems to occur infrequently in the animal kingdom. The same holds true for another type of reproduction that involves differentiation controlled exclusively by the maternal genes of the unfertilized but activated egg (parthenogenesis). Nevertheless, each event can happen spontaneously or be initiated experimentally in a number of vertebrates and insects (described in Beatty 1967; White 1978). Although different in origin, these uniparental animals with either a hemizygous-haploid or homozygous-diploid genome are of considerable interest for analysis of gene function during early development.

Various attempts to generate uniparental mice experimentally from fertilized and unfertilized eggs were limited to embryonic and early postimplantation stages. Similarly, most recent efforts using microsurgical techniques either to remove one pronucleus shortly after fertilization (Modlinski 1975) or to bisect the fertilized egg at the pronuclear stage (Tarkowski 1977) rarely resulted in the production of haploid blastocysts. Because mouse embryos with only one set of chromosomes are apparently nonviable, full restoration of the genome by diploidizing haploid embryos in the presence of cytochalasin B has recently been attempted. However,

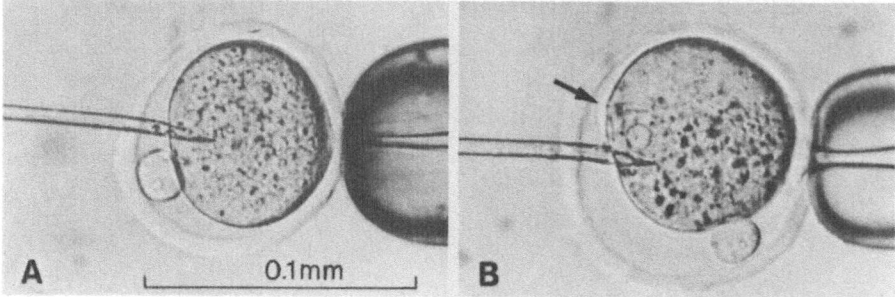

Fig. 1A,B. Microsurgical production of homozygous-diploid mice after removal of the female **(A)** or male **(B)** pronucleus from fertilized mouse eggs. About 9 h after sperm entrance, both pronuclei are located at the egg periphery, the smaller female pronucleus near the polar body and the larger male pronucleus in the opposite position. **A** After attaching the egg to a blunt holding pipet, a sharply tipped micropipet is pushed through the zona pellucida and plasma membrane above the polar body into the egg. Subsequently, the female pronucleus is gently withdrawn from the egg. **B** The micropipet has been inserted into the egg near the male pronucleus. Note the sperm tail *(arrow)*. After microsurgery, the haploid eggs are cultured overnight in the presence of cytochalasin B to restore a diploid set of chromosomes. The resulting blastocysts are then transferred surgically to the uterus of a pseudopregnant foster mother to allow development to term. Live-born mice possess exclusively the genome from either the father (androgenesis) or the mother (gynogenesis) depending on whether the female or male pronucleus had been removed from the egg, respectively (Hoppe and Illmensee 1977)

only blastocytes have been obtained, and no genetic markers were introduced to allow testing for homozygosity at the preimplantation stage (Markert and Petters 1977).

Recently, we reported that viable homozygous-diploid mice can indeed be generated by microsurgically removing one pronucleus from the fertilized egg and diploidizing the residual one in the presence of cytochalasin B (Fig. 1). The resulting blastocytes are then implanted into the uterus of a pseudopregnant female in order to allow development to term. The isogenic mice are expected to be females, because diploidized eggs left with only a Y chromosome bearing pronucleus should eventually develop into Y/Y embryos which usually die during early cleavage (Tarkowski 1977). The X/X embryos, however, which grow normally will give rise to females. The uniparental mice carry either the maternal or paternal genome, depending on whether the female or male pronucleus has been retained in the egg. Confirmation of such a process seems to be absolutely essential at the genetic level and, in fact, ought to be considered as *onus probandi* in order to reveal the gynogenetic or androgenetic origin of the experimental mice. Homozygosity for a number of gene loci positioned on different chromosomes has adequately been demonstrated in all the isogenic females. Some of the uniparental females proved to be fertile and delivered healthy offspring (Hoppe and Illmensee 1977).

In addition to using coat color as a genetic marker for identifying the isogenic progeny, the blood of these mice was analyzed for glucose-phosphate isomerase, hemoglobins, plasma proteins and esterases, and carbonic anhydrase; the urine was screened for the major urinary protein complex. All of the isogenic mice exhibited

Fig. 2. *Agouti* BL10SJLF₁ foster mother with her "adopted offspring" showing a nonmanipulated regular *agouti* BL10SJLF₁ mouse *(left)*, a gynogenetic *black* BL10 female *(right)* and an androgenetic *albino* SJL female *(middle)*. The *black* mouse derived from the maternal genome of a fertilized egg from which the male pronucleus had been removed microsurgically, whereas the *albino* mouse derived from the paternal genome of an egg after removing its female pronucleus. Therefore, genetically speaking, both experimental mice had only one parent (Hoppe and Illmensee 1977)

only the enzymatic form or protein profile of one of the inbred parents. The occurrence of pure strain-specific variants of these different gene products clearly demonstrated that, after microsurgical removal of one pronucleus, the initial F_1 hybrid egg was still able to develop normally with the genome of the residual pronucleus. Furthermore, all seven uniparental mice showed normal diploid karyotypes with two X chromosomes, thus substantiating their diploid genetic constitution at the chromosomal level. Karyotypic analysis also revealed that X chromosome inactivation had occurred normally as in regular females, irrespective of whether the X chromosomes had been inherited paternally or maternally (Eicher et al. 1980).

The development to term of females which obtained the genome only from their father was comparable to that of females with genes inherited only from their mother indicating that the paternal genome alone is fully capable of initiating and promoting normal development (Fig. 2). The paternal genes must have also been active in extraembryonic tissues which usually show a preferential inactivation of the paternal X chromosome (Takagi 1978; West et al. 1978). The apparently normal X-activation-inactivation process in our gynogenetic females as well as in parthenogenetic embryos (Kaufman et al. 1978) and human female patients (Latt et al. 1976) contradicts two current hypotheses (Cooper 1971; Chandra and Brown 1975) that would require two active X chromosomes in these individuals, which have not been observed. Furthermore, these hypotheses would demand inactive X chromosomes for our androgenetic females as well as for XO females which both inherited the X chromosome from their father and therefore should not be viable.

The normal survival of uniparental mice raises again the question about the true origin of the prenatal developmental arrest during parthenogenesis frequently discussed as a result of deleterious genes or gene combinations being uncovered in these gynogenetic embryos (reviewed by Graham 1974). The microsurgical production of homozygous-diploid mice provides a new means of analyzing the reasons and causes for the developmental failure of parthenotes, and may also be useful in studying the phenotypic consequences of homozygosity for recessive mutations, lethal or X-linked genes, and maternal versus paternal gene expression during mammalian differentiation. It now should be possible to produce homozygous-diploid daughters from the uniparental mothers by repeating the microsurgical removal of the male pronucleus in the next generation. In this way, all the daughters would be genetically identical among each other (like monozygotic twins) and with their isogenic mother. Such a "family" would certainly be of great value for immunological studies and clonal propagation of a given cellular phenotype.

III. Xenogeneic Gene Expression in Chimeric Mice

Genetic manipulation in vivo can be accomplished not only by removing genetic material from the organism but also by introducing cells, chromosomes, nuclei or eventually genes into the early mouse embryo in order to study their phenotypic expression during in situ differentiation (Fig. 3).

Recently it was found that, after microinjection into genetically different blastocysts, malignant cells of a particular tumor, a teratocarcinoma (described by Stevens 1975; Graham 1977), lost their neoplastic properties in the embryonic environment, participated in normal development and clonally contributed to virtually all major adult tissues in chimeric mice including those not seen before in the solid tumors (reviewed by Illmensee 1978). It therefore appears as if the initially malignant teratocarcinoma cells remain developmentally totipotent and are able to express their genetic repertoire in an orderly sequence of differentiation into somatic and germ-line tissues. The reversion of malignancy was obviously a stable process, because when pieces of the teratocarcinoma-derived mosaic tissues were grafted under the skin of syngeneic mice, they never formed tumors. One could still argue that the teratocarcinoma cell population was heterogeneous with respect to malignant properties, so that one cell might be able to differentiate normally whereas another cell might be destined to form a tumor. An alternative explanation of the experimental results is that a single teratocarcinoma cell can either

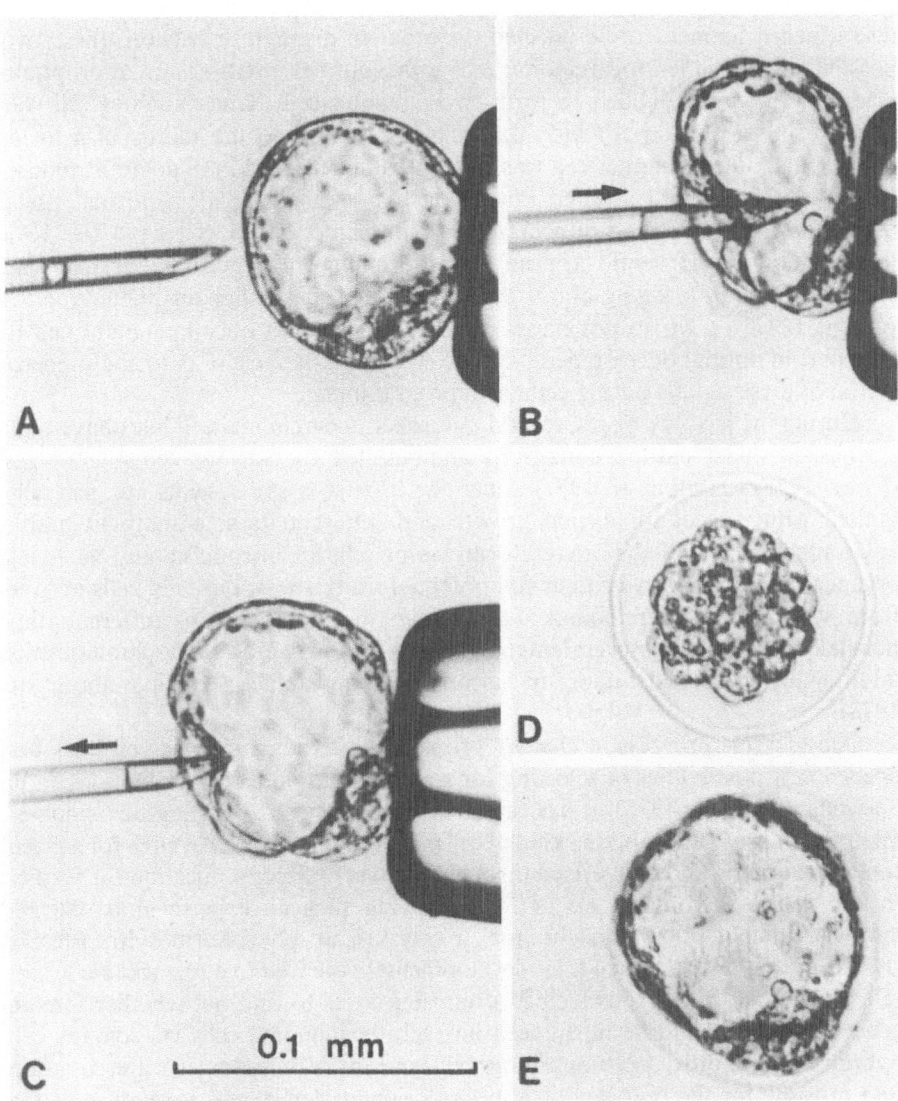

Fig. 3A-E. Microinjection of a single teratocarcinoma cell into a mouse blastocyst. After holding the blastocyst in position **(A)**, a micropipet containing a cell is pushed through the zona pellucida and trophoplast into the blastocoel **(B)**. The cell is injected near the inner cell mass, the presumptive embryonic region. After the injection pipet is withdrawn **(C)**, the blastocyst collapses **(D)**, and soon expands again to its normal size **(E)**. The injected cell is still attached to the inner cell mass. Subsequently, the blastocyst is surgically implanted into the uterus of a pseudopregnant foster mother to allow further development. The resulting mouse can then be analyzed for any tissue contributions derived from the injected cell (Illmensee 1978)

differentiate normally or give rise to a tumor, depending on the microenvironment into which it happens to be injected. In order to distinguish between these two possibilities, single teratocarcinoma cells were cultured in vitro in an appropriate medium, and some divided to form two daughter cells. One daughter cell was injected into a blastocyst which was then transferred to the uterus of a foster mother; the other daughter cell was implanted under the skin of an adult mouse. In some cases the cell injected into a blastocyst participated in normal tissue differentiation whereas its subcutaneously implanted sister cell gave rise to a teratocarcinoma. It would appear that the embryonic environment plays an important role in bringing about the reversion of the malignant phenotype. At present, however, we do not know what signals cause a once-malignant cell to take part in normal development, nor do we have any insight as to the mode of action of these signals during cellular reprogramming.

During the past few years, several mouse teratocarcinoma cell lines have been established under culture conditions and selected for various cell phenotypes (Evans 1975; Chung et al. 1977). Since the in vitro assay systems are generally limited with respect to normal growth and differentiation, a more favorable environment is provided when teratocarcinoma cells are introduced into the living organism in order to reveal their full potential. In this way, cultured cells derived from various teratocarcinomas were shown to be capable of differentiating normally in the coat and several internal organs, although most of the chimeric mice additionally developed tumors in various anatomical sites (Papaioannou et al. 1978).

Considerable progress in *clonally* propagating teratocarcinoma cell lines has opened new possibilities of selecting for somatic mutations in vitro (described in Sherman and Solter 1975). It has therefore been proposed that teratocarcinomas might provide us with a unique kind of cell which can be selected in vitro for a given somatic mutation and then cycled through mice via blastocyst injection for further in situ analysis (Mintz et al. 1975). Following such an experimental scheme, teratocarcinoma cells deficient for hypoxanthine phosphoribosyltransferase (HPRT) apparently retained their developmental potential to a remarkable extent (Dewey et al. 1977). It thus seemed promising to us to find out whether foreign genetic material could be introduced into teratocarcinoma cells via somatic cell hybridization in order to study xenogeneic gene expression during differentiation and to assay for the ontogenetic appearance, coexistence, and regulation of the foreign gene products in various tissues during development.

A. Human × Mouse Hybrid Cells

In collaboration with Dr. C. Croce (Wistar Institute), the combination of cell hybridization techniques with our biological system has recently opened up a new line of research. In his laboratory, cultured mouse teratocarcinoma cells first were selected for thymidine kinase (TK) deficiency. After microinjection into mouse blastocysts, the TK deficient cells became integrated during normal organogenesis and contributed substantially to several internal tissues, thus demonstrating their suitability as a gene carrier. The TK-deficient mouse teratocarcinoma cells where then fused in vitro with HPRT-deficient human fibrosarcoma cells using inactivated sendai virus. Only the interspecific hybrid cells can grow in hypoxanthine/aminopterin/thymidine (HAT) selective medium. On the contrary, TK- and HPRT-deficient parental cells, which lack the enzymes required for the incorporation of thymidine and hypoxanthine respectively, do not survive in this medium. Under these selective conditions, the viable hybrid cells, which quickly lose human but not mouse chromosomes, retain at least human chromosome 17 that carries the locus for TK. This particular chromosome also carries a second known gene that is closely linked to TK and coded for galactokinase (GLK). The latter enzyme, with its characteristic electrophoretic mobility quite different from the equivalent mouse enzyme, serves as another useful biochemical marker for detecting the presence and normal expression of the human gene product in the hybrid cells.

After subcutaneous implantation into athymic *nude* mice, the human × mouse hybrid cells formed predominantly undifferentiated tumors. In contrast, when injected into genetically marked mouse blastocysts, the malignant hybrid cells became integrated during embryogenesis and participated in orderly differentiation of the coat and seven internal organs (Fig. 4). Although the hybrid cells contributed up to 60% in some tissues of chimeric mice as judged from enzyme analysis, the human-specific gene product of GLK has only been detected in the heart of one chimera and the kidneys of another; both tissues, by the way, showed the highest participation of hybrid cells in the enzyme tests (Illmensee et al. 1978). The failure to disclose human GLK in the other mosaic organs might have resulted from, among other possibilities, the loss of human chromosome 17 during development in the absence of any selective pressure to retain it.

B. Rat × Mouse Hybrid Cells

In order to facilitate a more extensive analysis of foreign gene expression in vivo, it would be desirable to utilize a hybrid cell line containing several xenogeneic chromosomes, thereby allowing the search for a number of different gene products. In this respect, interspecific hybrid cells between TK-deficient mouse teratocarcinoma and HPRT-deficient rat hepatoma that had retained almost all of the mouse chromosomes and various numbers of rat chromosomes, seemed ideal for blastocyst injections. While the cell hybrids usually formed malignant tumors in athymic *nude* mice (Fig. 5), at least some of these tumor cells were still capable of reverting to a normal phenotype in chimeric mice (Illmensee and Croce 1979). In contrast to previous results, tissue contributions of the rat × mouse hybrid cells remained limited to the liver and a few other organs of endomesodermal origin, probably due to the rat hepatoma cell parent. Reversion of malignancy appeared more surprising because, on the one hand, the hybrid cells formed undifferentiated tumors in adult hosts and, on the other hand, differentiated normally into liver, lung, kidney, gut, and fat pad during in situ organogenesis. A comparable situation occurred with the human × mouse hybrid cells, which produced undifferentiated tumors in *nude* mice but contributed normally to a number of different tissues in chimeric mice. It therefore seems as if the prospective potential of the malignant hybrid cells remains limited in the adult host and will only be fully revealed in the environment of the embryo.

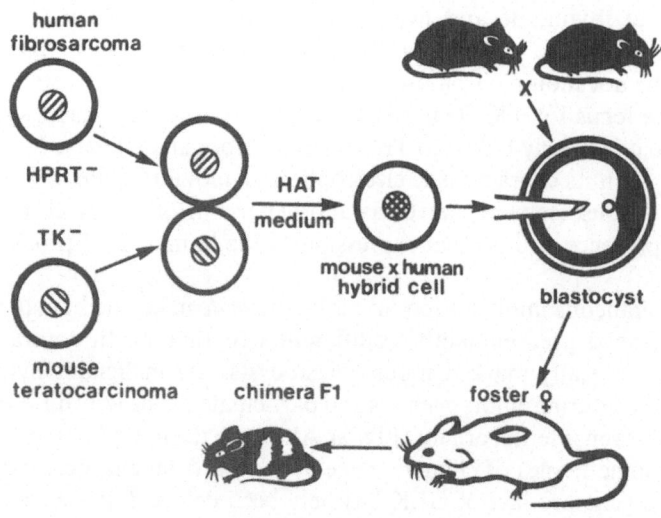

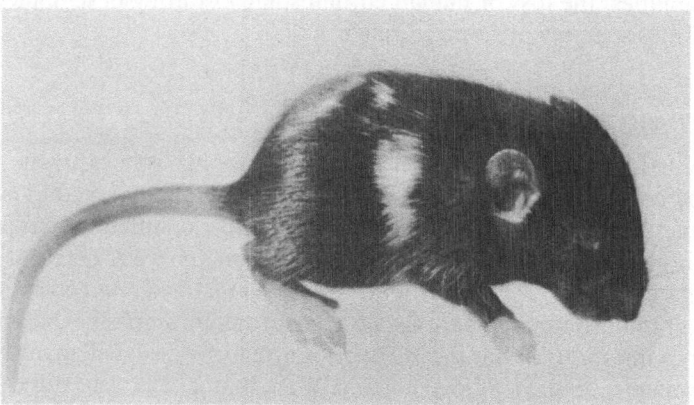

Fig. 4. Experimental scheme of cycling human × mouse hybrid cells through mice via blastocyst injection. Malignant hybrid cells between human fibrosarcoma, deficient in hypoxanthine phosphoribosyltransferase (HPRT⁻), and mouse teratocarcinoma, deficient in thymidine kinase (TK⁻), are selected in vitro and cultured in hypoxanthine/aminopterin/ thymidine (HAT) medium. A single hybrid cell is then injected into a C57BL/6 blastocyst carrying many genetic markers in order to detect any in vivo tissue differentiation derived from the injected cell. Shortly after micromanipulation, the blastocyst has to be implanted into the uterus of a foster mother to allow developmental to term. The live-born chimeric mouse shows normal hybrid-cell contributions in the coat, the white patches, which contrast to the black coat of the recipient and in seven internal tissues (Illmensee et al. 1978)

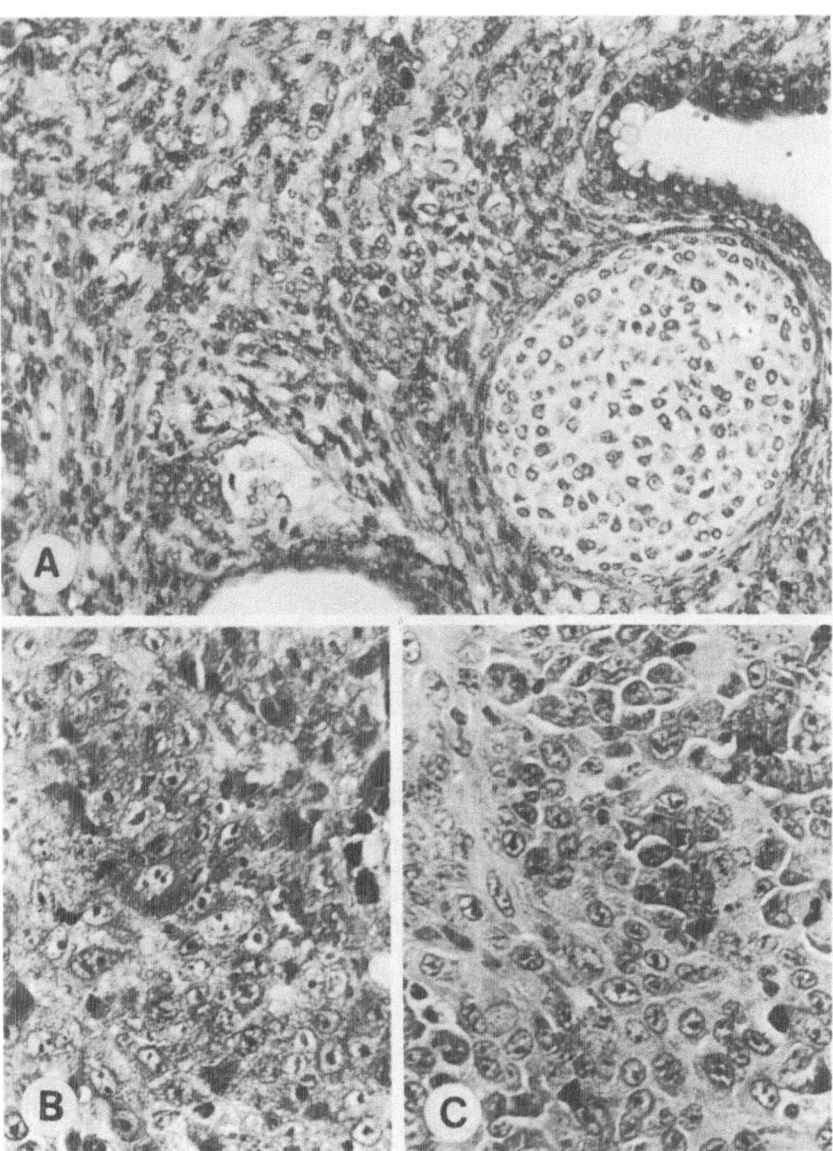

Fig. 5A-C. Malignant tumors derived from in vitro cultured cells of TK deficient mouse teratocarcinoma (**A**), HPRT deficient rat hepatoma (**B**), and mouse × rat hybrid (**C**) after injection into the kidney capsule of athymic *nude* mice. **A** Pluripotent teratocarcinoma shows cartilage, glandular epithelium, mesenchymal tissue, and embryonal carcinoma cells. **B** Rat hepatoma contains pycnotic hepatocytes. **C** Mouse × rat hybrid resembles morphologically the rat hepatoma parent to some extent but also contains clusters of embryonal carcinoma cells

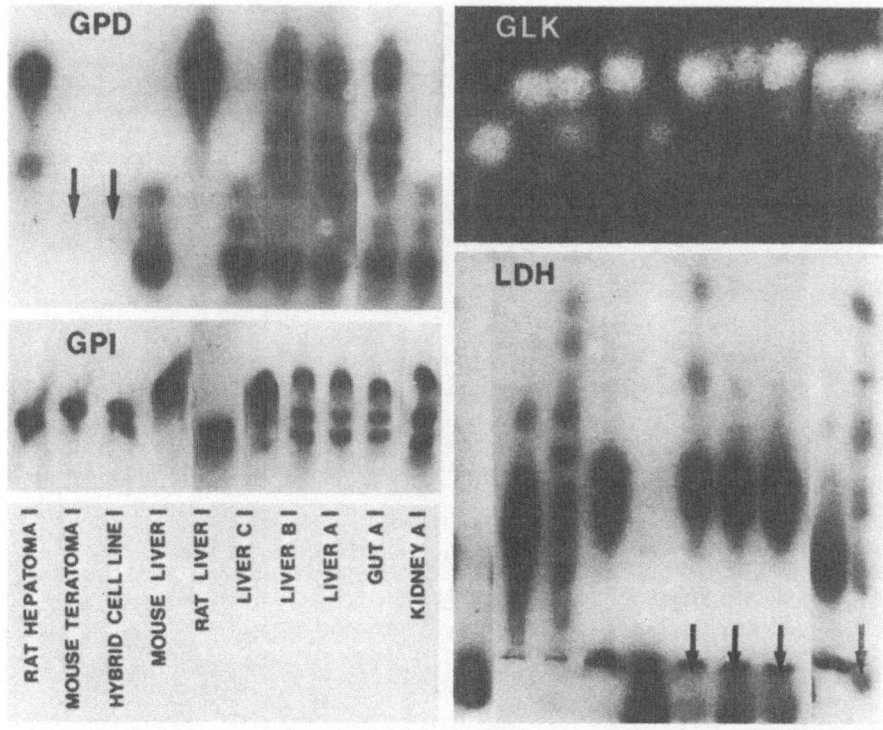

Fig. 6. Starch gel electrophoresis of rat- and mouse-specific enzyme variants of glycerolphosphate dehydrogenase (GPD), glycosephosphate isomerase (GPI), galactokinase (GLK), and lactate dehydrogenase (LDH). Cell extracts from liver, gut, and kidneys of chimeric mice A, B and C reveal rat × mouse hybrid cell contributions. Note the absence of GPD activity *(arrow)* in the hybrid cells and mouse teratoma and its appearance in the liver B and A as well as in gut A. The formation of rat-mouse heteropolymers is these mosaic tissues documents functional cooperation of the interspecific gene products. The appearance of adult rat LDH-5 *(arrow)*, not detectable in the hybrid cells, indicates differential modulation of the xenogeneic genes during in vivo development (Illmensee and Croce 1979)

But what happened to the rat chromosomal genes in the chimeric mice? Because the rat × mouse hybrid cells retained several rat chromosomes, it was possible to detect nine different rat-specific enzymes in the various mosaic tissues (Illmensee and Croce 1979). The appearance in vivo of rat gene products not detectable in the hybrid cells in vitro and the formation of heteropolymers with the corresponding mouse enzymes indicate a functional expression of the rat genes during mouse development. The synthesis of adult-specific enzyme variants further demonstrates the proper modulation of the xenogeneic genes during organogenesis (Fig. 6). Extending this kind of analysis to other genes (including X chromosomal ones) should reveal their developmental stage-specificity and ultimately give insight into the processes that control gene activity during mammalian differentiation.

IV. Prospectives

Recent progress in experimental embryology has provided new approaches to the genetic manipulation of the mammalian embryo. The production of uniparental mice following microsurgical removal of one of the two pronuclei from a fertilized egg will eventually allow us to study the expression of mutant genes, to compare maternal and paternal gene activity during development, to further elucidate the problem of X chromosome inactivation, and to gain more insight into the unsolved process of parthenogenesis in mammals.

The injection of teratocarcinoma cells into mouse blastocysts has led to the striking observation that the malignant stem cells of these tumors contribute to normal tissue differentiation in chimeric mice and has therefore drawn attention to the fact that the neoplastic state of a cell can sometimes be reversed under appropriate conditions. In a manner of speaking, a teratocarcinoma resembles a mammalian embryo to some extent (reviewed by Jacob 1975). In our investigations we exploit this phenomenon to study a central problem in development: how does a single cell, the fertilized egg, differentiate to form all the cell types and tissues of the organism? Since the teratocarcinoma cell is malignant and yet gives rise to normal tissues, we also expect to learn something from these tumors about the nature of neoplastic transformation. Now that it is possible to hybridize different kinds of cells with mouse teratocarcinoma cells and thereby make them carriers of foreign genetic material for integration into an excellent bioassay system—the tissues of a developing animal—the path should be open for new advances in the study of mammalian differentiation and neoplasia.

Acknowledgments. I should like to thank Drs. C.M. Croce, P.C. Hoppe, and L.C. Stevens for continuous collaboration. Our current research is supported by grants FN 3.183–0.77 and 3.183–1.77 from the Swiss National Science Foundation and by an appropriation from the G. and A. Claraz Foundation.

References

Beatty, R.A.: Parthenogenesis in vertebrates. In: Fertilization (eds. C.B. Metz, A. Monroy), Vol. I, pp. 413-440. New York-London: Academic Press 1967

Chandra, H.S., Brown, S.W.: Chromosome imprinting and the mammalian X-chromosome. Nature (London) 253, 165-168 (1975)

Chung, A.E., Estes, L.E., Shinozuka, H., Braginski, J., Lorz, C., Chung, C.A.: Morphological and biochemical observations on cells derived from the in vitro differentiation of the embryonal carcinoma cell line PCC4-F. Cancer Res. 37, 2072-2081 (1977)

Cooper, D.W.: A direct genetic change model for X-chromosome inactivation in eutherian mammals. Nature (London) 230, 292-294 (1971)

Dewey, M.J., Martin, D.W., Martin, G.R., Mintz, B.: Mosaic mice with teratocarcinoma-derived mutant cells deficient in hypoxanthine phosphoribosyltransferase. Proc. Natl. Acad. Sci. USA 74, 5564-5568 (1977)

Eicher, E.M., Washburn, L.L., Hoppe, P.C., Illmensee, K.: X-chromosome inactivation in gynogenetically and androgenetically derived female mice. In preparation (1980)

Evans, M.J.: Studies with teratoma cells in vitro. In: The Early Development of Mammals (eds. M. Balls, A.E. Wild), pp. 265-284. London: Cambridge University Press 1975

Graham, C.F.: The production of parthenogenetic mammalian embryos and the use in biological research. Biol. Rev. 49, 399-422 (1974)

Graham, C.F.: Teratocarcinoma cells and normal mouse embryogenesis. In: Concepts in Mammalian Embryogenesis (eds. M.I. Sherman, C.F. Graham), pp. 315-394. Cambridge: MIT Press 1977

Hoppe, C.P., Illmensee, K.: Microsurgically produced homozygous-diploid uniparental mice. Proc. Natl. Acad. Sci. USA 74, 5657-5661 (1977)

Illmensee, K.: Reversion of malignancy and normalized differentiation of teratocarcinoma cells in chimeric mice. In: Genetic Mosaics and Chimeras in Mammals (ed. L.B. Russell), pp. 3-25. New York: Plenum 1978

Illmensee, K., Croce, C.M.: Xenogeneic gene expression in chimeric mice derived from rat-mouse hybrid cells. Proc. Natl. Acad. Sci. USA 76, 879-883 (1979)

Illmensee, K., Hoppe, P., Croce, C.M.: Chimeric mice derived from human-mouse hybrid cells. Proc. Natl. Acad. Sci. USA 75, 1914-1918 (1978)

Jacob, F.: Mouse teratocarcinomas as a tool for the study of the mouse embryo. In: The Early Development of Mammals (eds. M. Balls, A.E. Wild), pp. 233-241. London: Cambridge University Press 1975

Kaufman, M.H., Guc-Cubrilo, M., Lyon, M.F.: X chromosome inactivation in diploid parthenogenetic mouse embryos. Nature (London) 271, 547-549 (1978)

Latt, S., Willard, H.F., Gerald, P.S.: BrdU-33258 Hoechst analysis of DNA replication in human lymphocytes with supernumerary or structurally abnormal X chromosomes. Chromosoma 57, 135-153 (1976)

Markert, C.L., Petters, R.M.: Homozygous mouse embryos produced by microsurgery. J. Exp. Zool. 201, 295-302 (1977)

Mintz, B., Illmensee, K., Gearhart, J.D.: Developmental and experimental potentialities of mouse teratocarcinoma cells from embryoid body cores. In: Teratomas and Differentitation (eds. M.I. Sherman, D. Solter), pp. 59-82. New York-London: Academic Press 1975

Modlinski, J.A.: Haploid mouse embryos obtained by microsurgical removal of one pronucleus. J. Embryol. Exp. Morphol. 33, 897-905 (1975)

Papaioannou, V.E., Gardner, R.L., McBurney, M.W., Babinet, C., Evans, M.J.: Participation of cultured teratocarcinoma cells in mouse embryogenesis. J. Embryol. Exp. Morphol. 44, 93-104 (1978)

Sherman, M.I., Solter, D. (eds.): Teratomas and Differentiation. New York-London: Academic Press 1975

Stevens, L.C.: Teratocarcinogenesis and spontaneous parthenogenesis in mice. In: The Developmental Biology of Reproduction (eds. C.L. Markert, J. Papaconstantinou), pp. 93-106. New York-London: Academic Press 1975

Takagi, N.: Preferential inactivation of the paternally derived X chromosome in mice. In: Genetic Mosaics and Chimeras in Mammals (ed. L.B. Russell), pp. 341-360. New York: Plenum 1978

Tarkowski, N.K.: In vitro development of haploid mouse embryos produced by bisection of one-cell fertilized eggs. J. Embryol. Exp. Morphol. 38, 187-202 (1977)

West, J.D., Papaioannou, V.E., Frels, W.I., Chapman, V.M.: Preferential expression of the maternally derived X chromosome in extraembryonic tissues of the mouse. In: Genetic Mosaics and Chimeras in Mammals (ed. L.B. Russell), pp. 361-377. New York: Plenum 1978

White, M.: Modes of Speziation. San Francisco: Freeman 1978

The Analysis of Cell Differentiation by Hybridization of Somatic Cells

M.C. WEISS

Centre de Génétique Moléculaire du C.N.R.S., 91190 Gif-Sur-Yvette, France

I. Introduction

This paper will give a very brief review of some of the major observations made on somatic hybrid cells concerning the expression of differentiated functions. Only a few examples will be cited since exhaustive reviews may be found in the literature (Ephrussi 1972; Davidson 1974; Ringertz and Savage 1976; Weiss 1977). The six principal interactions observed in hybrid cells will be described and in a few cases illustrated, and their relevance to our understanding of cell differentiation will be considered.

Hybridization of somatic cells permits the confrontation of two genomes in different functional states, and from the properties of the resulting hybrid cells conclusions can be drawn concerning first, the "dominant" or "recessive" nature of the expressing and nonexpressing states, and second, the heritability of the commitment of a genome to carry out a given pattern of gene expression.

Cell hybridization is usally performed with cells of permanent lines, often derived from tumors. We are therefore dealing with abnormal cells that have altered karyotypes. Some permanent lines are characterized by the expression of functions specific to the tissue of origin, and these are refered to as "differentiated" or "well-differentiated" cells. For many lines, we are not aware of the production of tissue-specific proteins, but this may be because we do not know what to look for or have not used sufficiently sensitive assays. In any case, all permanent lines of fetal or adult somatic cells must be considered as committed cells, and should not be referred to as undifferentiated. I will speak of expressing and nonexpressing cells with reference to the production of a specific protein or set of proteins characteristic of a given tissue type.

The expressing and nonexpressing cells used in crosses are most often of different histogenetic origin. In a few cases, the cells are of the same tissue (and even cell line) origin, one parental line being a stable variant or "mutant" that no longer expresses differentiated functions.

As will be seen below, cell hybridization has permitted us to document several types of interactions that occur between expressing and nonexpressing cells, to establish what parameters are critical in determining these interactions, and to deduce which aspects of the expression of the differentiated state involve diffusible substances, or reflect an autonomous heritable mechanism. It is only when cell hybridization is used in conjunction with the appropriat biochemical analyses that one can hope to identify the specific mechanisms involved, and to determine their roles in normal development and differentiation.

II. Interactions in Hybrid Cells

We will turn now to a consideration of some of the interactions observed in hybrid cells formed by crossing parental cells which do and do not express specific differentiated functions.

Extinction of differentiated functions is nearly always observed in crosses between expressing and nonexpressing cells of approximately equivalent ploidy (Ephrussi 1972). This was first described for melanogenesis in hybrids between hamster melanoma cells and mouse fibroblasts (Davidson et al. 1966). The examples of extinction in hybrid cells are by now too numerous to list, but it can be added that (1) this phenomenon is observed for essentially all of the functions of a given cell type (Weiss et al. 1975; Conscience et al. 1977); (2) extinction may be only partial (Peterson 1974); (3) it may be observed when the nonexpressing parent is a variant of the same line as the expressing parent (Deschatrette and Weiss 1975), and its basis appears to be pretranslational since mRNA for an extinguished protein cannot be detected in hybrid cells (Deisseroth et al. 1975). These results suggest that cells not expressing a given function produce diffusible substances, whose final effects are negative, that prevent expression of the function (Davidson 1974).

Gene Dosage Effects refer to the fact that extinction may not be observed in crosses where the ploidy of the expressing partner is much higher than that of the nonexpressing parent. It is even possible to demonstrate that for a given cross the inclusion in the hybrid of a single set of chromosomes of the expressing parent results in extinction, but a double set of chromosomes from the same parent may lead to continued expression in the hybrid cell (Davidson 1972; Fougère et al. 1972). When extinction is not observed in hybrids owing to gene dosage effects, one generally observes expression of the entire set of functions of the expressing parent (Brown and Weiss 1975). These observations suggest that the factors responsible for extinction (and perhaps also of expression) are produced in relatively small amounts.

Activation is the synthesis, directed by the nonexpressing genome, of one or more of the differentiated proteins characteristic of the expressing parent. Activation of "silent" genes to express liver specific proteins has been observed in crosses of hepatoma cells with fibroblasts (Peterson and Weiss 1972), lymphoblasts (Malawista and Weiss 1974), lymphocytes (Darlington et al. 1974), and melanoma cells (Fougère and Weiss 1978), all cell types that would normally never produce these proteins. Activation can be observed (1) in hybrids where extinction fails to occur owing to gene dosage effects, (2) in hybrids where extinction is only partial (Malawista and Weiss 1974; Brown and Weiss 1975), or (3) in hybrids showing reexpression (see below) of a previously extinguished protein (Bertolotti 1977; Fougère and Weiss 1978). The fact that activation occurs suggests that differentiated cells produce diffusible substances capable of inducing the expression of a previously silent gene.

Reexpression is the reappearance of a previously extinguished function, apparently correlated with loss from the hybrid cells of chromosomes of the nonexpressing parent (Klebe et al. 1970; Weiss and Chaplain 1971), and the reexpression of the various functions of a given cell type appears to occur independently (Sparkes and Weiss 1973; Bertolotti and Weiss 1974). Reexpression demonstrates that extinction imposes merely a transitory block to the final expression of differentiated functions without altering the potential of the genome to reestablish its original functional activity.

Retention of Pluripotentiality has been observed in hybrids between pluripotent mouse teratoma cells and normal mouse thymocytes (Miller and Ruddle 1976) or nearly diploid Friend leukemia cells (Miller and Ruddle 1977). By contrast, hybrids between such teratoma cells and hyperdiploid mouse L cells (Finch and Ephrussi 1967; Jami et al. 1973) showed extinction of pluripotentiality. This is the first (and so far only) instance encountered among the numerous kinds of hybrids studied up until now where a modification in determination of one of the parental genomes has been observed. The retention of pluripotentiality of these hybrids implies that the genome of the somatic thymocyte or Friend cell has been reversed to an "undetermined" state; however, other interpretations remain possible and further experimentation is required to clarify this question.

Exclusion of the simultaneous expression of functions characteristic of two different differentiations has been observed in mouse melanoma-rat hepatoma hybrids. While mutual extinction of pigment and albumin production is observed in these hybrids, reexpression of both of these functions can be observed. However, in a given cell, only one of the two parental functions is expressed at a time, and it has been possible to demonstrate successive and exclusive shifts of phenotype (flip-flop) in hybrid subclones (Fougère and Weiss 1978). Thus, hybrid cells can conform to one of the fundamental rules of normal development, the law of exclusivity formulated by Paul Weiss (1939).

III. Interpretations and Conclusions

Having seen the basic interactions observed in hybrid cells concerning the expression of differentiated functions and the heritability of determination, we must now consider what these results tell us about cell differentiation.

Extinction, dosage effects and activation provide insight into the problem of the regulation of overt expression of differentiated functions, where the final effect is either negative (extinction) or positive (activation). Since these interactions occur when two genomes in different functional states are simply confronted, we are forced to conclude that diffusible substances are involved, although we have as yet no idea of their chemical nature nor of their sites of action. In the case of activation however, it appears highly probable that the substance is acting at the gene level (Brown and Weiss 1975).

Reexpression of previously extinguished functions by segregated hybrid cells demonstrates that the determination of somatic cells can be inherited in the absence of overt expression. We must therefore conclude that determination is due to an autonomous mechanism that is not altered or lost in cell hybrids. However, we have already seen above that diffusible substances can act in cell hybrids to alter drastically the pattern of proteins that is actually produced, including cases of activation of a genome to produce a protein foreign to its own determination. This apparent contradiction is most clearly illustrated by the case of mouse melanoma-rat hepatoma hybrids (Fougère and Weiss 1978) cited above to illustrate the phenomenon of exclusion in hybrid cells. In these cells, only pigment *or* albumin is produced at any one time; however, when albumin is produced, most of it is of the mouse type. The albumin producing hybrid cells retain the potential to synthesize pigment, for pigmented subclones can be isolated from the progeny of albumin-producing cells. It is therefore clear that both parental determinations coexist in the hybrid nucleus, in spite of the fact that the mouse melanoma genome has been activated to participate in the synthesis of a liver protein.

From these various observations of interactions in hybrid cells, it appears clear that there are at least two fundamentally different levels of regulation in mammalian somatic cells, that concerned with *determination* or genomic commitment, and that responsible for *expression* of tissue-specific proteins. The former is characterized by an extraordinary degree of stability and is not lost when cells of different determinations are hybridized; the latter appears to be regulated by *trans*-acting substances that can have a final effect that is negative (in the case of extinction) or positive (activation), without causing loss of the original determination of the cell's genome.

One of the first questions we must pose concerning these conclusions is whether the interactions observed in hybrid cells reflect normal control mechanisms that play a role in development (Ephrussi 1972). Although there is no direct answer to this question, there are four indirect arguments in favor of the idea that they do. The first series of arguments is concerned with the general properties of somatic hybrid cells. Such cells, whether intraspecific or interspecific, are perfectly viable: polymeric enzymes are composed of subunits from both parents and show normal activity, and cell cycle regulation occurs normally, implying that regulatory signals involved in these events do not show genomic or species specificity (Ephrussi 1972). Secondly, reexpression of the multiple differentiated functions characteristic of a given tissue type occurs independently, demonstrating that extinction is not due to a nonspecific messing up of regulation; if the latter were the case, reexpression would be coordinate. Thirdly, melanoma-hepatoma hybrids conform to the law of exclusivity in the expression of parental differentiations. Finally, it has been observed that many of the same genomic interactions that are observed in hybrids formed between cells of different histotypes and different species are encountered when cells of the same histotype and same species are crossed (Fourquignon and Ephrussi, in preparation). For all of these reasons, it appears reasonable to conclude that hybrid cells are not simply abnormal monsters, but rather valid experimental models for exploring problems of regulation in cells of higher organisms.

While a final answer to the question of the mechanisms involved in these interactions and their normal or abnormal nature will only be resolved by further and more detailed analyses, it is clear that the existence of pure populations of cells, for which it is possible to alter gene expression experimentally in a predictable fashion, will provide precious material for the molecular analysis of the differentiation of normal and malignant cells.

Acknowledgments. The author is grateful to Prof. Boris Ephrussi and to many colleagues in the laboratory, numerous discussions with whom have led to the interpretations given in this paper.

References

Bertolotti, R.: Expression of differentiated functions in hepatoma cell hybrids: selection in glucose-free media of hybrid cells which reexpress gluconeogenic enzymes. Somat. Cell Genet. 3, 579-602 (1977)

Bertolotti, R., Weiss, M.C.: Expression of differentiated functions in hepatoma cell hybrids V. Re-expression of aldolase B in vitro and in vivo. Differentiation 2, 5-17 (1974)

Brown, J.E., Weiss, M.C.: Activation of production of mouse liver enzymes in rat hepatoma-mouse lymphoid cell hybrids. Cell 6, 481-494 (1975)

Conscience, J.F., Miller, R.A., Henry, J., Ruddle, F.H.: Acetylcholinesterase, carbonic anhydrase and catalase activity in Friend erythroleukemia cells, non-erythroid mouse cell lines and their somatic hybrids. Exp. Cell Res. 105, 401-412 (1977)

Darlington, G.J., Bernhard, H.P., Ruddle, F.H.: Human serum albumin phenotype activation in mouse hepatoma-human leukocyte cell hybrids. Science 185, 859-862 (1974)

Davidson, R.L.: Regulation of melanin synthesis in mammalian cells: effect of gene dosage on the expression of differentiation. Proc. Natl. Acad. Sci. USA 69, 951-955 (1972)

Davidson, R.L.: Gene expression in somatic cell hybrids. Annu. Rev. Genetics 8, 195-218 (1974)

Davidson, R.L., Ephrussi, B., Yamamoto, K.: Regulation of pigment synthesis in mammalian cells, as studied by somatic hybridization. Proc. Natl. Acad. Sci. USA 56, 1437-1440 (1966)

Deisseroth, A., Burk, R., Picciano, D., Minna, J., Anderson, W.F., Neinhuis, A.: Hemoglobin synthesis in somatic cell hybrids: globin gene expression in hybrids between mouse eryhtroleukemia and human marrow cells or fibroblasts. Proc. Natl. Acad. Sci. USA 72, 1102-1106 (1975)

Deschatrette, J., Weiss, M.C.: Extinction of liver-specific functions in hybrids between differentiated and dedifferentiated rat hepatoma cells. Somat. Cell Genet. 1, 279-292 (1975)

Ephrussi, B.: Hybridization of Somatic Cells. Princeton, NJ: Princeton University Press 1972

Finch, B.W., Ephrussi, B.: Retention of multiple developmental potentialities by cells of a mouse testicular teratocarcinoma during prolonged culture in vitro and their extinction upon hybridization with cells of permanent lines. Proc. Natl. Acad. Sci. USA 57, 615-621 (1967)

Fougère, C., Weiss, M.C.: Phenotypic exclusion in mouse melanoma-rat hepatoma hybrid cells: pigment and albumin production are not reexpressed simultaneously. Cell 15, 843-854 (1978)

Fougère, C., Ruiz, F., Ephrussi, B.: Gene dosage dependence of pigment synthesis in melanoma × fibroblast hybrids. Proc. Natl. Acad. Sci. USA 69, 330-334 (1972)

Jami, J., Failly, C., Ritz, E.: Lack of expression of differentiation in mouse teratoma-fibroblast somatic hybrids. Exp. Cell Res. 76, 191-199 (1973)

Klebe, R.J., Chen, T., Ruddle, F.H.: Mapping of a human regulator element by somatic cell genetic analysis. Proc. Natl. Acad. Sci. USA 66, 1220-1227 (1970)

Malawista, S., Weiss, M.C.: Expression of differentiated functions in hepatoma cell hybrids: high frequency of induction of mouse albumin production in rat hepatoma-mouse lymphoblast hybrids. Proc. Natl. Acad. Sci. USA 71, 927-931 (1974)

Miller, R.A., Ruddle, F.H.: Pluripotent teratocarcinoma-thymus somatic cell hybrids. Cell 9, 45-55 (1976)

Miller, R.A., Ruddle, F.H.: Teratocarcinoma × Friend erythroleukemia cell hybrids resemble their pluripotent embryonal carcinoma parent. Dev. Biol. 56, 157-173 (1977)

Peterson, J.A.: Discontinuous variability, in the form of a geometric progression, of albumin production in hepatoma and hybrid cells. Proc. Natl. Acad. Sci. USA 71, 2062-2066 (1974)

Peterson, J.A., Weiss, M.C.: Expression of differentiated functions in hepatoma cell hybrids: induction of mouse albumin production in rat hepatoma-mouse fibroblast hybrids. Proc. Natl. Acad. Sci. USA 69, 571-575 (1972)

Ringertz, N., Savage, R.E.: Cell Hybrids. New York-London: Academic Press 1976

Sparkes, R.S., Weiss, M.C.: Expression of differentiated functions in hepatoma cell hybrids: VII Alanine aminotransferase. Proc. Natl. Acad. Sci. USA 70, 377-381 (1973)

Weiss, M.C.: The use of somatic cell hybridization to probe the mechanisms which maintain cell differentiation. In: Human Genetics (eds. S. Armendares, R. Lisker), pp. 284-292. Amsterdam-Oxford: Excerpta Medica 1977

Weiss, M.C., Chaplain, M.: Expression of differentiated functions in hepatoma cell hybrids. III. Re-expression of tyrosine aminotransferase inducibility. Proc. Natl. Acad. Sci. USA 68, 3026-3030 (1971)

Weiss, M.C., Sparkes, R.S., Bertolotti, R.: Expression of differentiated functions in hepatoma cell hybrids: IX. Extinction and re-expression of liver-specific enzymes in rat hepatoma-chinese hamster fibroblast hybrids. Somat. Cell Genet. 1, 27-40 (1975)

Weiss, P.: Principles of Development. New York: Holt 1939

Chromosomal Changes Associated with Premalignancy and Cancer

J. Cervenka

Division of Oral Pathology and Oral and Human Genetics
School of Dentistry, Department of Medicine
Medical School University of Minnesota, Minneapolis, MN 55455, USA

I. Introduction

In this overview I will concentrate on the data derived from studies in human patients. In certain cases it is only with difficulty that we allow ourselves the diagnosis of "malignancy". The transitional states of the disease, the intermediate steps of the malignant transformation are only too well known to the pathologist, and criteria of cancer or leukemia are frequently based on arbitrarily set limits of cellular qualities and quantities. The transition from "dysplasia" to "carcinoma in situ" to "carcinoma" manifests itself as a continuous process which, out of necessity, we attempt to categorize into three stages. In cytogenetics it appears likely that the most informative data on the malignant transformation and etiologic agents involved would be obtained by studying the very early stages of this process or possibly even by the study of events preceding any detectable changes of cellular morphology and physiology related to cancer. In "preneoplasia" or "premalignancy" we frequently have the opportunity to study the subtle anomalies of cell differentiation, the biological characteristics and chromosomal constitutions which precede the apparent abnormality of the cancer cell.

The term "premalignancy", however, can be understood in several ways. The broader definition would include any state of an organism (patient) in a situation where the risk of developing cancer is increased in comparison with that of a random population: an increased age of the individual; a predisposition to cancer due to the occurrence of malignancy in siblings and in the ascendant line of the pedigree; preceding consanguineous unions and inbreeding (Schull 1977); and also the exposure of the individual to an environment known to increase the mutation rate; for instance the intense solar radiation in the near-equatorial regions in relation to skin cancers (Segi 1963; Cervenka et al. 1979).

More narrowly defined, premalignancy would include disorders which, not having characteristic of malignant diesease per se, predispose the patient to development of leukemia or solid neoplasia. There are a great number of these disorders, both congenital and acquired, inherited and nongenetic. This review will be limited to only those which are characterized by detectable chromosomal changes: (1) *congenital chromosomal anomaly syndromes,* (2) *preleukemia with abnormal clones,* (3) *chromosomal instability syndromes,* and (4) *other situations with chromosomal aberrations.*

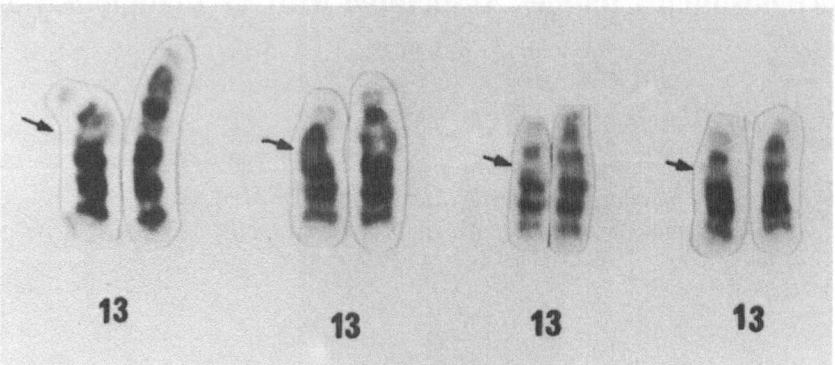

Fig. 1. Four pairs of chromosome 13 from four lymphocytic mitosis from a patient with congenital 13q- anomaly and bilateral retinoblastoma. *Arrows* indicate interstitial deletion of the band q13 and part of q14 (pter q12 q14 qter)

II. Congenital Chromosomal Anomaly Syndromes

Since the original observation by Krivit and Good (Krivit and Good 1956, 1957), it has been confirmed repeatedly that in the trisomy 21 Down's syndrome the prevalence of leukemia is about 20 times higher than in the general population, and also it occurs with earlier onset (Miller 1970). Further, it has been observed that other trisomies predispose the individual to cancers and leukemias: XXY Klinefelter's syndrome, +13, +18, XXX, and trisomy 8 mosaicism (Harnden et al. 1969, 1971; Nevin et al. 1972; Opitz and Herrmann 1977). Other chromosomal anomalies, such as balanced translocations and deletions, have been implicated (Blattner et al. 1976). Of chromosomal deletions, the best known is the interstitial deletion of the long arm of chromosome 13 (Fig. 1). Congenital deletion of probably band q14 of No. 13 causes mental retardation, failure-to-thrive in some, anomalies of the heart and eye, prominent frontal and nasal bones (Orye et al. 1974; Yunis and Chander 1977). The preliminary data show that in about 25–40% of these patients retinoblastoma will develop (Francois et al. 1975; Niebuhr 1977). On the other hand, in about 10% of all patients with retinoblastoma, this deletion anomaly will be found.

In analogy with 13q-syndrome, the triad of *aniridia, genito-urinary anomalies*, and *mental retardation* results, in some cases, from interstitial deletion of the short arm of chromosome 11. The association of Wilms' tumor and 11p- has been found in three out of nine patients with 11p- syndrome recently by Riccardi et al. (1978). Wilms' tumor is an embryonal nephroblastoma of heterogeneous etiology; about one-third of cases show autosomal dominant transmission. It has be noted that other chromosomal anomalies were found to be associated with retinoblastoma and Wilms' tumor. Both tumors also occur without any association with chromosomal anomalies, either as an autosomal dominant trait or sporadically.

The analysis of data on retinoblastoma, Wilms' tumor, and other dominant neoplasias formed the basis of Knudson's two-mutation hypothesis of oncogenesis

(Knudson 1971, 1974). Kundson postulates that in familial cases, first germinal mutation occurs and later a second, somatic mutation triggers the malignant transformation. In sporadic cases, both mutations occur in the somatic cell. However, a delayed mutation model has been favored by some (Herrmann 1977).

III. Preleukemias with Abnormal Clones

Well-defined and less well-defined disorders in this category include for instance polycythemia vera, pancytopenia (non-Fanconi's anemia), myelofibrosis, and other less frequent hematopoietic deficiencies. Clones with chromosomal translocations, monosomies, and deletions have been described in these disorders; trisomies include mainly those of chromosomes 1, 8, 9, 19, and 20. Nowell's comprehensive review (Nowell et al. 1976) of the cytogenetics of "preleukemia" patients summarizes data on 106 patients of whom 30 had chromosomally abnormal clones of cells from peripheral blood or bone marrow. Later, half of these 30 patients developed true leukemia. Of the remaining 76 patients with normal chromosomal constitution, 14 (18.4%) developed leukemia.

It appears that the emergence of a chromosomally abnormal clone in a patient with a hematopoetic disorder enhances the risk of leukemia, especially in nonspecific pancytopenia. Detection of this tendency, it is hoped, can be exploited in preventive therapy.

IV. Chromosomal Instability Syndromes

Sometimes designated as chromosomal breakage syndromes, this category includes a number of genetic disorders characterized by either elevated chromosomal breakage rate, increased rate of sister chromatid exchanges (SCE) and/or a defect in DNA repair. All are autosomal recessive, have an impaired immunologic system, and predispose to neoplasias. Well established are Fanconi's anemia, Bloom's syndrome, Louis-Bar syndrome, and a series of disorders grouped under the term xeroderma pigmentosum; the latter has a normal frequency of chromosomal breaks and SCE. The mechanism of chromosomal aberrations is not known; however, the high frequency of breaks is known to lead to frequent chromosomal rearrangements, to the emergence of abnormal clones, and eventually to malignancy (Hecht et al. 1973). For instance, a rise in pseudo-diploid clones, mostly with translocations involving chromosome 14, has been well described in Louis-Bar syndrome (ataxia telangiectasia) (Hecht et al. 1973; McCaw et al. 1975). Still, other possible mechanisms of oncogenesis could be postulated: an increased tendency to malignant transformation by viruses has been demonstrated in Fanconi's anemia cells in vitro (Todaro et al. 1966). The striking increase of SCE's in Bloom's syndrome from the normal rate of 6–10 SCE/mitosis to around 100 SCE/mitosis (Chaganti et al. 1974) reflects chromosomal instability which is not clearly understood in its relationship to oncogenesis.

There are a number of other congenital syndromes where both the chromosomal instability and frequent occurrence of malignancy have been demonstrated. These include scleroderma (Emerit and Housset 1973), incontinentia pigmenti (Cantu et al. 1973), dyskeratosis congenita (Burgdorf et al. 1977b), and nevoid basal cell carcinoma syndrome (Bazopoulou-Kyrkanidou and Cervenka, unpubl. data; Horland et al. 1975). However, in all of these conditions additional cytogenetic data are still desirable.

V. Other Situations with Chromosomal Aberrations

This category cannot be either exhaustively covered or dismissed. It includes a variety of conditions and familial tendencies where chromosomal defects coincide with cancer, thus suggesting an etiologic relationship. A number of pedigrees were reported where several members were affected by leukemia or myeloma and their unaffected relatives carried chromosomal aberrations, such as +F, +21, +B, translocations (B,G), (D,G), and (C,G) (Cervenka et al. 1977). At least three "leukemia families" are known, where there was considerable increase of tetraradial and triradial chromosomal exchange figures in lymphocytes of first degree releatives of members affected with leukemia (Cervenka et al. 1977). A high frequency of SCE in lymphocytes of patients with skin cancers who were exposed to arsenic treatment has been reported (Burgdorf et al. 1977a). Further, the "natural" process of aging as a "premalignant" state with chromosomal anomalies should not be omitted, since it has been established that with increasing age, both the rate of chromosomal breaks (Littlefield and Goh 1973) and the frequency of cancers increase. A multitude of cancerogenic chemicals and viruses which have clastogenic potential (breaking chromosomes) have been uncovered and could be included here.

VI. Research of Sister Chromatid Exchanges (SCE)

It is evident that SCE occur 200 times more frequently in normal cultured lymphocytes and fibroblasts than chromosomal breaks (Fig. 2A). Thus, elevation of their frequency could be easily observed, and consequently SCE are potentially more sensitive indicators of mutagenesis than breakage or structural and numerical chromosomal aberrations. While there is no known causative correlation between the SCE frequency and frequency of breaks, both are elevated by the action of several cancerogenes (Perry and Evans 1975; Stetka and Wolff 1976). In premalignant disorders, the most striking example remains the Bloom's syndrome, where the SCE number for one diploid mitosis reaches 100 (Chaganti et al. 1974).

To illustrate briefly the variation of SCE frequency in different conditions, I shall present examples from our research. The following data in Tables 1 and 2 are mostly the work of K. Kurvink, W. Burgdorf, J. Wiencke, C. Yasis, and E. Bazopoulou-Kyrkanidou, of my laboratory at the University of Minnesota, and of colleagues from Niigata University in Japan and the University of Nigeria in Enugu.

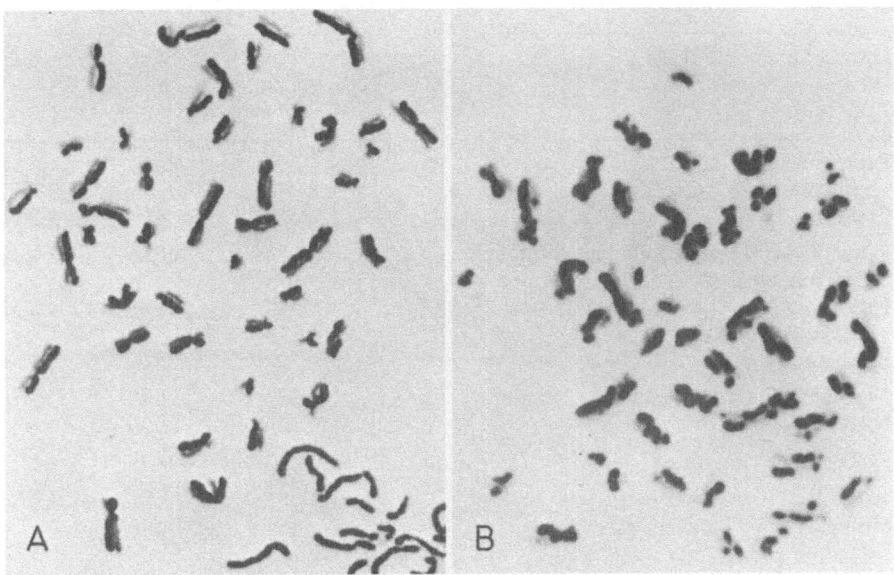

Fig. 2A,B. Sister chromatid differential staining of peripheral lymphocyte mitoses. **A** Normal mitosis with 7 SCE. **B** Mitosis of lymphocyte exposed in vitro to cis-2-2-diaminodichloroplatinum with manifold increase to 50 SCE

The interpretation of the presented data could not be done without considering separately each group of disorders studied. Significant were our observation of difference between matched controls and patients with chronic exposure to arsenic, who all later developed skin cancers; with tuberous sclerosis; with deSanctis-Cacchione type of xeroderma pigmentosum treated with high doses of the somewhat mysterious Murayama vaccine; and patients with untreated Hodgkin's and non-Hodgkin's lymphomas. In most of the listed disorders our studies continue. With the exception of Bloom's syndrome, the use of SCE analysis for diagnostic purposes or for monitoring therapy is still not practical in the majority of premalignancies and malignancies, while its use in uncovering genetic toxicity in a number of chemical substances has been demonstrated clearly by several authors (Perry and Eans 1975; Stetka and Wolff 1976; Wolff 1977).

VII. Chromosomes in Solid Tumors and Leukemias

Most information on the role of chromosomal rearrangements in malignancy has come not from observations on humans, but from cell hybridization experiments and studies of induced tumors and leukemias in experimental animals. Studies of heterokaryons show that the addition of a specific chromosome or a specific translocation either maintains the malignant character of the cell or suppresses it (Harris 1971; Hitotsumachi et al. 1972; Codish and Paul 1974). Also, it has been suggested that a specific cancerogene induces malignancy with a specific

Table 1. SCE frequencies in peripheral lymphocytes of patients with diverse disorders

Disorder or condition	Number of subjects	Number of diploid mitoses	SCE/ mitosis	SD
Patients with chronic arsenic exposure plus cutaneous cancers	6	196	14.0	±3.12
Controls	44	1564	6.0	±2.1
Patients with chronic cadmium exposure (Itai Itai disease)	14	283	10.66	2.34
Patients with chronic mercury exposure (Minamata disease)	11	140	11.43	2.36
Japanese controls	6	90	9.6	1.56
Nigerian albinos with tendency to cancer	9	137	6.43	
Nigerian controls	3	49	5.20	
Patients with viral disease	17	552	8.7	±2.9
Controls	44	1564	6.0	±2.1
Multiple nevoid basal cell carcinoma syndrome	6	164	14.73	
Controls	6	184	8.12	
Sezary syndrome	3	126	11.3	
Fanconi's anemia	5	245	7.6	3.6
Tuberous sclerosis	1	79	15.6	5.8
Dyskeratosis congenita	1	106	12.2	3.35
Aplastic anemia (non-Fanconi)	1	22	13.2	
Xeroderma pigmentosum deSanctis Cacchione syndrome treated by "Murayama vaccine"	1	20	17.0	3.6
Seckel dwarfs	2	56	8.4 and	8.0

Table 2. SCE frequencies in peripheral lymphocytes of patients with untreated malignancies[a]

Type of malignancy	Number of subjects	Number of diploid mitoses	SCE per mitosis	SD
Malignant lymphoma	13	292	12.7	0.9
Wilm's tumor	2	65	7.9	3.3
Seminoma	2	116	9.0	3.6
Ovarian carcinoma	3	43	7.2	2.3
Controls	40	1540	6.0	0.3

[a] Unpublished data by Dr. Karen Kurvink

chromosomal aberration (Mitelman et al. 1972). However, generalization from experimental data to all cancers is not tenable since the chromosomal aberration has not been found to be a prerequisite of malignant cells. A number of acute leukemias show normal diploid karyotypes as determined by present methods. Further, there are no asurances that experimental cancerogenesis is directly comparable to the transformation in vivo, especially concerning its very early stages.

A vast amount of cytogenetic data is available on human tumors and leukemias (Cervenka and Koulischer 1973;, Codish and Paul 1974). They indicate the potential importance of cytogenetics in diagnosis, treatment, and the search for etiologic factors. Because of the space limitation, I would summarize that there are only three human malignancies recognized with specific chromosomal aberrations, and even these include exceptions. *Chronic myeloid leukemia* blastic cells contain, in over 90% of cases, Ph_1 chromosomes, that is No. 22 of which part of the long arm has been deleted and translocated to the long arm of No. 9 (Rowley 1973). *Burkitt lymphoma* and derived cell lines are characterized in about 70% of cases by translocation of a portion of 8q to the long arm of No. 14 (Zech et al. 1976). However, the marker 14q+ as an unbalanced translocation has been observed in non-Burkitt lymphomas, too (Mark 1977). Finally, evidence has been accumulated that the loss of chromosome 22 is the most consistent anomaly in human *meningiomas* (Mark et al. 1972; Mark 1977). Recently, attention has been paid to the clonal evolution of leukemias and solid tumors and nonrandom chromosomal patterns have been detected. Preferential involvement of certain chromosomes, duplication of markers, and loss of sex chromosomes have been observed.

In summary, the causative relationship of gross chromosomal mutation and human neoplasia is not well understood and remains a hypothesis, however persuasive. Enough data have been accumulated which show definite association of specific chromosomal aberration with either specific types of malignancy or an effect of a certain cancerogenic agent. At this stage of knowledge, we are thus not capable of answering the basic question whether chromosomal aberration could be the primary triggering event in malignant transformation. However, it appears that in specific instances this might well be the case.

References

Bazopoulou-Kyrkanidou, E., Cervenka, J.: Unpublished data

Blattner, W.A., Whang-Peng, J., Kistenmacher, M.: Unpublished data (1976)

Burgdorf, W., Kurvink, K., Cervenka, J.: Elevated sister chromatid exchange rate in lymphocytes of subjects treated with arsenic. Hum. Genet. 36, 69-72 (1977a)

Burgdorf, W., Kurvink, K., Cervenka, J.: Sister chromatid exchange in dyskeratosis congenita lymphocytes. J. Med. Genet. 14, 256-257 (1977b)

Cantu, J.M., del Castillo, V., Jiminez, M., Ruiz-Barquin, E.: Chromosomal instability in incontinentia pigmenti. Ann. Genet. 16, 117-119 (1973)

Cervenka, J., Koulischer, L.: Chromosomes in Human Cancer. Springfield, Ill.: Thomas 1973

Cervenka, J., Anderson, R.S., Nesbit, M.E., Krivit, W.: Familial leukemia and inherited chromsomal aberration. Int. J. Cancer 19, 783-788 (1977)

Cervenka, J., Witkop, C.J., Jr., Okoro, A.N., King, R.A.: Chromosome breaks and sister chromatid exchanges in albinos in Nigeria. Clin. Genet.,15, 17–21 (1979)

Chaganti, R.S.K., Schonberg, S., German, J.: A manyfold increase in sister chromatid exchanges in Bloom's syndrome lymphocytes. Proc. Natl. Acad. Sci. USA 71, 4508-4512 (1974)

Codish, S.D., Paul, B.: Reversible appearance of a specific chromosome which suppresses malignancy. Nature (London) 252, 610 (1974)

Emerit, I., Housset, E.: Chromosomal breakage in scleroderma. Possible presence of a breakage factor in the serum of patients. Ann. Genet. 16, 135-138 (1973)

Francois, J., Matton, M.T., DeBie, S., Tanaka, Y., Vandenbulcke, D.: Genesis and genetics of retinoblastoma. Ophthalmologica 170, 405-425 (1975)

German, J. (ed.): Chromosomes and Cancer. New York-London: Wiley 1974

Harnden, D.G., Langlands, A.D., McBeath, S., O'Riordan, M., Faed, M.J.X.: The frequency of constitutional chromosome abnormalities in patients with malignant disease. Eur. J. Cancer 5, 605 (1969)

Harnden, D.G., Maclean, N., Langlands, A.O.: Carcinoma of the breast and Klinefelter's syndrome. J. Med. Genet. 8, 460-461 (1971)

Harris, H.: Cell fusion and the analysis of malignancy. Proc. R. Soc. London 179, 1-20 (1971)

Hecht, F., McCaw, B.K., Koler, R.D.: Ataxia-telangiectasia—Clonal growth of translocation lymphocytes. New Engl. J. Med. 289, 286 (1973)

Herrmann, J.: Delayed mutation model: Carotid body tumors and retinoblastoma. In: Genetics of Human Cancer (eds. J.J. Mulvihill, R.W. Miller, J.F. Fraumeni, Jr.), pp. 417-427. New York: Raven Press 1977

Hitotsumachi, S., Rabinowitz, Z., Sachs, L.: Chromosomal control of reversion in transformed cells. Nature (London) 231, 511 (1972)

Horland, A.A., Wolman, S.R., Cox, R.P.: Cytogenetic studies in patients with the basal cell nevus syndrome and their relatives. Am. J. Hum. Genet. 27, 47A (1975)

Knudson, A.G., Jr.: Mutation and cancer: Statistical study of retinoblastoma. Proc. Natl. Acad. Sci. USA 68, 820-823 (1971)

Knudson, A.G., Jr.: Heredity and human cancer. Am. J. Pathol. 77, 77-84 (1974)

Krivit, W., Good, R.A.: The simultaneous occurrence of mongolism and leukemia. Am. J. Dis. Child 91, 289 (1956)

Krivit, W., Good, R.A.: The simultaneous occurrence of leukemia and mongolism. Am. J. Dis. Child. 94, 289 (1957)

Littlefield, L.G., Goh, K.O.: Cytogenetic studies in control men and women. Cytogenet. Cell Genet. 12, 17-34 (1973)

Mark, J.: Chromosomal abnormalities and their specificity in human neoplasms: An assessment of recent observations by banding techniques. In: Advances in Cancer Research, Vol. 24 (eds. J. Klein, S. Weinhouse), pp. 165-222. New York: Academic Press 1977

Mark, J., Levan, G., Mitelman, F.: Identification by fluorescence of the G chromosome lost in humnan meningiomas. Hereditas 71, 163-168 (1972)

McCaw, B.K., Hecht, F., Harnden, D.G., Teplitz, R.L.: Somatic rearrangement of chromosome 14 in human lymphocytes. Proc. Natl. Acad. Sci. USA 72, 2071-2075 (1975)

Miller, R.W.: Neoplasia and Down's syndrome. Ann. N.Y. Acad. Sci. 171, 637-644 (1970)

Mitelman, F., Mark, J., Levan, G., Levan, A.: Tumor etiology and chromosome pattern. Science 176, 1340-1341 (1972)

Nevin, N.C., Dodge, J.A., Allen, J.V.: Two cases of trisomy D associated with adrenal tumors. J. Med. Genet. 9, 119-122 (1972)

Niebuhr, E.: Partial trisomies and deletions of chromosome 13. In: New Chromosomal Syndromes (ed. J.J. Yunis), pp. 273-299. New York-London: Academic Press 1977

Nowell, P., Jensen, J., Gardner, F., Murphy, S., Chaganti, R.S.K., German, J.: Chromosome studies in "preleukemia". III. Myelofibrosis. Cancer 38, 1873-1881 (1976)

Opitz, J.M., Herrmann, J.: Clinical genetics and cancer. In: Genetics of Human Cancer (eds. J. Mulvihill, R.W. Miller, J.F. Fraumeni, Jr.), pp. 465-473. New York: Raven Press 1977

Orye, E., Delbeke, M.J., Verhaaren, H.: Retinoblastoma and long arm deletion of chromosome 13. Attempts to define the deleted segment. Clin. Genet. 5, 457-464 (1974)

Perry, P., Evans, H.J.: Cytological detection of mutagen-carcinogen exposure by sister chromatid exchange. Nature (London) 258, 121-125 (1975)

Riccardi, V.M., Sujansky, E., Smith, A.C., Francke, U.: Chromosomal imbalance in the aniridia-Wilms' tumor association: 11p interstitial deletion. Pediatrics 61, 604-610 (1978)

Rowley, J.D.: A new consistent chromosomal abnormality in chronic myelogenous leukemia identified by quinacrine flueorescence and Giemsa staining. Nature (London) 243, 290-293 (1973)

Schull, W.J.: Cancer and inbreeding. In: Genetics of Human Cancer (eds. J.J. Mulvihill, R.W. Miller, J.F. Fraumeni, Jr.), pp. 15-18. New York: Raven Press 1977

Segi, M.: World incidence and distribution of skin cancer. Natl. Cancer Inst. Monogr. No. 10, 245-255 (1963)

Stetka, D., Wolff, S.: Sister chromatid exchange as an assay for genetic damage induced by mutagen carcinogens. I and II. Mutat. Res. 41, 333-350 (1976)

Todaro, G.J., Green, H., Swift, M.R.: Human diploid fibroblasts transformed with SV40 or hybrid Adeno-7 x SV40. Science 153, 1252-1254 (1966)

Wolff, S.: Sister chromatid exchange. Annu. Rev. Genet. 11, 183-201 (1977)

Yunis, J.J., Chandler, M.E.: High-resolution chromosome analysis in clinical medicine. In: Progress in Clinical Pathology, Vol. 7 (eds. M. Stefanini, A. Hossaini), pp. 261-288. New York: Grune & Stratton 1977

Zech, L., Haglund, U., Nilsson, K., Klein, G.: Characteristic chromosomal abnormalities in biospies and lymphoid-cell lines from patients with Burkitt and non-Burkitt lymphomas. Int. J. Cancer 17, 47-56 (1976)

Chromosomes and Tumor Progression

P.C. NOWELL

Department of Pathology, School of Medicine, University of Pennsylvania
Philadelphia, PA 19104, USA

Neoplasms frequently become more aggressive during the course of their development, often with concomitant loss of specific differentiated properties. This is "tumor progression", and my purpose is to suggest how cytogenetic data may provide clues to the nature of the process.

It is important to stress at the outset that chromosome studies represent a relatively crude approach to the investigation of genetic changes in cell populations. They can indicate certain alterations taking place within tumor cells, but there may be a wide variety of mutational events which are not recognizable at the chromosomal level.

I. Chromosome Changes in Tumors

Sufficient data are currently available from both human and experimental neoplasms to permit several generalizations to be made:

1. Most tumors have cytogenetic abnormalities, although chromosome changes are not essential for the neoplastic state. In general, the more advanced a malignancy, the more extensive the chromosome alterations.

2. In a tumor, the cells frequently all show the same karyotypic abnormality, or changes which are demonstrably related.

3. Very rarely, however, do all tumors of a given type show the same cytogenetic alteration, although enough nonrandom changes are now being identified with banding techniques to indicate a number of specific chromosomal sites where genes important in the development of neoplasia are apparently located (Levan et al. 1977).

II. Clonal Evolution in Tumors

These observations on tumor chromosomes, as well as other data on isoenzyme patterns and immunoglobulin products, have led to a concept of tumor

development which has been termed "clonal evolution" (Nowell 1976; Fialkow 1977). This hypothesis has two major components:
1. Most tumors are clones, that is, derived from a single cell of origin.
2. Tumor progression results from genetic instability in the neoplastic population which permits sequential selection, over time, of increasingly mutated subpopulations within the original clone.

It is postulated that a single cell initially acquires a carcinogen-induced selective growth advantage, often without visible chromosome change, and the progeny of that cell then begin to grow as a neoplasm. Subsequently, as a result of genetic lability within the clone, mutants are generated, and occasionally one arises which has an additional selective advantage. The progeny of this cell are then able to from a predominant subpopulation.

The basis for the apparent enhanced genetic instability of the neoplastic cells is not understood. Rarely, there may be an inherited defect in all the cells of the body, as in the "chromosome breakage syndromes" which are described elsewhere in this conference (Cervenka this vol.); but usually it appears to be an acquired defect, limited to the tumor, resulting perhaps from activation of a "mutator gene" or from continued presence of a mutagenic carcinogen (Nowell 1976).

In any event, over a time course which may be quite variable, new mutants continue to appear within the neoplastic clone, and selection of more and more abnormal subpopulations, often marked by visible chromosome changes, continues to go forward. As a result both of the nature of the mutation and of the environment in which the cells are proliferating, there is continuing selection for those cellular properties most conducive to growth and "malignant" behavior of the tumor cells, and often selection against specific specialized properties of the tissue of origin. This appears to be the basis for the biological phenomenon of tumor progression, with increasing malignancy and associated "loss of differentiation" by tumor cells, which has been long recognized both in human and experimental material.

In the common malignancies of man, the solid epithelial cancers, we are generally seeing a very late stage of this sequential process, and as a result, the predominant population is typically composed of cells which are very extensively altered genetically (and cytogenetically), as well as biologically, often with considerable loss of specialized characteristics.

The postulated sequence of events leading to this end stage has been documented in a number of human and experimental systems and reviewed elsewhere (Fialkow 1977, Nowell 1976). I would like to discuss briefly just two examples, both of which involve hematopoietic neoplasms. These frequently appear not to be as far advanced in clonal evoluation as the typical solid malignanicies, but they do permit additional speculation concerning possible mechanisms underlying "dedifferentiation' during tumor progression.

III. Chronic Granulocytic Leukemia

There is perhaps more information concerning the relationship between chromosomal change and tumor progression in this disorder than in any other human neoplasm. In its early stages, chronic granulocytic leukemia (CGL) is clinically a benign diesease in which the relatively slowly expanding neoplastic clone consists mostly of differentiated myeloid elements. After several years, the clinical picture often changes dramatically. The well-differentiated cells are replaced by a rapidly expanding population of undifferentiated myeloblasts, filling the blood and bone marrow, and leading to the death of the patient. Tumor progression in this instance, with associated loss of specialized properties of the neoplastic cells, appears to be due to a marked shift within the tumor population in the direction of more immature forms.

Both the early and late stages of CGL are usually accompanied by chromosomal abnormalities. In the benign phase, the neoplastic cells typically show the Philadelphia chromosome as the only cytogenetic change. In the accelerated phase, cells with karyotypic abnormalities in addition to the Ph chromosome often become predominant. These additional changes are not consistent from case to case, but frequently involve one or more of several specific alterations (a second Ph, iso-17, trisomy 8) (Levan et al. 1977). It is thus possible to speculate that in CGL the disease is initiated by a visible genetic change in a marrow stem cell (the Ph chromosome), and that tumor progression results from a second mutation in a cell of the original clone, also recognizable cytogenetically, allowing a more aggressive subpopulation to develop and overwhelm the patient.

The specific genes and gene products involved in these stages are not known, but these profound clinical derangements could result from relatively minor quantitative alterations in the neoplastic cells. For instance, the mutation represented by the Philadelphia chromosome might produce, on a marrow stem cell and its progeny, slight changes in the density and/or distribution of membrane receptors for some local growth-regulatory factor (chalone). The subsequent altered response to growth regulation could produce a minor imbalance between proliferation and differentiation in the Ph-positive population, resulting in gradual expansion of the stem cell pool within the neoplastic clone, and ultimately the increased numbers of partially differentiated and well-differentiated myeloid elements observed in the chronic phase of CGL.

The second mutation within the clone might further alter membrane receptors so that nearly all of the cells with the additional chromosome change remained as blasts, with very few undergoing terminal differentiation. This would lead to rapid expansion of the stem cell pool and the clinical picture of the "blast crisis".

This postulated mechanism of altered growth regulation in myeloid leukemia obviously represents only one of many ways in which the Philadelphia chromosome, and other chromosome abnormalities, might exert their effects. Such considerations are discussed in more detail by Sachs, McCulloch, and by others in this volume. It is perhaps worth emphasizing, however, that the visible cytogenetic changes and their disastrous clinical consequences might well be mediated by relatively minor alterations at the cellular level. Furthermore, in theory at least, even in the advanced stage of tumor progression epitomized by the blast crisis in CGL, the undifferentiated neoplastic population might be forced back into a normal balance between proliferation and differentiation with effective reversal of the neoplastic state, if the dysfunctional membrane receptors for growth regulation could be successfully bypassed. Loss of differentiation in a neoplastic population need not necessarily reflect an irreversible "block" in maturation, particularly if the acquired genetic changes within the cells are not extensive.

IV. Chronic Lymphocytic Leukemia

As a second example of changing cytogenetic patterns in an evolving tumor, I would like to describe a patient with the rare T cell variant of chronic lymphocytic leukemia (CLL). In this instance, the loss of certain specific differentiated properties by the neoplastic cells has not yet been accompanied by the aggressive biological behavior that is usually implied in the term tumor progression.

The patient has had an unexplained marked lymphocytosis for over three years, and this has been the basis for the diagnosis of CLL, although she has had no other clinical signs or symptoms referable to her lymphoproliferative disorder (Nowell et al. 1979). The circulating cells have consistently been well-differentiated, morphologically normal, small lymphocytes. When first studied, 77% of the lymphocytes formed rosettes with sheep red blood cells (E rosettes), a T cell marker, and only 8% formed complement rosettes (EAC rosettes), a B cell marker. The cells also gave a normal proliferative response to several T cell mitogens (PHA, Con A, A23187). On the basis of these findings, the patient's disease was considered a T cell neoplasm, although in most instances CLL is a B cell disorder. The cells proliferating in the mitogen-stimulated cultures did not initially show any chromosome changes.

In following the patient for nearly three years, we have observed no change in her clinical status, nor has there been any alteration in the number or morphology of the circulating lymphocytes. However, during this time, a clone of cells with multiple chromosome changes has gradually replaced the original diploid neoplastic population. These cells have 46 chromosomes with several translocations, including one which has produced the $14q^+$ marker which has previously been described frequently in lymphoproliferative disorders. In our most recent study of this patient, this chromosomally abnormal clone constituted more than 90% of the metaphases in several cultures.

This evidence of cytogenetic progression has been accompanied by loss of certain T cell properties by many of the circulating lymphocytes. Cells forming E rosettes have fallen as low as 16% (without an increase in EAC rosettes), and there has been an associated gradual loss in responsiveness to T cell mitogens.

It is tempting to suggest that the loss of these specialized T cell characteristics is the result of the chromosomally altered subline having largely replaced the original diploid neoplastic cells. The ability to form E rosettes and to respond to T cell mitogens both appear to depend on specific membrane receptors (Nowell et al. 1979). In this instance, the chromosome changes in the mutant subpopulation may have caused alterations in these receptors without any important change in growth-regulatory pathways, either at the cell membrane or elsewhere within the cell. The observations in this patient suggest that clonal evolution within a neoplastic population can on occasion result in loss of specific differentiated properties without major alteration in growth kinetics.

In general, the currently available chromosome data in tumors, as well as the specific evidence of these two examples, support the concept of a microevolutionary basis of tumor progression, with sequential selection of mutant subpopulations. The cytogenetic findings do not usually provide any direct explanation for the genetic instability underlying this process, except in the rare instances where there is an inherited defect causing chromosomal fragility (Cervenka, this vol.). Nor do the karyotypic data serve to identify individual genes and gene products involved in neoplastic development, although they are increasingly suggesting specific chromosomal sites where such genes may be sought (Levan et al. 1977). Cytogenetics can generate leads to the mechanisms underlying tumor progression and associated loss of differentiation, but finer techniques are needed to substantiate these hypotheses and develop more complete understanding.

V. Summary

Chromosome studies suggest that most neoplasms are unicellular in origin (i.e., "clones"), and that biological progression, typically associated with loss of "differentiated" properties, results from acquired genetic instability in the neoplastic population leading to sequential appearance of increasingly mutant subpopulations with greater selective growth advantage. In these circumstances, acquisition by the tumor of an apparent "defect in maturation" could reflect either abnormal response to external growth regulators or loss ability to synthesize certain specific differentiation products. Karyotypic data from a variety of mammalian neoplasms support both possibilities. The terminal accelerated phase of chronic granulocytic leukemia, for instance, in which undifferentiated blast cells replace a previously well-differentiated population, seems to represent emergence of a subline, often recognizable cytogenetically, in which an additionally altered response to growth regulation further shifts the balance between proliferation and differentiation at the stem cell level, so that undifferentiated (proliferating) cells come to predominate. In a case of chronic T cell leukemia, on the other hand, we have observed loss of certain specific T cell characteristics (membrane markers, response to mitogens) in conjunction with emergence of a chromosomally abnormal subclone, without alteration in the "mature" morphology of the neoplastic lymphocytes or any apparent change in their proliferation kinetics in vivo. Presumably, loss of these differentiated characteristics reflects alterations in specific gene functions within the neoplastic subpopulation, rather than any significant expansion in the undifferentiated stem cell pool. Cytogenetic studies may provide clues to the genetic basis of both types of "dedifferentiation".

References

Cervenka, J.: Chromosomal changes associated with premalignancy and cancer (this vol.)
Fialkow, P.: Clonal origin and stem cell evolution of human tumors. Prog. Cancer Res. Ther. 3, 439-453 (1977)
Levan, A., Levan, G., Mitelman, F.: Chromosomes and cancer. Hereditas 86, 15-29 (1977)
McCulloch, E.: Cellular heterogeneities in acute myeloblastic leukemia (this vol.)
Nowell, P.: The clonal evolution of tumor cell populations. Science 194, 23-28 (1976)
Nowell, P., Rowlands, D., Daniele, R., Berger, B., Guerry, D.: Changes in membrane markers and chromosome patterns in chronic T cell leukemia. Clin. Immunol. Immunopathol. 12, 323-330 (1979)
Sachs, L.: Activation of normal differentiation genes and the origin and development of myeloid leukemia (this vol.)

Surface Antigens in Early Embryonic Development

R. Kemler, D. Morello, Ch. Babinet, and F. Jacob

*Service de Génétique Cellulaire du Collège de France
et de l'Institut Pasteur, 75015 Paris, France*

I. Introduction

For over ten years now, several authors have proposed a model in which cell surface structures play a key role in embryonic development (Glueckson-Waelsch and Erickson 1970; Bennett et al. 1971; Jacob 1978). It has been thought that during development surface antigens are in some way implicated in the unfolding of the genetic program. The experimental study of this hypothesis raised many difficulties because of the scarcity of material, and of the rapid differential transition of the cells through successive states of differentiation. This hypothesis implies that not only are there defined patterns of surface structures during successive stages of differentiation which might be quite easy to study in a descriptive way, but it also suggests that during development, at least some surface antigens play a functional role which demands a rather new experimental technology. Some of these difficulties have been overcome with the help of teratocarcinoma cells. Teratocarcinoma are tumors, first described by Stevens (Stevens 1967), from which many stable cell lines have been established. Tumor stem cells, the embryonal carcinoma (EC) cells, have the potential to differentiate into derivatives of the three germ layers.

With the help of antisera against EC cells, several surface antigens have been described with different cell type distributions (Jacob 1977). These antigens are mainly expressed on early embryonic cells and cells of the male germ line. Hence the teratocarcinoma system allows the characterization of cell surface markers in embryonic development.

A first attempt to investigate the physiological role of surface antigens during embryonic development has been done with antisera against an EC cell (Kemler et al. 1977), the F9 line, and will be discussed in more detail.

II. The F9 Line

The F9 cell is a nullipotent EC cell with slight endodermal differentiation under certain in vitro conditions (Sherman and Miller 1978). The cells have been used to immunize syngeneic mice and rabbits.

Table 1. Effect of antisera on the in vitro development of two-cell embryos

Line	Culture in	Total number of embryos	Blastocysts after 66 h in culture EGT = 102		Number of grape-like structure
			(N)	(%)	
1	Whitten's medium	215	167	76	0
2	Mouse anti-F9 1/20	53	35	66	0
3	Rabbit anti-F9 1/20	41	28	68	0
4	Rabbit anti-F9 Fab 1/20[a]	253	0	0	215[b]
5	1/40	20	0	0	16[b]
6	Rabbit anti-lymphocytes Fab 1/10	28	11	39	0
7	1/20	58	48	82	0
8	Rabbit anti-liver Fab 1/10	59	41	69	0
9	1/20	42	33	78	0
10	Rabbit anti-brain Fab 1/10	59	42	71	0
11	1/20	42	26	61	0
12	Rabbit anti-F9 Fab absorbed on F9 1/20	100	68	68	0
13	succ-con A (50 µg/ml)	60	39	65	0

[a] All Fab preparations have been adjusted to 3.4 mg protein/ml
[b] Early arrested and fragmented embryos account for the difference between the total number of embryos and the number of either blastocyst or grape-like structures scored in this table. For details, see Kemler et al. (1977)

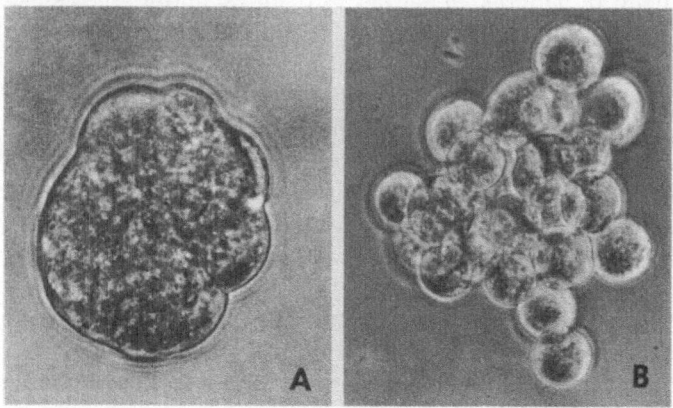

Fig. 1A,B. Two-cell embryos were grown in Whitten's medium. **A** Control. **B** Addition of rabbit anti-F9, IgG; Fab fragments (3.4 mg/ml at 1/20 dilution)

Hyperimmunization of syngeneic mice with F9 cells produces antisera which allow the detection of a surface antigen(s) called the F9 antigen. Extensive studies have shown that the F9 antigen has a very restricted cell distribution (Jacob 1977). Beside EC cells, it is only expressed on early embryonic cells and cells of the male germ line. On embryonic cells, F9 antigen seems to be most highly expressed at the morula stage.

The effect of embryonal growth and differentiation was studied by adding various antisera or monovalent IgG fragments (Fab) to cultured two-cell embryos.

As can be seen from Table 1, only the monovalent fragment of a rabbit anti-F9 antibody effects morulae compaction and blastocyst formation (Fig. 1). Although addition of anti-F9 Fab fragments does not alter cleavage, it prevents formation of compact morulae, a prerequisite for blastocyst differentiation (Ducibella and Anderson 1975). These experiments suggest that rabbit anti-F9 IgG detects a cell surface structure which is involved in embryonic development, a result which supports the hypothesis mentioned above. However, the rabbit anti-F9 serum, even after appropriate absorptions, consists of a rather heterogeneous antibody population.

A more detailed characterization of the specificity responsible for the effect on embryonic development is of course of great interest. For this reason, we have used hybrid cell lines producing monospecific antibodies, according to the technique described by Köhler and Milstein (Köhler and Milstein 1975).

III. Monoclonal Antibodies Against F9 Cells

Experimental details of this work are described elsewhere (Kemler et al. 1978). Briefly, mouse myeloma cells (X63-Ag8, obtained from Dr. C. Milstein, MRC, Cambridge, England) were fused with spleen cells of rats previously immunized with F9 cells, using polyethylene-glycol (PEG, MG : 4,000) as a fusion agent. Hybrid cell lines were tested against F9 cells in a radioactive binding-assay and cytotoxic test.

Out of 93 hybrids 28 were found to produce anti-F9 antibodies. Five of these were chosen for successive subclonings and subsequent serological screening. They are referred to as ECMA 1 to 5 (Embryonal Carcinoma Monoclonal Antibody). Four produce IgM and only one IgG antibodies. In a cytotoxic assay, concentrated culture supernatants from each ECMA have high titers (between 10^{-4} and 10^{-6}) on F9 cells tested. Each ECMA activity was then tested (by absorption or indirect immunofluorescence tests) on various EC cells, their differentiated derivatives, mouse lymphocytes and sperm, as well as preimplantation embryos. Table 2 summarizes the results of these experiments. It gives the preliminary distributions of the antigens detected by each ECMA. As can be seen, each ECMA detects an antigen with a different distribution. ECMA 5 is not listed in Table 2. Its activity is absorbed by mouse lymphocytes, sperm, and sheep red blood cells. The antigen ECMA 5 detects might be similar to the one described by Stern et al. (1978) which appears to be a glycolipid.

The presence of antigens detected by ECMA 1 to 4 was studied during in vitro differentiation of EC cells (PCC3/A/1). ECMA 4 labels 100% of the cells at any time during differentiation. ECMA 1 labels only 2–4% of the cell population at any time during differentiation.

On the contrary, ECMA 2 and 3 label 70% of EC cells. This percentage decreases rapidly as differentiation of EC cells proceeds. It drops about 1% at day 13. On preimplantation embryos ECMA 4 does not react with any stage. ECMA 3 begins to react only at the blastocyst stage and sometimes only a few trophectoderm cells are positive. In contrast, ECMA 2 and 3 label from two-cell stage on. They

110 R. Kemler et al.

Table 2. Surface antigens detected by ECMA on various cell types[a]

Cell type	ECMA			
	Embryo-nal 1	Carci-noma 2	Mono-clonal 3	Anti-body 4
EC, multipotential from 129/Sv origin	+	+	+	+
C₃H origin	−	+	+	n.t.
LT origin	−	+ (40[b])	+ (15)	n.t.
Differentiated derivatives from teratocarcinoma cells[c]				
PYS-2: parietal yolk sac carcinoma cell	−	−	−	+ (100)
3/A/1-D-3: mesenchymal-like cell	+	+	−	+
3/TM 1: trophoblastoma cell line	−	+ (70)	−	n.t.
PCD/2: myoblast-like cell	−	+ (25)	+ (5)	+ (100)
PCD/3: fibroblast-like cell	−	+	−	n.t.
Mouse lymphocytes (129/Sv mice)	+	+	−	−
sperm (129/Sv mice)	+	+	+	−
Mouse preimplantation embryos (129/Sv mice)				
non fertilized egg (129/Sv mice)	n.t.	−	−	n.t.
two-cell embryo (129/Sv mice)	n.t.	+	+	n.t.
morula (129/Sv mice)	−	+	+	−
blastocyst (129/Sv mice)	+	+	+	−
inner cell mass (129/Sv mice)	+	+	+	−

[a] Tests were done by absorption or indirect immunofluorescence
[b] Percentage of labeled cells
[c] For more detailed description see Nicolas et al. (1976)

differ however because the antigen detected by ECMA 2, but not that detected by ECMA 3 is also expressed on a small subpopulation of adult lymphocytes (D. Morello, unpublished observations).

IV. Conclusion

Teratocarcinoma cell lines allow a study of cell surface structures expressed on embryonic cells. Antisera against embryonal carcinoma cells provide tools to detect surface antigens at successive differentiation stages of the embryo. With such antisera, it was also possible to show that a particular surface structure of the embryo is involved in development.

Introducing the myeloma fusion technique into studies of embryonic cell surface structures seems to be a great advantage.

Monoclonal antibodies against embryonic carcinoma antigens (ECMA) can be readily produced and no doubt allow a more precise analysis than crude antisera. The technique can be used to obtain a battery of monoclonal antibodies, each detecting a different membrane structure. Clear-cut and unambiguous results following differential expression of surface antigens during embryonic development can thus be obtained. This promising experimental approach should also facilitate a study of the functional role of precisely defined structures during embryonic development.

Acknowledgments. This work was supported by grants from the Centre National de la Recherche Scientifique (LA 269), the Institute National de la Santé et de la Recherche Médicale (C.R.A.T. n° 76.4.311), the Délégation Générale à la Recherche Scientifique et Technique (A.C.C. n° 77.7.0966), the National Institutes of Health (CA 16355) and the André Meyer Foundation.

References

Bennett, D., Boyse, E.A., Old, L.J.: Cell surface immunogenetics in the study of morphogenesis. In: Cell Interactions. Third Lepetit Symp. London (ed. L.G. Silvestri), pp. 247-263. Amsterdam: North-Holland 1971

Ducibella, T., Anderson, E.: Cell shape and membrane changes in the eight-cell mouse embryo: pre-requisite for morphogenesis of the blastocyst. Dev. Biol. 47, 45-58 (1975)

Glueckson-Waelsch, S., Erickson, R.P.: The T locus of the mouse: implications for mechanisms of development. In: Current Topics in Developmental Biology, pp. 281-316. New York-London: Academic Press 1970

Jacob, F.: Mouse teratocarcinoma and embryonic antigens. Immunol. Rev. 33, 3-32 (1977)

Jacob, F.: Mouse teratocarcinoma and mouse embryo. (The Leeuwenhoek Lecture, 1977). Proc. R. Soc. London Ser. B 201, 249-270 (1978)

Kemler, R., Babinet, C., Eisen, H., Jacob, F.: Surface antigen in early differentiation. Proc. Natl. Acad. Sci. USA 74, 4449-4452 (1977)

Kemler, R., Morello, D., Jacob, F.: Properties of monoclonal antibodies against mouse embryonal carcinoma cells. In: Cell Lineage, Stem Cells and Cell Determination. INSERM Symposium No 10. (ed. N. Le Douarin). Elsevier/North Holland Biomedical Press 1979

Köhler, G., Milstein, C.: Continuous cultures of fused cells secreting antibody of predefined specificity. Nature (London) 256, 495-497 (1975)

Nicolas, J.F., Avner, P., Gaillard, J., Guénet, J.L., Jacob, H., Jacob, F.: Cell lines derived from teratocarcinomas. Cancer Res. 36, 4224-4231 (1976)

Sherman, M.I., Miller, R.A.: F9 embryonal carcinoma cells can differentiate into endoderm-like cells. Dev. Biol. 63, 27 (1978)

Stern, P., Willison, K., Lennox, E., Galfré, G., Milstein, C., Secher, D., Ziegler, A.: Monoclonal antibodies as probes for differentiation and tumor associated antigens: a Forssman specificity on teratocarcinomas and early mouse embryos. Cell 14, 775-783 (1978)

Stevens, L.C. The biology of teratomas. Adv. Morphog. 6, 1-31 (1967)

The Role of Fibronectin in Cellular Behavior

R.O. Hynes

*Department of Biology and Center for Cancer Research
E17-227, M.I.T., Cambridge, MA 02139, USA*

I. Introduction

Fibronectin, which we have previously called LETS protein (Hynes 1973, 1976; Hynes et al. 1978), is a large glycoprotein found on the external surfaces of many normal cell types, in particular fibroblasts. It has a subunit molecular weight of 230,000 daltons ($\pm 10\%$) but is disulfide-bonded into dimers and higher aggregates (Hynes and Destree 1977; Ali and Hynes 1978b). It can be isolated in a membrane-free fraction apparently identical with the cell surface coat or glycocalyx (Graham et al. 1975, 1978) and immunofluorescent microscopy shows that it occurs in fibrillar arrays around the cells (Hedman et al. 1977; Mautner and Hynes 1977; Furcht et al. 1978).

The levels of fibronectin on transformed cells are greatly reduced (see Hynes 1976, for review) and a good correlation has been reported between tumorigenicity and loss of fibronectin (Chen et al. 1976). Addition of purified fibronectin to transformed cultures produces striking effects on their behavior (Yamada et al. 1976a,b; Ali et al. 1977). The cells show increased adhesion to the substratum, increased spreading, reduced overlapping, elongation and alignment of cells, and a decrease in surface ruffling. These effects are all likely to be secondary to the increase in cell-substratum adhesion. All the alterations produced by addition of fibronectin to transformed cells are in the direction of normality and it is easy to argue that these aspects of the transformed phenotype which are affected are secondary consequences of reduced adhesivity which is in turn a consequence of the reduced levels of fibronectin. This argument will be pursued further later in this paper.

Another effect of fibronectin on transformed cells is on the cytoskeletal structures inside the cells. Many electron and immunofluorescent microscopic analyses of the arrangement of microfilaments in normal cells show them to be organized in submembranous sheaths and in bundles traversing the cytoplasm. These arrays of microfilaments are much less well-organized in transformed cells. However, addition of fibronectin leads to reappearance of the microfilament bundles (Ali et al. 1977; Willingham et al. 1977).

Because of these pleiotropic effects of fibronectin on the behavior of transformed cells, we have recently been analyzing the distribution of fibronectin in normal cells in a variety of states in order to analyze in more detail the role of this protein in normal cell functions. The aim of this paper will be to review and discuss some of these results and develop hypotheses arising from them as to the way in which cell surface proteins and protein complexes might be involved in cell behavior.

II. Techniques and Data

The methods used in these studies and most of the results will not be described in detail. Details can be found in several recent papers. Procedures for immunofluorescence are described by Mautner and Hynes (1977) and for double-label immunofluorescence by Hynes and Destree (1978). Methods for cell culture are described in the same papers and the analysis of cell migration was by the technique of Albrecht-Buehler (1977) and Ali and Hynes (1978a). Methods for biochemical analysis of cell surface proteins are described by Hynes and Destree (1977) and by Ali and Hynes (1977, 1978b). Purification of fibronectin and removal of fibronectin from serum was by several methods (Yamada et al. 1976a; Ali et al. 1977; Engvall and Ruoslahti 1977; Hynes et al. 1978).

III. Results and Discussion

A. Fibronectin is Involved in Spreading and Migration

Fibronectin clearly promotes cell adhesion and spreading when added to transformed cells which lack it. A factor which promotes cell spreading has been isolated from serum (Grinnell 1976; Grinnell et al. 1978) and it has recently been shown that this factor is identical with a plasma and serum protein which is antigenically and biochemically similar to cellular fibronectin (Grinnell and Hays 1978). This protein is known as cold-insoluble globulin (CIg) or plasma fibronectin (Mosesson et al. 1975). It is not clear that the cell surface and plasma forms are identical but they are very similar and both appear to promote cell spreading.

It was therefore instructive to investigate the arrangement of fibronectin in spreading normal cells. It appears first as a ring of punctate patches beneath the cell; at later times this pattern is replaced by short fibrils which increase in length with time (Hynes and Destree 1978). These patterns of fibronectin appear even when cells are seeded in serum depleted of fibronectin but do not appear when, in addition, protein synthesis is inhibited by cycloheximide (unpublished data). These results suggest that the fibronectin beneath spreading cells is predominantly produced by the cells, not derived from the serum. This is consistent with, but does not prove, a role for cell-derived fibronectin in cell spreading.

Spreading cells are surrounded by a circular ruffling lamella (Goldman and Knipe 1972; Witkowski and Brighton 1972; Bragina et al. 1976; Vasiliev and Gelfand 1976). In many respects this lamella resembles that at the leading edge of a migrating cell (Abercrombie et al. 1971; Wessells et al. 1973). Fibronectin is also found beneath the leading lamellae of migrating cells (Fig. 1; Hynes et al. 1978), and

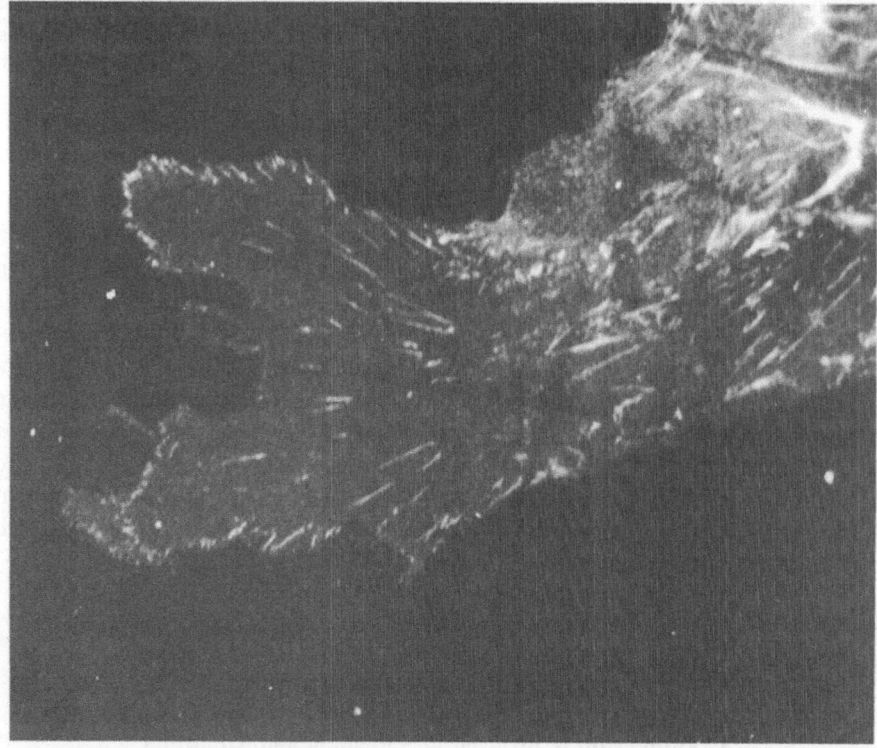

Fig. 1. Fibronectin beneath spreading lamella of fibroblast. Figure shows leading lamella of a cell stained with anti-fibronectin antiserum. Note striae of fibronectin

furthermore, addition of fibronectin to cultures of normal or transformed cells leads to greatly increased migration (Ali and Hynes 1978a). Again, these results suggest a role for fibronectin in cell migration.

B. Fibronectin Interacts with the Cytoskeleton

Added fibronectin causes appearance of microfilament bundles in transformed cells (Ali et al. 1977). Reciprocally, treatment of normal cells with cytochalasin B, a drug which causes dissociation of microfilaments in cells, causes release of cell surface fibronectin and prevents its reattachment (Ali and Hynes 1977). These results suggest some sort of interaction between the two sets of filaments, fibronectin fibrils outside the cell and microfilaments inside the cell. Similar interactions with other cytoskeletal components were not detected; transformed cells do not lack microtubules or intermediate filaments, and addition of fibronectin leads merely to their realignment consistent with the changes in cell shape (Ali et al. 1977; Hynes et al. 1978). Furthermore, colchicine, which disassembles microtubules and causes intermediate filaments to rearrange into a

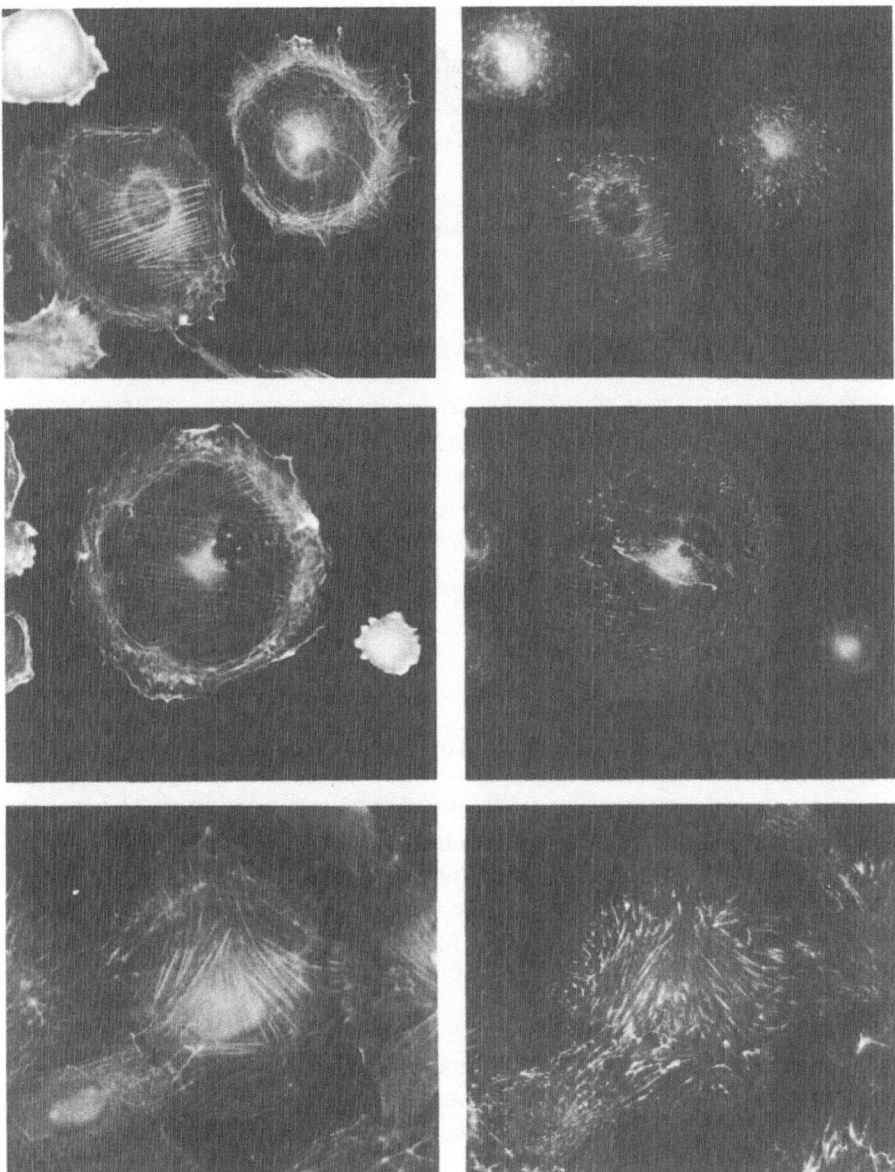

Fig. 2. Codistribution of fibronectin and actin in NIL 8 hamster cells. Figure shows pairs of photographs of cells stained for actin *(top)* and fibronectin *(bottom)* showing correspondences between the two arrays of fibrils. Note that while the patterns are clearly related, they are not identical, showing that codistribution of the two stains is not merely due to cross-reaction of the antibodies

coil, has no effect on cell surface fibronectin (Ali and Hynes 1977; Mautner and Hynes 1977).

To investigate possible connections between actin microfilaments and fibronectin, arrangements of the two proteins were studied in the same cells using double-label immunofluorescence (Hynes and Destree 1978). Striking correspondences were seen between the two patterns (Fig. 2). The distributions were not identical, but at many points the two sets of filaments lie over one another, clearly indicating some sort of transmembrane interaction.

C. Disulfide Bonding is Important at the Cell Surface

As mentioned earlier, fibronectin exists on the cell surface in disulfide-bonded complexes, dimers and higher aggregates (Hynes and Destree 1977). These disulfide bonds appear to be important in retention of fibronectin at the cell surface, since reduction with dithiothreitol leads to rapid loss of fibronectin from the cells without loss of cell viability (Ali and Hynes 1978b). The action of the spreading factor studied by Grinnell, which is presumably plasma fibronectin, is inhibited by sulfhydryl blocking reagents (Grinnell 1976), suggesting that its function may also require disulfide bonding.

D. Adhesion or Attachment Plaques

The appearance of fibronectin beneath cells during spreading and migration, its involvement in adhesion, and particularly the correspondence between the arrangement of fibronectin and of microfilament bundles, all suggest the hypothesis that fibronectin may be involved in the formation of "adhesion or attachment plaques". These are the regions of closest approach between cells and substrata and have been studied by electron microscopy (Abercrombie et al. 1971; Brunk et al. 1971) and interference reflection microscopy (Abercrombie and Dunn 1975; Izzard and Lochner 1976; Heath and Dunn 1978). In fact, the pattern of attachment plaques seen beneath cells by interference is strikingly similar to some of the patterns of fibronectin observed by immunofluorescence. Adhesion plaques are frequently at the termini of microfilament bundles or at intervals along them (Abercrombie et al. 1971; NcNutt et al. 1971, 1973; Heaysman and Pegrum 1973; Revel and Wolken 1973; Abercrombie and Dunn 1975; Bragina et al. 1976; Izzard and Lochner 1976; Heath and Dunn 1978). Very similar patterns have been observed by double-label immunofluorescence (Hynes and Destree 1978). Adhesion plaques are also observed in regions of cell-cell contact (Heaysman and Pegrum 1973) and fibronectin is frequently observed in a similar location (Chen et al. 1976; Mautner and Hynes 1977). Adhesion plaques have been proposed to be fixed points of attachment against which tension develops and is converted into cell movement (Abercrombie and Dunn 1975; Izzard and Lochner 1976).

E. Working Hypotheses and Speculations

The attachment plaque hypothesis would be consistent with many of the results described earlier. Thus, one could propose that fibronectin functions in adhesion by forming focal plaques. Development of increasing numbers of plaques and perhaps their individual extension could promote extension of lamellae during spreading and migration. At the same time, these plaques could function as nucleation centers for the development or organization of microfilament bundles. The latter proposal requires a transmembrane interaction. The bulk of the evidence is against a transmembrane arrangement for either actin or fibronectin. Actin is not detected from the outside of intact cells either by surface labeling methods or by antibodies. Fibronectin appears to be external to the lipid bilayer. It can be separated from the plasma membrane after nitrogen cavitation (Graham et al. 1975, 1978). Fibronectin does not appear to be mobile in the plane of the membrane and does not insert into phospholipid vesicles (Schlessinger et al. 1977). Hence one would have to propose the existence of one or more linker proteins between the two.

One could also propose that the role of the disulfide bonding is to hold the various molecules together in the plaques. This hypothesis is consistent with the known propensity of fibronectin to form disulfide-bonded aggregates and for the rapid release of fibronectin on reduction. If one proposes, as we did above, that fibronectin interacts with actin via other proteins, some of which must be integral membrane proteins, one could further propose that binding of individual fibronectin molecules to individual integral membrane proteins is followed by lateral diffusion of the latter and disulfide-bonding between fibronectin molecules. Alternatively, formation of fibronectin aggregates could be the first event, followed by lateral diffusion and binding of the integral membrane proteins. In either case, the end result would be a patch of integral membrane proteins held together by fibronectin outside the membrane. This plaque of integral membrane proteins could form the attachment point for the microfilaments on the inside of the membrane. The plaque of fibronectin on the outside could form a "stick patch" which would form an adhesive bond with the substratum. The latter is actually more precisely a film of bound serum proteins (Brunk et al. 1971; Revel and Wolken 1973) which includes serum fibronectin (Grinnell 1976; Grinnell et al. 1978; our unpublished data). Fibronectin is known to show binding affinities for collagen (Engvall and Ruoslahti 1977), fibronogen or fibrin (Mosher 1975; Ruoslahti and Vaheri 1975), and certain proteoglycans (Stathakis and Mosesson 1977). Any or all of these might also be involved in cell-substratum adhesion mediated by fibronectin. Culp has shown that the substrate adhesion area beneath cells contains proteoglycans and fibronectin as well as other proteins (Terry and Culp 1974; Roblin et al. 1975; Culp 1976), supporting the hypothesis that they may cooperate in cell-substratum adhesion. This working hypothesis is consistent with most of the data discussed earlier and is a testable proposal.

One set of results which is not immediately readily reconciled with the model proposed concerns the effects of cytochalasin B. The proposal outlined above suggests that the fibronectin interacts with integral membrane proteins and that this in turn organizes the microfilaments. That is, the stream of causality is outside-to-inside. The drug cytochalasin B has been proposed to act on microfilaments causing their disassembly (Wessells et al. 1971). On this interpretation of the drug's effects, cytochalasin B would affect the microfilaments and this would lead to the observed release of fibronectin (Ali and Hynes 1978b). In this case the train of causality is inside-to-outside. It is possible to propose a reciprocal dependence between the two proteins. However, another possibility is that the cytochalasin B does not interact with the microfilaments themselves. In fact, a certain amount of evidence supports this viewpoint. It has been shown that multiple binding sites for various cytochalasins to membranes exist and these include probable glucose transport molecules (Lin and Spudich 1974; Jung and Rampal 1977; Lin and Snyder 1977; Atlas and Lin 1976). Furthermore, Weber et al. (1976) showed that in glycerol-extracted cell ghosts, which lack a membrane, cytochalasin B has no effect on the microfilament bundles, which do not dissociate. If ATP is restored to the glycerinated cells, then the microfilament bundles condense (contract?) into small actin-rich clumps. The simplest interpretation of these results is that, in intact cells, cytochalasin B causes detachment of the microfilaments from membrane attachment sites, releasing them to contract. The contraction would then produce the "arborized" appearance of cytochalasin-treated cells and the clustered or clumped appearance of the actin. A similar hypothesis for the action of cytochalasin B was proposed some time ago (Spooner 1973). Pursuing this line of argument, one can then propose that the target for cytochalasin B is somewhere in the membrane protein complex containing the fibronectin. Binding of cytochalasin B to its target would also release the fibronectin.

F. Fibronectin and the Transformed Phenotype

Let us now consider how these hypotheses can be applied to the various aspects of the transformed phenotype which may be related to loss of fibronectin.

It has been shown that transformed chicken cells synthesize fibronectin at a reduced rate (Olden and Yamada 1977) reflecting a lower level of translatable mRNA (Adams et al. 1977). Transformed hamster cells also show a reduced rate of fibronectin synthesis (Hynes et al. 1977, 1978). However, in neither system is the reduced rate of synthesis sufficient to account completely for the reduction in surface levels. Some other factor must also contribute. It has been shown that transformation leads to an increased rate of turnover of fibronectin (Robbins et al. 1974; Hynes and Wyke 1975). Furthermore, fibronectin binds less well to transformed cells than to normal ones (Hynes et al. 1978). These results suggest some other alteration in the cell surface that interferes with binding of the reduced amount of fibronectin which is made. We will return later to possible hypotheses for this alteration.

Whatever the nature of any further alteration, it is clear that fibronectin will bind to transformed cells when added exogenously. Moreover, this binding has marked effects on the phenotype leading to increased adherence and spreading, elongation and alignment of cells, reduction in cell overlapping, and decreases in surface ruffles and microvilli (Yamada et al. 1976a,b; Ali et al. 1977). It is easy to see how increased adhesion could lead to increased spreading of lamellae. Diversion of membrane into spreading could readily produce a reduction in surface ruffles and microvilli. The reduced cell overlapping and alignment of cells are characteristic of normal cells and are usually considered to arise from contact inhibition of movement (Abercrombie 1970). In this phenomenon, one cell contacting another is inhibited in its forward movement, movement commences in another direction, and the cell turns away. When cells are elongated, as many motile cells are, this leads eventually to parallel alignment of the cells. Transformed cells show contact inhibition of movement with much less frequency; they tend rather to continue their forward motion, underlapping the cell they contact. This leads to the criss-crossing or overlapping of cells characteristic of transformed cultures. Harris (1973a,b) and Bell (1977) have shown that the decision between contact inhibition of movement and underlapping depends on whether the cell being contacted is well attached to the substratum at the point of contact. The leading lamella of a motile cell is always well attached. If it contacts another well-attached region of cell periphery, underlapping is impossible and the cell turns away. This happens much more frequently with normal cells which are strongly adherent than with less adherent transformed cells. It follows from this line of argument that increasing adhesion by addition of fibronectin should lead to increased contact inhibition of movement, reduced overlapping (actually underlapping) and alignment of the cells. Hence, one can see that many of the alterations in phenotype brought about by addition of fibronectin could follow directly from a single effect on adhesion. Conversely, reduction of fibronectin levels would be expected to reduce adhesion and the other properties would follow from that. While this is undoubtedly an oversimplified view, it is a useful working hypothesis.

The effects of fibronectin on microfilament arrangement could also be argued to follow from cell spreading, since in general well-spread cells have better-developed microfilament bundles. However, the evidence for a close relationship between fibronectin and actin suggests that the effect might be more direct than that, as discussed earlier. This leads us back to a consideration of the concept of protein complexes which we have suggested here and elsewhere (Hynes 1974, 1976) may play a role in the function of fibronectin. As was discussed earlier for the effects of cytochalasin B, it is conceivable that transformation could affect one of the other components in this putative complex, leading to a reduced binding capacity for fibronectin and perhaps also for actin microfilaments, leading to the disruption of both the internal cytoskeleton and the external network of fibrils containing fibronectin. For instance, disruption of the complex by absence or modification of one of the other components could explain many of the effects observed, including the reduction in surface levels of fibronectin below that expected from the biosynthetic rate. Given the recent results indicating that the product of the *src* gene of avian sarcoma viruses may be a protein kinase (Collett and Erickson 1978) it is extremely tempting to speculate that the modification might involve phosphorylation. Proteolysis by transformation-specific proteases (Unkeless et al. 1973; Chen and Buchanan 1975; Quigley 1976; Mahdavi and Hynes 1979) would be another possibility, and many others can be envisaged.

IV. Conclusions

From the forgoing discussion it is clear that the cell surface protein fibronectin and its interactions are likely to be involved in multiple aspects of cell behavior. While many of the specific proposals being discussed are likely to prove incorrect, they provide testable working hypotheses and suggest experiments to be tried. The possibility that pleiotropic effects could follow from the alteration in levels of a single protein is perhaps not so surprising when one considers that a single gene of a transforming virus triggers the whole phenomenon of transformation; pleiotropic effects must be expected. The phenomena we have been considering represent only a part of the phenotype which is altered on transformation. Other parameters (nutrient transport, cyclic nucleotides, and growth control) appear to be unrelated to the complex of properties related to adhesion which we have discussed (Yamada et al. 1976a; Ali et al. 1977; Willingham et al. 1977; Beug et al. 1978) and yet they, too, are affected by the action of the *src* gene.

Owing to a lack of space and an even greater lack of data, we have not considered any of the aspects of in vivo behavior of tumor cells, although an implicit assumption behind the interest in transformation in vitro is that it has some bearing on the in vivo situation. The presence of molecules related to fibronectin in connective tissues, basal lamellae, and plasma throughout the body provide ample material for speculation on the role of this protein in vivo, but that will have to wait for another time and a lot more data.

References

Abercrombie, M.: Contact inhibition in tissue culture. In Vitro 6, 128-142 (1970)

Abercrombie, M, Dunn, G.A.: Adhesions of fibroblasts to substratum during contact inhibition observed by interference reflection microscopy. Exp. Cell Res. 92, 57-62 (1975)

Abercrombie, M., Heaysman, J.E.M., Pegrum, S.M.: The locomotion of fibroblasts. IV. Electron microscopy of the leading lamella. Exp. Cell Res. 67, 359-367 (1971)

Adams, S.L., Sobel, M.E., Howard, B.H., Olden, K., Yamada, K.M., de Crombrugghe, B., Pastan, I.: Levels of translatable mRNAs for cell surface protein, collagen precursors and two membrane proteins are altered in Rous sarcoma virus-transformed chick embryo fibroblasts. Proc. Natl. Acad. Sci. USA 74, 3399-3403 (1977)

Albrecht-Buehler, G.: The phagokinetic tracks of 3T3 cells. Cell 11, 395-404 (1977)

Ali, I.U., Hynes, R.O.: Effects of cytochalasin B and colchicine on attachment of a major surface protein of fibroblasts. Biochim. Biophys. Acta 471, 16-24 (1977)

Ali, I.U., Hynes, R.O.: Effects of LETS glycoprotein on cell motility. Cell 14, 439-446 (1978a)

Ali, I.U., Hynes, R.O.: Role of disulfide bonds in the attachment and function of large, external, transformation-sensitive glycoprotein at the cell surface. Biochim. Biophys. Acta 510, 140-150 (1978b)

Ali, I.U., Mautner, V.M., Lanza, R.P., Hynes, R.O.: Restoration of normal morphology, adhesion and cytoskeleton in transformed cells by addition of a transformation-sensitive surface protein. Cell 11, 115-126 (1977)

Atlas, S.J., Lin, S.: High affinity cytochalasin B binding to normal and transformed BALB/3T3 cells J. Cell Physiol. 89, 751-756 (1976)

Bell, P.B.: Locomotory behavior, contact inhibition, and pattern formation of 3T3 and polyoma virus-transformed 3T3 cells in culture. J. Cell Biol. 74, 963-982 (1977)

Beug, H., Claviez, M., Jockusch, B., Graf, T.: Differential expression of Rous sarcoma virus specific transformation parameters in enucleated cells. Cell 14, 843-156 (1978)

Bragina, E.E., Vasiliev, L.M., Gelfand, I.M.: Formation of bundles of microfilaments during spreading of fibroblasts on the substrate. Exp. Cell Res. 97, 241-248 (1976)

Brunk, U., Ericsson, J.L.E., Ponten, J., Westermark, B.: Specialization of cell surfaces in contact-inhibited human glia-like cells in vitro. Exp. Cell Res. 67, 407-415 (1971)

Chen, L.B., Buchanan, J.M.: Plasminogen-independent fibrinolysis by proteases produced by transformed chick embryo fibroblasts. Proc. Natl. Acad. Sci. USA 72, 1132-1136 (1975)

Chen, L.B., Gallimore, P.H., McDougall, J.K.: Correlation between tumor induction and the large external transformation sensitive protein on the cell surface. Proc. Natl. Acad. Sci. USA 73, 3570-3574 (1976)

Collett, M.S., Erickson, R.L.: Protein kinase activity associated with the avian sarcoma virus src gene product. Proc. Natl. Acad. Sci. USA 75, 2021-2024 (1978)

Culp, L.A.: Electrophoretic analysis of substrate-attached proteins from normal and virus-transformed cells. Biochemistry 15, 4094-4104 (1976)

Engvall, E., Ruoslahti, E.: Binding of soluble form of fibroblast surface protein fibronectin to collagen. Int. J. Cancer 20, 1-5 (1977)

Furcht, L.T., Mosher, D.F., Wendelschafer-Crabb, G.: Immunocytochemical localization of fibronectin (LETS protein) on the surface of L6 myoblasts: light and electron microscopic studies. Cell 13, 263-271 (1978)

Goldman, R.D., Knipe, D.M.: Functions of cytoplasmic fibers in non-muscle cells. Cold Spring Harbor Symp. Quant. Biol. 37, 523-534 (1972)

Graham, J.M., Hynes, R.O., Davidson, E.A., Bainton, D.F.: The location of proteins labeled by the ^{125}I-lactoperoxidase system in the NIL8 hamster fibroblast. Cell 4, 353-365 (1975)

Graham, J.M., Hynes, R.O., Rowlatt, C., Sandall, J.K.: Cell surface coat of hamster fibroblasts. Ann. N.Y. Acad. Sci. 312, 221-239 (1978)

Grinnell, F.: Cell spreading factor: occurrence and specificity of action. Exp. Cell Res. 102, 51-62 (1976)

Grinnell, F., Hays, D.G.: Cell adhesion and spreading factor: identity with cold insoluble globulin in human serum. Exp. Cell Res. 115, 221-229 (1978)

Grinnell, F., Hays, D.G., Minter: Cell adhesion and spreading factor: partial purification and properties. Exp. Cell Res. 110, 175-190 (1978)

Harris, A.: Behavior of cultured cells on substrata of variable adhesiveness. Exp. Cell Res. 77, 285-297 (1973a)

Harris, A.K.: Location of cellular adhesions to solid substrata. Dev. Biol. 35, 97-114 (1973b)

Heath, J.P., Dunn, G.A.: Cell to substratum contacts of chick fibroblasts and their relation to the microfilament system. A correlated interference-reflexion and high voltage electron-microscope study. J. Cell Sci. 29, 197-212 (1978)

Heaysman, J.E.M., Pegrum, S.M.: Early contacts between fibroblasts. An ultrastructural study. Exp. Cell Res. 78, 71-78 (1973)

Hedman, K., Vaheri, A., Wartiovaara, J.: Cell surface protein, fibronectin, of human fibroblast cultures has a membrane-associated and a predominant percellular matrix form. J. Cell Biol. 76, 748-760 (1977)

Hynes, R.O.: Alteration of cell-surface proteins by viral transformation and by proteolysis. Proc. Natl. Acad. Sci. USA 70, 3170-3174 (1973)

Hynes, R.O.: Role of surface alterations in cell transformation: the importance of proteases and surface proteins. Cell 1, 147-156 (1974)

Hynes, R.O.: Cell surface proteins and malignant transformation. BBA Reviews on Cancer 458, 73-107 (1976)

Hynes, R.O., Destree, A.T.: Extensive disulfide bonding at the mammalian cell surface. Proc. Natl. Acad. Sci. USA 74, 2855-2859 (1977)

Hynes, R.O., Destree, A.T.: Relationship between fibronectin (LETS protein) and actin. Cell 15, 875-886 (1978)

Hynes, R.O., Wyke, J.A.: Alterations in surface proteins in chicken cells transformed by temperature-sensitive mutants of Rous sarcoma virus. Virology 64, 492-504 (1975)

Hynes, R.O.:, Destree, A.T., Mautner, V., Ali, I.U.: Synthesis, secretion, and attachment of LETS glycoprotein in normal and transformed cells. J. Supramol. Struct. 7, 397-408 (1977)

Hynes, R.O., Ali, I.U., Destree, A.T., Mautner, V., Perkins, M.E., Senger, D.R., Wagner, D.D., Smith, K.K.: A large glycoprotein lost from the surface of transformed cells. Ann. N.Y. Acad. Sci. 312, 317-342 (1978)

Izzard, C.S., Lochner, L.R.: Cell-to-substrate contacts in living fibroblasts: an interference reflexion study with an evaluation of the technique. J. Cell Sci. 21, 129-159 (1976)

Jung, C.Y., Rampal, A.L.: Cytochalasin B binding sites and glucose transport carrier in human erythrocyte membranes. J. Biol. Chem. 252, 5456-5463 (1977)

Lin, S., Snyder, C.E.: High affinity cytochalasin B binding to red cell membrane proteins which are unrelated to sugar transport. J. Biol. Chem. 252, 5464-5471 (1977)

Lin, S., Spudich, J.: Biochemical studies on the mode of action of cytochalasin B. Cytochalasin binding to red cell membrane in relation to glucose transport. J. Biol. Chem. 249, 5778-5783 (1974)

Mahdavi, V., Hynes, R.O.: Proteolytic enzymes in normal and transformed cells. Biochem. Biophys. Acta 583, 167-178 (1979)

Mautner, V., Hynes, R.O.: Surface distribution of LETS protein in relation to the cytoskeleton of normal and transformed cells. J. Cell Biol. 75, 743-758 (1977)

McNutt, N.S., Culp, L.A., Black, P.H.: Contact-inhibited revertant cell lines isolated from SV-40-transformed cells. II. Ultrastructural study. J. Cell Biol. 50, 691-708 (1971)

McNutt, N.S., Culp, L.A., Black, P.H.: Contact-inhibited revertant cell lines isolated from SV-40-transformed cells. IV Microfilament distribution and cell shape. J. Cell Biol. 56, 413-428 (1973)

Mosesson, M.W., Chen, A.B.: Huseby, R.M.: The cold-insoluble globulin of human plasma: studies of its essential structural features. Biochim. Biophys. Acta 386, 509-524 (1975)

Mosher, D.F.: Cross-linking of cold-insoluble globulin by fibrin-stabilizing factor. J. Biol. Chem. 250, 6614-6621 (1975)

Olden, K., Yamada, K.M.: Mechanisms of the decrease in the major cell surface protein of chick embryo fibroblasts after transformation. Cell 11, 957-969 (1977)

Quigley, J.P.: Association of a protease (plasminogen activator) with a specific membrane fraction from transformed cells. J. Cell Biol. 71, 472-486 (1976)

Revel, J.P., Wolken, K.: Electron microscope investigations of the underside of cells in culture. Exp. Cell Res. 78, 1-14 (1973)

Robbins, P.W., Wickus, G.G., Branton, P.E., Gaffney, B.J., Hirschberg, C.B., Fuchs, P., Blumberg, P.M.: The chick fibroblast cell surface following transformation by Rous sarcoma virus. Cold Spring Harbor Symp. Quant. Biol. 39, 1173-1180 (1974)

Roblin, R.O., Albert, S.O., Gelb, N.A., Black, P.H.: Cell surface changes correlated with density-dependent growth inhibition. Glycosaminoglycan metabolism in 3T3, SV3T3 and canA-selected revertant cells. Biochemistry 14, 347-357 (1975)

Ruoslahti, E., Vaheri, A.: Interaction of soluble fibroblast surface antigen with fibrinogen and fibrin. J. Exp. Med. 141, 497-501 (1975)

Schlessinger, J., Barak, L.S., Hammes, G.G., Yaonada, K.M., Pastan, I., Webb, W.W., Elson, E.L.: Mobility and distribution of a cell surface glycoprotein and its interaction with other membrane components. Proc. Natl. Acad. Sci. USA 74, 2909-2913 (1977)

Spooner, B.S.: Cytochalasin B.: toward an understanding of its mode of action. Dev. Biol. 35, 813-818 (1973)

Stathakis, N.E., Mosesson, M.W.: Interactions among heparin, cold-insoluble globulin, and fibrinogen in formation of the heparin-precipitable fraction of plasma. J. Clin. Invest. 60, 855-865 (1977)

Terry, A.H., Culp, L.A.: Substrate-attached glycoproteins from normal and virus-transformed cells. Biochemistry 13, 414-425 (1974)

Unkeless, J.C., Tobia, A., Ossowski, L., Quigley, J.P., Rifkin, D.B., Reich, E.: An enzymatic function associated with transformation of fibroblasts by oncogenic viruses. I. Chick embryo fibroblast cultures transformed by avian RNA tumor viruses. J. Exp. Med. 137, 85-111 (1973)

Vasiliev, J.M., Gelfand, I.M.: Effects of colcemid on morophogenetic processes and locomotion of fibroblasts. In: Cell Motility (eds. R. Goldman, T. Pollard, J. Rosenbaum), pp. 279-304. New York: Cold Spring Harbor Laboratory 1976

Weber, K., Rathke, P.C., Osborn, M., Franke, W.W.: Distribution of actin and tubulin in cells and in glycerinated cell models after treatment with cytochalasin B. Exp. Cell Res. 107, 285-297 (1976)

Wessells, N.K., Spooner, B.S., Ash, J.F., Bradley, M.O., Luduena, M.A., Taylor, E.L., Wrenn, J.T., Yamada, K.M.: Microfilaments in cellular and developmental processes. Science 171, 135-143 (1971)

Wessells, N.K., Spooner, B.S., Luduena, M.A.: Surface movements, microfilaments and cell locomotion. CIBA Symp. 14, 53-77 (1973)

Willingham, M.C., Yamada, K.M., Yamada, S.S., Pouyssegur, J., Pastan, I.: Microfilament bundles and cell shape are related to adhesiveness to substratum and are dissociable from growth control in cultured fibroblasts. Cell 10, 375-380 (1977)

Witkowski, J.A., Brighton, W.. D.: Influence of serum on attachment of tissue cells to glass surfaces. Exp. Cell Res. 70, 41-48 (1972)

Yamada, K.M., Yamada, S.S., Pastan, I.: Cell surface protein partially restores morphology, adhesiveness, and contact inhibition of movement to transformed fibroblasts. Proc. Natl. Acad. Sci. USA 73, 1217-1221 (1976a)

Yamada, K.M., Ohanian, S.H., Pastan, I.: Cell surface protein decreases microvilli and ruffles on transformed mouse and chick cells. Cell 9, 241-245 (1976b)

Desmin and Intermediate Filaments in Muscle Cells

E. LAZARIDES

Division of Biology, California Institute of Technology
Pasadena, CA 91125, USA

I. Different Classes of Intermediate (100Å) Filaments

With electron microscopy, a morphological class of filaments with an average diameter of 100 Å has been identified in many vertebrate cells. These filaments were initially referred to as intermediate filaments since their diameter is intermediate to those of the 60 Å actin filaments and the 250 Å microtubules in nonmuscle cells and intermediate to those of the 60 Å actin filaments and the 150 Å myosin filaments in muscle cells. Presently, intermediate filaments are commonly referred to as neurofilaments in neurons, glial filaments in glial cells, 10 nm filaments in fibroblastic cells, tonofilaments or keratin filaments in epiderimal and epithelial cells, and intermediate filaments in skeletal, cardiac, and smooth muscle cells. Due to their wide distribution and morphological similarity, it was originally believed that the major subunit of the intermediate filaments from these various sources would be highly conserved in its amino acid sequence as is the case with actin and tubulin. The recent isolation of intermediate filaments from a variety of avian or mammalian cell types has enabled investigators to characterize the subunit proteins of these different filament preparations and to determine the extent of their homology. By a number of criteria including peptide mapping, antibody cross-reactivity, isoelectric point and molecular weight, it was shown that the major protein subunit of the intermediate filaments from muscle cells (desmin), fibroblasts, glial cells, neurons, and epidermal and epithelial cells (keratins) was distinct (Benitz et al. 1976; Dahl and Bignami 1976; Davison et al. 1977; Schachner et al. 1977; Schlaepfer 1977; Franke et al. 1978; Lazarides 1978; Lazarides and Balzer 1978; Milstone and McGuire 1978; Schlaepfer and Freeman 1978; Starger et al. 1978; Sun and Green 1978).

The predominance of a particular class of intermediate filament in a given cell type suggested that each class of filaments performs a function tailored to the differentiated phenotype of the cell type. However, it has not been demonstrated that only one class of intermediate filament may be present in a specific cell type. The simultaneous presence of two kinds of intermediate filaments in the same cell wold seem to imply that the filaments serve complementary functions. Although the exact function of intermediate filaments has not been established, they are believed to play a cytoskeletal role in cells. With the exception of keratin filaments (Sun and Green 1978), they also have the remarkable ability to aggregate into perinuclear caps in cells exposed to the antimitotic drug colcemid.

We have been particularly concerned with the cytoskeletal function performed by the desmin-containing intermediate filaments of muscle cells. We have used fluorescent antibody methods to define the intracellular localization of desmin and have begun the biochemical characterization of desmin. We hope thereby to determine the control mechanisms for the assembly of desmin into cytoskeletal structures. Finally, we have been interested in the possibility that desmin or desmin filaments might also be present in nonmuscle cells.

II. Desmin and Intermediate Filaments in Smooth Muscle

We chose to focus our attention on the intermediate filaments in muscle rather than nonmuscle cells because the biochemistry and structure of contractile filaments in muscle cells are well characterized thus facilitating the investigation of new structural relationships between the contractile and intermediate filaments. The cytoplasm of developing and adult smooth muscle is permeated by an interconnecting network of intermediate filaments which terminates together with actin filaments at numerous cytoplasmic and membrane-bound electron dense structures; these structures can be thought of as the smooth muscle analogues of skeletal muscle Z lines (Cooke and Chase 1971; Uehara et al. 1971). These well-defined morphological associations then provide a structural background for investigating the biochemistry of any putative actin-intermediate filament-membrane interactions.

Cooke was the first to observe that intermediate filaments from smooth muscle are insoluble in high concentrations of KCl (0.6–1.0 M) that render the majority of actin and myosin filaments soluble (Cooke and Chase 1971; Cooke 1976). We found that chicken gizzard smooth muscle cells extracted with 0.6 M KI remain as ghosts that contain a cytoskeleton composed almost entirely of intermediate filaments. However, electrophoretic analysis of these ghosts revealed the presence of two major proteins in approximately equal amounts: actin and a novel 50,000 mol.wt. protein that, for reasons discussed below, we have named desmin (Lazarides and Hubbard 1976). Desmin is soluble only under a variety of fairly extreme conditions. It can be solubilized in 1 M acetic acid (Small and Sobieszek 1977), ethylene diamine, 4–8 M urea, and 3 M guanidine-HCl or by the anionic detergents Sarkosyl NL-97 and sodium dodecyl sulfate. It cannot be solubilized by sodium deoxycholate, Triton X-100, or Nonidet NP-40 (Hubbard and Lazarides 1979). Thus, unless cell disruption and KI extraction render desmin insoluble, the vast majority of desmin appears to exist as an insoluble polymer under physiological conditions. We have taken advantage of these properties to purify desmin from chicken gizzard by repeated cycles of solubilization in 1 M acetic acid and precipitation at pH 4 (Hubbard and Lazarides 1979). In all cases, a small proportion of the total actin in smooth muscle cells remains insoluble along with desmin during high salt extraction and copurifies with it through the cycles of solubilization and precipitation.

Although desmin precipitates when acetic and acid solutions are neutralized with base, it frequently remains in solution when the acetic acids is instead removed by dialysis against distilled water. These dialyzed solutions are metastable and exposure to either 10–$100\,\mu$ ionic strengths or glass results in a rapid gelation and subsequent syneresis of the gel into a tight mass. This syneresis phenomenon is ATP-independent. Electron microscopy reveals the gels to be composed of a highly intertwined network of filaments with average diameters of 120 Å (Hubbard and Lazarides 1979). Immunofluorescence with antidesmin and antiactin antibodies has suggested that actin and desmin copolymerize in vitro to form the 120 Å filaments. It should be noted that the polymerization properties of the actin-desmin mixture are totally unlike those of the conventional actin that is solubilized along with myosin at high salt concentrations.

These results strongly implied that actin may be specifically associated with desmin. We have therefore investigated the existence of an actin-desmin complex under milder extraction conditions that avoid the use of acetic acid or detergents as a solvent for desmin. We observed that extration of chicken gizzard with Tris-EGTA at low ionic strength (10 mM Tris-HCl; 10 mM EGTA, pH 7.5) slowly solubilizes a number of proteins, including actin and desmin. At least 50% of the total desmin can be extracted in this way, presumably by the dissociation of a calcium dependent structure. Two different types of experiments demonstrate that in these extracts desmin and actin exist in the form of a tenacious complex: (1) Actin has a high affinity ($K_d = 10^{-11}$) for pancreatic deoxyribonuclease I. DNase-I affinity chromatography of the Tris-EGTA extract demonstrates that actin and desmin coelute from the column with 3 M guanidine-HCl, conditions previously shown to elute the most tightly bound actin from DNase-I (Lazarides and Lindberg 1974). (2) Indirect immunoprecipitation of the Tris-EGTA extract demonstrated that both actin and desmin coprecipitate with antibodies specific for either desmin or actin.

The exact stoichiometry of the actin-desmin complex is unclear because it varies with the different extraction procedures; it can be approximately 7:1 (D:A) in acetic acid purified desmin, 1:1 in the antibody precipitated desmin or actin and 1:5 (D:A) in the 3 M guanidine HCl fraction of the DNase-I column.

The actin-desmin complexes isolated by the Tris-EGTA extraction procedure may include the cytoplasmic dense bodies and membrane-bound plaques of smooth muscle. It has been hypothesized that actin and desmin interact at these sites in vivo; both of these structures are removed by treating smooth muscle with Tris-EGTA (Schollmeyer et al. 1976).

From this work, we have drawn the following conclusions: (1) Desmin is the major component of smooth muscle intermediate filaments. (2) Desmin exists as a stable polymer under physiological conditions. (3) Desmin forms a tenacious complex with actin (4) Desmin confers upon the small fraction of conventional smooth muscle actin with which it interacts its own characteristic insolubility and polymerization properties.

III. Desmin in Striated Muscle

The small size of smooth muscle cells and their complicated morphology makes in situ studies difficult, and for this reason we have focused our cytological inquiries on desmin in skeletal muscle. Electrophoretic, isoelectric, and immunological criteria all indicate that smooth muscle and striated muscle desmin are very closely similar molecules (Lazarides and Hubbard 1976; Izant and Lazarides 1977; Lazarides 1978; Lazarides and Balzer 1978). Using immunofluorescence, we observed that desmin is a component of both skeletal and cardiac muscle Zlines and also of regions where the Z line is apposed with the plasma membrane (intercalated discs) (Lazarides and Hubbard 1976; Lazarides 1978; Lazarides and Balzer 1978). Desmin is thus the third protein to be localized at the Z line, the other two being α-actinin and actin. These localizations were performed on longitudinal sections of myofibrils, however, and did not yield any information on the spatial relationships of these proteins within the plane of a single Z line.

Before proceeding, we should define some terminology. The sarcomeres of a single myofibril are connected longitudinally by individual Z-discs. The Z-disc is thus confined to a single myofibril and is the smallest unit that we will be considering. In a bundle of myofibrils, the Z-discs of each fibril are in lateral register and define a Z-plane. When a myofibril bundle is viewed laterally (side view), the Z-discs in each plane superimpose to form the phase dense Z line band seen by phase contrast light microscopy. Until now, Z-planes have been viewed face on (surface view) only with electron microscopy in embedded and sectioned material.

We have been able to pursue the question of how actin, α-actinin, and desmin are spatially related with the development of a technique for performing immunofluorescence on the surface of a single Z-plane (Granger and Lazarides 1978). When a muscle fiber is extracted with 50% glycerol and then sheared in a blender, it is cleaved along its long axis to produce small bundles of myofibrils. Separation occurs between laterally associated Z-discs. The actomyosin in these bundles is structurally and functionally intact. We found, however, that if glycerol extraction is followed instead by 0.6 M KI extraction, most of the actomyosin is removed to yield stacked arrays of KI-insoluble Z-planes which are still tethered together by residual longitudinal fibrils. When these stacked arrays are sheared, cleavage now occurs between Z-planes, not between adjacent myofibrils, and individual Z-sheets result (Granger and Lazarides 1978). These are easily observed in the light microscope and have a honeycomb appearance (Fig. 1).

In chicken skeletal muscle Z-planes, the closely packed Z-discs are separated from each other by a small boundary. This boundary region is permeated by the transverse tubular (T) system, an elaborate finger-like invagination of the plasma membrane. The lumen of the T system freely communicates with the extracellular medium (for complete references, see Lazarides and Granger 1978). The Z-planes thus provide an ideal system for investigating the cytological basis of the actin-desmin-membrane interaction that the smooth muscle biochemistry has implied.

The use of specific probes (didansyl ornithine, didansyl lysine) that are fluorescent only in hydrophobic environments indicates the presence of T tubule membrane remnants in the Z-disc boundary region of isolated Z-sheets (Fig. 1). This is confirmed by direct electron microscopic observation which reveals the presence of vesicular material in the Z-disc boundary region; little or no membrane is present in the Z-discs themselves (Granger and Lazarides 1978; Lazarides and Granger 1978).

Immunofluorescent antibody studies of the isolated Z-sheets indicate that α-actinin and desmin have a complementary distribution: α-actinin is confined to the Z-discs while desmin is only present in the boundary region between the discs. Thus,

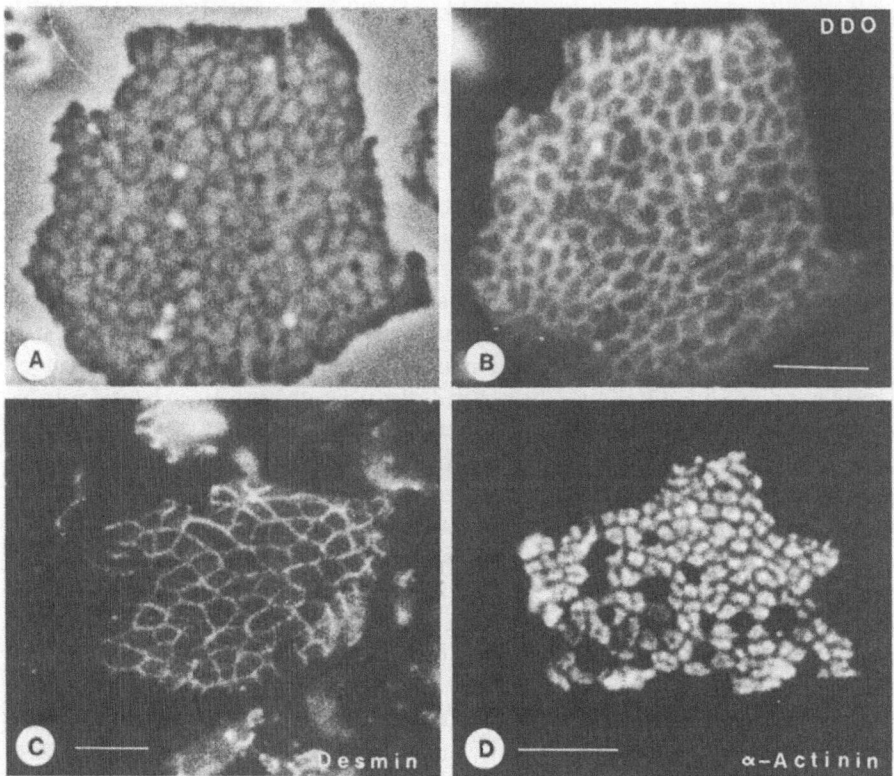

Fig. 1A-D. Distribution of didansyl ornithine (DDO), desmin and α-actinin in Z-disc lattices. **A, B** Lattices treated with DDO. **A** Phase-contrast image; **B** fluorescence image; **C** Indirect immunofluorescence using antibodies to desmin; **D** Indirect immunofluorescence using antibodies to α-actinin. *Bars* = 5 μm. (Reprinted from Lazarides and Granger 1978)

the distribution of desmin (which is a hydrophobic protein) and T membranes coincide (Fig. 1). Actin appears to be uniformly distributed through the Z-discs and their boundaries (Granger and Lazarides 1978). The immunofluorescent distribution of actin provides cytological evidence for the biochemical observations that actin interacts with both desmin and α-actinin.

Several possible functions of desmin emerge from this work. First, desmin may serve to integrate all the actin filaments that terminate in the Z-disc and to link individual myofibrils laterally by binding together the individual Z-discs in a Z-plane. The result of this would be the mechanical unification of the contractile actions of all of the separate myofibrils. Thus, in the biogenesis and assembly of myofibrils, desmin may play an important role in the generation of the striated phenotype of muscle by bringing the Z-disc into lateral register. Secondly, desmin may function in the biogenesis of the T tubule system. Some mechanism is required to direct the localization of the invaginating plasma membrane to the Z-disc boundary region of the Z-sheets. A protein with both membrane and Z-disc affinities, like desmin, could play a vital role in this process. Preliminary chemical

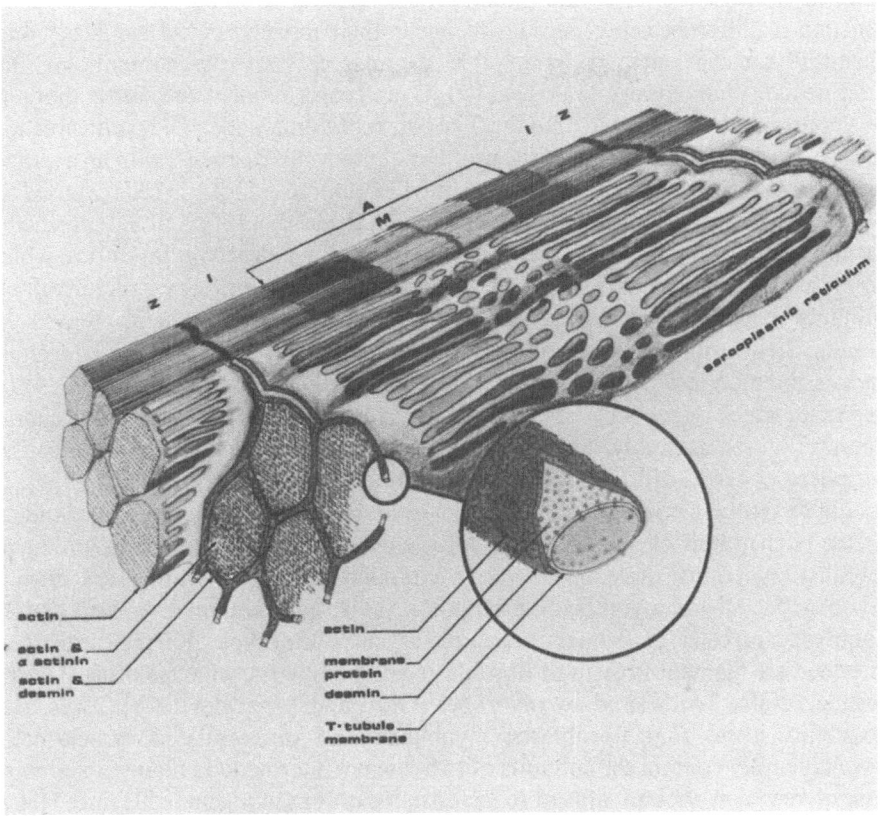

Fig. 2. Schematic representation of the distribution of desmin, α-actinin, and the transverse tubular membrane in chicken skeletal myofibril Z-discs

characterization of desmin indicates that the amino acid distribution in the molecule is asymmetric with a majority of the neutral and hydrophobic amino acids occurring near one terminus. This part of the molecule may be responsible for the hydrophobic properties of desmin and for interacting with the plasma membrane. Finally, desmin may function in linking actin to the plasma membrane (T system, intercalated discs). Our thoughts on the linking functions of desmin in the Z-discs of muscle cells are summarized in Fig. 2. It is to illustrate these linking roles for desmin that we have named it from the Greek noun δεσμός, which means link or bond (Lazarides and Hubbard 1976).

IV. Conclusions and Prospects for the Future

The results summarized above clearly indiate that intermediate filaments may perform a very specialized cytoplasmic structural function and, in retrospect, it is not a big surprise that the subunit of intermediate filaments are heterogeneous. We can only conclude that such a heterogeneity reflects their specific cytoplasmic function in different cell types. To reconcile their morphological similarity and chemical heterogeneity, it is possible to imagine that the subunits of the intermediate filaments are analogous to IgG molecules, namely, they are composed of a constant and a variable domain. The constant domain may be responsible for the conserved polymerization of the protein subunits into filaments with an average diameter of 100 Å and their aggregation into filamentous bundles in cells exposed to colcemid. The variable domain may confer upon the molecule functional specificity for the cell it is part of. This question will be answered clearly in the future when amino acid sequencing data will be available comparing the primary structure of the subunits from the different classes of intermediate filaments. With regard to desmin, these results raise some interesting questions: (1) The insolubility of desmin and its interaction with actin leads us to ask whether there is a soluble desmin precursor which is incapable of interacting with actin. This is part of a more general question on the assembly steps desmin goes through before it is deposited at the periphery of the Z-disc. (2) If desmin is indeed specific for muscle cells, then it should be synthesized predominantly after myoblast fusion into myotubes. Studies on the biosynthesis of desmin during muscle differentiation in tissue culture have revealed something quite unexpected. Although desmin is synthesized during myoblast fusion and myotube differentiation, it also exist both in myoblasts and in nonmyogenic cells in primary cultures of chick embryos. Furthermore, the intermediate filament protein of fibroblasts and myoblasts, which is distinct from desmin, is also synthesized by myotubes (Gard and Lazarides 1979). Thus, the possibility arises that fibroblasts, myoblasts, and differentiated muscle cells simultaneously contain the subunits of two distinct intermediate filament systems. This observation does not appear to be restricted only to myogenic cells since HeLa cells have been found recently to contain, in addition to the intermediate filament subunit of fibroblasts normally found in these cells (in the author's nomenclature, vimentin), a small number of keratin filaments (Franke et al. 1978).

Such observations raise a whole set of new biological and chemical questions about the structure and cytoplasmic function of intermediate filaments which will be undoubtedly elucidated in the near future.

Acknowledgments. Supported by a grant from the Muscular Dystrophy Association of American and NIH grant PHS-GM 06965-19.

References

Benitz, W.E., Dahl, D., Williams, K.W., Bignami, A.: The protein composition of glial and nerve fibers. FEBS Lett. 66, 285-289 (1976)

Bennett, G.S., Fellini, S.A., Croop, J.M., Otto, J.J., Bryan, J., Holtzer, H.: Differences among 100 Å filament subunits from different cell types. Proc. Natl. Acad. Sci. USA 75, 4364-4368 (1978)

Cooke, P.: A filamentous cytoskeleton in vertebrate smooth muscle fibers. J. Cell Biol. 68, 539-556 (1976)

Cooke, P.H., Chase, R.H.: Potassium chloride-insoluble myofilaments in vertebrate smooth muscle cells. Exp. Cell Res. 66, 417-425 (1971)

Dahl, D., Bignami, A.: Isolation from peripheral nerve of a protein similar to the glial fibrillary acidic protein. FEBS Lett. 66, 281-284 (1976)

Davison, P.F., Hong, B.-S., Cooke, P.: Classes of distinguishable 10 nm cytoplasmic filaments. Exp. Cell Res. 109, 471-474 (1977)

Franke, W.W., Schmid, E., Osborn, M., Weber, K.: Different intermediate-sized filaments distinguished by inmunofluorescence microscopy. Proc. Natl. Acad. Sci. USA 75, 5034-5038 (1978)

Gard, D.L., Lazarides, E.: Specific fluorescent labeling of chicken mayofibril Z-line proteins catalyzed by guinea pig liver transglutaminase. J. Cell Biol. 81, 336-347 (1979)

Granger, B.L., Lazarides, E.: The existence of an insoluble Z-disc scaffold in chicken skeletal muscle. Cell 15, 1253-1268 (1978)

Hubbard, B.D., Lazarides, E.: The copurification of actin and desmin from chicken smooth muscle and their co-polymerization in vitro to intermediate filaments. J. Cell Biol. 80, 166-182 (1979)

Izant, J.G., Lazarides, E.: Invariance and heterogeneity in the major structural and regulatory proteins of chick muscle cells revealed by two-dimensional gel electrophoresis. Proc. Natl. Acad. Sci. USA 74, 1450-1454 (1977)

Lazarides, E.: Distribution of desmin (100 Å) filaments in primary cultures of embryonic chick cardiac cells. Exp. Cell Res. 112, 265-273 (1978)

Lazarides, E., Balzer, D.B, Jr.: Specificity of desmin to avian and mammalian muscle cells. Cell 14, 429-438 (1978)

Lazarides, E., Granger, B.L.: Fluorescent localization of membrane sites in glycerinated chicken skeletal muscle fibers and the relationship of these sites to the protein composition of the Z disc. Proc. Natl. Acad. Sci. USA 75, 3683-3687 (1978)

Lazarides, E., Hubbard, B.D.: Immunological characterization of the subunit of the 100 Å filaments from muscle cells. Proc. Natl. Acad. Sci. USA 73, 4344-4348 (1976)

Lazarides, E., Lindberg, U.: Actin is the naturally occurring inhibitor of deoxyribonuclease I. Proc. Natl. Acad. Sci. USA 71, 4742-4746 (1974)

Milstone, L.M., McGuire, J.S.: Comparisons of proteins that form 8–10 nm filaments in two keratinizing epithelia and endothelial cells in culture. J. Cell Biol. 79, (2, Pt. 2) 30a (1978)

Schachner, M., Hedley-Whyte, E.T., Hsu, D.W., Schoonmaker, G., Bignami, A.: Ultrastructural localization of glial fibrillary acidic protein in mouse cerebellum by immunoperoxidase labeling. J. Cell Biol. 75, 67-73 (1977)

Schlaepfer, W.W.: Immunological and ultrastructural studies of neurofilaments isolated from rat peripheral nerve. J. Cell Biol. 74, 226-240 (1977)

Schlaepfer, W.W., Freeman, L.A.: Neurofilament proteins of rat peripheral nerve and spinal cord. J. Cell Biol. 78, 653-662 (1978)

Schollmeyer, J.E., Furcht, L.T., Goll, D.E., Robson, R.M., Stromer, M.H.: Localization of contractile proteins in smooth muscle cells and in normal and transformed fibroblasts. In: Cell Motility. Cold Spring Harbor Conferences on Cell Proliferation, Book A (eds. R.D. Goldman, T.D. Pollard, J. Rosenbaum), pp. 364-388. Cold Spring Harbor, N.Y.: Cold Spring Harbor Laboratories 1976

Small, J.V., Sobieszek, A.: Studies on the function and composition of the 10 nm (100 Å) filaments of vertebrate smooth muscle. J. Cell Sci. 23, 243-268 (1977)

Starger, J.M., Brown, W.E., Goldman, A.E., Goldman, R.D.: Biochemical and immunological analysis of rapidly purified 10 nm filaments from baby hamster kidney (BHK-21) cells. J. Cell Biol. 78, 93-109 (1978)

Sun, T.T., Green, H.: Immunofluorescent staining of keratin fibers in cultured cells. Cell 14, 469-476 (1978)

Uehara, Y., Campbell, G.R., Burnstock, G.: Cytoplasmic filaments in developing and adult vertebrate smooth muscle. J. Cell Biol. 50, 484-497 (1971)

The Microtubule Cytoskeleton
in Normal and Transformed Cells in Vitro

B.R. Brinkley[1], L.J. Wible[1], B.B. Asch[1], D. Medina[1]
M.M. Mace[1], P.T. Beall[2], and R.M. Cailleau[3]

Department of Cell Biology[1] and Department of Pediatrics[2]
Baylor College of Medicine
The University of Texas M.D. Anderson Hospital and Tumor Institute[3]
Houston, TX 77030, USA

I. Introduction

Cell transformation in vitro is usually accompanied by alterations in morphology and growth properties which may be related directly or indirectly to malignancy. Although the initial lesion of transformation probably involves the interaction of a carcinogen with the cell genome, subsequent expression of the mutagenic event is manifested in altered cell shape, alterations in motility, spreading, and adhesive properties, loss of density-dependent growth control, loss of anchorage dependence and changes in cell surface properties. Recent evidence has suggested that many aspects of transformation may relate to changes in a delicate system of cytoplasmic microfilaments and microtubules known collectively as the cytoskeleton (Edelman 1976; Nicolson 1976). Components of the cytoskeleton play a major role in the regulation of cell form (Porter 1966; Nicolson 1976), and both microtubules and microfilaments have been shown to be involved in cell motility (Porter 1966; Allison 1973) and substrate attachment (Goldman and Follett 1970; Pollack and Rifkin 1975; Miller et al. 1977). Also, the modulation of cell surface proteins is regulated to a large extent by submembranous assemblies of microtubules and microfilaments (Edelman 1976).

Evidence that the cytoskeleton becomes altered in transformed cells has been supported by both electron microscopy (McNutt et al. 1971; Fonte and Porter 1974; Goldman et al. 1974, 1976; Nicolson 1976) and immunofluorescence studies involving antibodies to actin (Pollack and Rifkin 1975; Pollack et al. 1975; Goldman et al. 1976) and tubulin (Brinkley et al. 1975, 1976; Fuller et al. 1975; Edelman and Yahara 1976; Puck 1977). The majority of such studies have utilized either early passage cell cultures or established lines usually derived from embryonic or neonatal mesenchymal tissues. In the present report, we will review the structure and distribution of the cytoplasmic microtubule complex in established cell lines, and present more recent investigations involving normal, preneoplastic and neoplastic cells of epithelial origin.

II. The Cytoplasmic Microtubule Complex (CMTC)

Following the development of glutaraldehyde as a fixative for electron microscopic studies by Sabatini et al. (1963), microtubules were found to be present in the cytoplasm of most eukaryotic cells (Porter 1966). The full extent of microtubule distribution in the cytoplasm was not apparent, however, until immunofluorescence procedures involving monospecific tubulin antibodies were developed (Fuller et al. 1975). As shown in Fig. 1, an extensive network of fine fluorescent filaments can be seen in well-spread cells stained by indirect immunofluorescence using tubulin antibody. Since the stained network is not present in cells treated with microtubule inhibitors such as colcemid (Fig. 2) and cold temperatures, we can safely conclude that it is composed of intact micro-tubules which has been termed the cytoplasmic microtubule complex (CMTC). Recently, Osborn et al. (1978) have utilized combined immunofluorescence and electron microscopy to show that each fluorescent filament in such preparations represents a single microtubule.

In many cells the CMTC appears to be organized around one or two central foci (the centrosphere), with microtubules radiating outward toward the cell periphery. They either terminate near the cell surface or bend and extent along the plasma membrane. Recently we have used a mild hypotonic pretreatment to gently expand Swiss mouse 3T3 cells and show that some microtubules appear to be physically attached to the cell surface (Brinkley et al. 1977). When nonconfluent cell monolayers were exposed to media made hypotonic with distilled water (1 part medium: 3 parts H_2O), they were reversibly swollen to various degrees. The extent of cell swelling could be monitored with Nomarski optics. When such cells were fixed and stained with tubulin antibodies, the only microtubules present were those with one end associated with the cell membrane (Figs. 3 and 4). Cytoplasmic microtubules which were not associated with the cell periphery, including those of the mitotic spindle, were depolymerized by the hypotonic treatment. We conclude from these experiments that at least two populations of microtubules exist in the cytoplasm: hypotonically stable microtubules which have one end associated with the plasma membrane, and hypotonically labile microtubules which are not associated with the cell surface. It is possible that the cell surface associated microtubules represent those tubules which are involved in the modulation of cell surface receptors (Edelman 1976).

III. The CMTC in Transformed Cells

When various transformed cell lines were examined with antitubulin immunofluorescence, the cytoplasmic staining pattern was found to be quite different from that of their nontransformed counterparts (Brinkley et al. 1975, 1976; Fuller et al. 1975; Edelman and Yahara 1976). In the established lines that we studied, the cytoplasm displayed a diffuse staining pattern. Microtubules were present but they appeared thinner, shorter and more densely packed. Also, many cells appeared to have fewer microtubules. We defined this staining pattern as the *Diminished CMTC*. Similar observations were made by other investigators using different cell lines (Edelman and Yahara 1976) and a direct correlation between the altered CMTC and increased capacity for growth in vitro can be demonstrated (Miller et al. 1977).

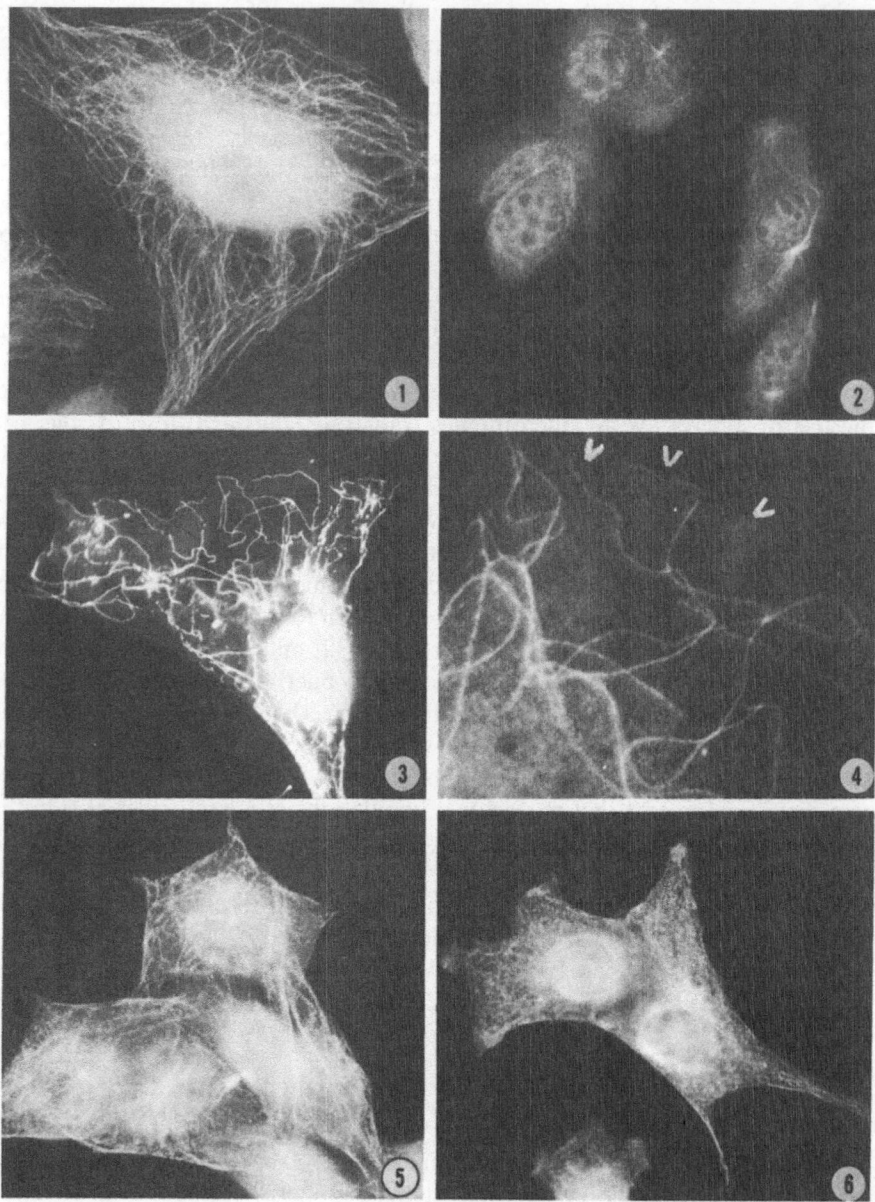

Fig. 1. Balbc/3T3 cell showing full CMTC. Note that individual microtubules can be observed in the cytoplasm. ×650

Fig. 2. 3T3 cell after colcemid treatment (0.06 µg/ml for 60 min). ×650

Fig. 3. 3T3 cell after hypotonic pretreatment (1 part medium: 3 parts H₂O). ×650

Fig. 4. Higher magnification of microtubule with free ends near the cell surface *(arrow heads)*. ×1300

Fig. 5. SV3T3 cells showing altered CMTC pattern. ×650

Fig. 6. Mouse hepatoma cell in vitro with diminished CMTC. ×650

Recently, several investigators using different fixation and staining regimens have concluded that few if any differences exist between the CMTC of normal and transformed cells (Osborn and Weber 1977; DeMey et al. 1978; Tucker et al. 1978). Although they generally agree that the tubulin staining pattern is different, they feel that it is due to the more rounded shapes of transformed cells. Clearly, most transformed cells have cytoplasmic microtubules, and there is little reason to suspect that a general defect in microtubule assembly exists in these cells. The relative abundance of polymerized microtubules and their organization in cells is difficult to assess by immunofluorescence microscopy. Quantitative estimations of the relative pool size of soluble and polymerized tubulin are needed in matched pairs of normal and transformed cells before this issue can be fully resolved. Nevertheless, it seems reasonable to conclude that the pronounced changes in cell shape which accompany transformation in many cells require a major alteration or restructuring of the CMTC. Such changes are clearly reflected in the tubulin staining pattern of many transformed cells as shown in Figs. 5 and 6.

IV. The CMTC in Malignant Cells of Epithelial Origin

The CMTC in tumor cells of epithelial origin was of interest to us for two major reasons. First, most of the human neoplasms are of epithelial origin and second, some epithelial tumor cells in vitro, such as those derived from mouse mammary epithelium, are indistinguishable on the basis of morphology and growth pattern (Das et al. 1974; Pickett et al. 1975; Voyles and McGrath 1976), adherence to substrates, and agglutination by lectins (Asch and Medina 1979). We wished to examine the CMTC in these types of cells to determine the relationship, if any, of microtubule organization to cell form and malignancy.

The details of the two in vitro epithelial systems used in the study will not be described here, and readers are referred to references Cailleau et al. (1974) and Medina (1973) for more information on culture conditions, morphology, and growth properties. The mouse mammary system has been described by Medina (1973), and consists of normal, preneoplastic, and neoplastic cells grown in vitro and in vivo. The system is particularly attractive because the tumorigenicity of cells in vitro can be easily assessed by injecting the cells into the mammary fat pads of female syngeneic mice. The human breast carcinoma cells used in this study were from long term continuous cell lines derived from patients with breast tumors at M.D. Anderson Hospital and Tumor Institute (Cailleau et al. 1974). The cells were derived from pleural effusions and brain biopsies of patients with metastatic tumors. Thirteen cell lines were investigated and in several instances tumorigenicity was confirmed in nude mice.

Tubulin staining patterns in normal, preneoplastic and neoplastic mouse mammary cells were essentially identical (Figs. 7 to 9). Variation in the expression of CMTC existed within each culture with some cells showing diminished CMTC patterns. However, the same variation existed in all three phenotypes with the full CMTC pattern being the predominant pattern. A study of actin immunofluorescence patterns revealed the same trend. It was concluded that no consistent differences existed in the cytoskeletons of normal, preneoplastic, and

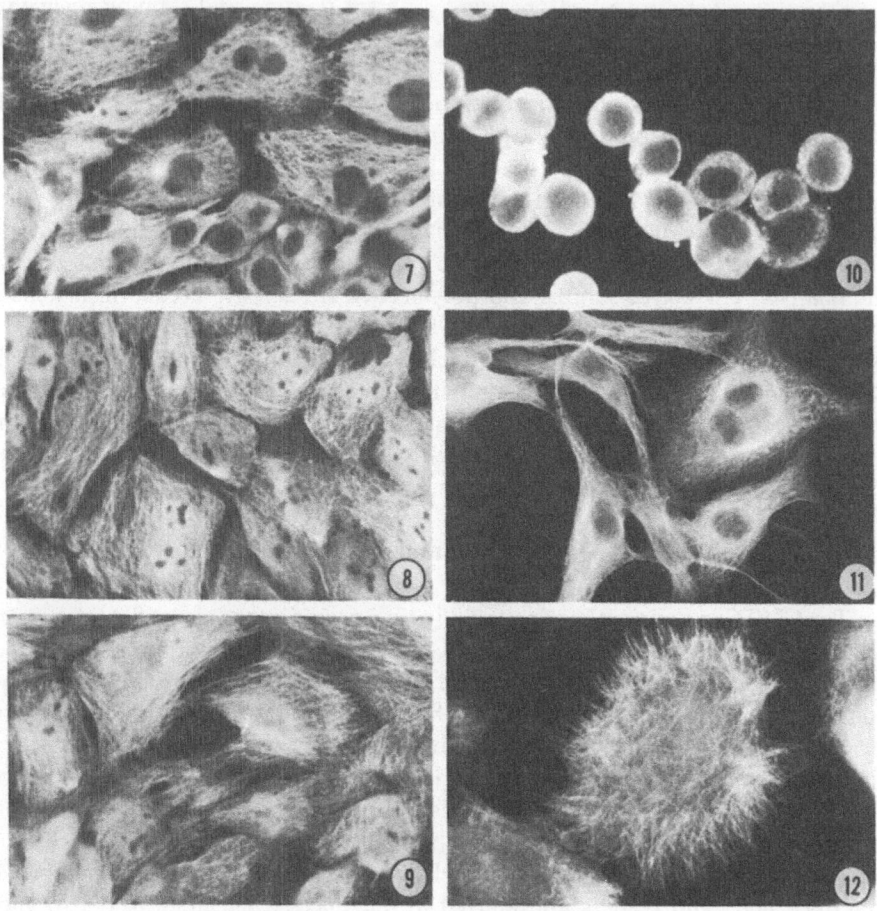

Fig. 7. Normal mouse epithelial mammary cells after immunofluorescence staining with antibodies to tubulin. × 400

Fig. 8. Preneoplastic nodule cell showing full CMTC. × 400

Fig. 9. Mouse mammary tumor cell showing CMTC like that of normal and preneoplastic cells above. × 400

Fig. 10. Human breast tumor cells in vitro. Note diffuse staining pattern and apparent absence of CMTC. × 400

Fig. 11. Human breast tumor cells in vitro. Note intermediate pattern of CMTC. × 400

Fig. 12. Human breast tumor cell in vitro. Note full CMTC. × 400

neoplastic mouse mammary cells in vitro (Asch et al. 1979). Thus, these cells fail to display any of the familiar properties which serve as "markers" for the in vitro transformation of established cell lines and, considerable caution should be used in correlating in vitro properties with malignancy in epithelial cells.

A different and considerably more interesting pattern was observed with respect to the organization of microtubules in human mammary carcinoma cells in vitro.

These cell lines displayed a wide range of stable morphological phenotypes (Figs. 10 to 12). Some lines were composed entirely of small rounded cells, while others were characterized by flattened epithelial cells which maintained extensive adherence to the substrate. The CMTC appeared diminished in the small rounded cells (Fig. 10), more extensive in the intermediate cells (Fig. 11) and fully extended in the flattened cells (Fig. 12).

V. Summary and Conclusions

Through the use of antitubulin immunofluorescence an extensive microtubule apparatus termed the CMTC can be demonstrated in interphase cells. The CMTC is present in both normal and transformed cells but appears modified in those cells which undergo major shape changes during transformation. Tumor cells in vitro which show little or no change in morphology after transformation display no apparent modification in the CMTC. Other tumor cell lines which display a wide range of cell morphologies exhibit a correspondingly wide range of CMTC patterns. Thus, the CMTC as visualized by tubulin immunofluorescence appears to be a reliable indicator of changes in cell morphology but should only be used in conjunction with a variety of other properties as a marker for transformation in vitro.

*Acknowledgments.*We wish to thank Ms. Jan Gibson for secretarial assistance and Ms. Susan Cox for proofreading the manuscript. This work was supported by NIH Grants CA-22610, CA-11944, and CA-21624 ONR C-0100 and C-0068.

References

Allison, A.C.: In: Locomotion of tissue cells CIBA Foundation Symposium, Vol. 14 (M. Abercrombie, ed.), p. 109. New York: Assoc. Scientific Publishers 1973

Asch, B.B., Medina, D.: Concanavalin A-induced agglutinability of normal, preneoplastic, and neoplastic mouse mammary cells. J. Natl. Cancer Inst., in press (1979)

Asch, B.B., Medina, D., Brinkley, B.R.: Microtubules and actin-containing filaments of normal preneoplastic and neoplastic mouse mammary epithelial cells. Cancer Res. 39, 893-907 (1979)

Brinkley, B.R., Fuller, G.M., Highfield, D.P.: Cytoplasmic microtubules in normal and transformed cells in culture: Analysis by tubulin antibody immunofluorescence. Proc. Natl. Acad. Sci. USA 73, 4981-4985 (1975)

Brinkley, B.R., Fuller, G.M., Highfield, D.P.: Tubulin Antibodies as Probes for Microtubules in Dividing and Nondividing Mammalian Cells. In: Cell Motility (R. Goldman, T. Pollard, J. Rosenbaum, eds.), pp. 435-456. Cold Spring Harbor: Cold Spring Harbor Lab. 1976

Brinkley, B.R., Marcum, J.M., Pepper, D.A.: Microtubule populations in cultured mammalian cells: Differential effects of hypotonic culture media. J. Cell Biol. 75, 296a (1977)

Cailleau, R., Mackay, B., Young, R.K., Reeves, W.J., Jr.: Tissue culture studies on pleural effusions from breast carcinoma patients. Cancer Res. 34, 801-809 (1974)

Das, N.K., Hosick, H.L., Nandi, S.: Influence of seeding sensity on multicellular organization and nuclear events in cultures of normal and neoplastic mouse mammary epithelium. J. Natl. Cancer Inst. 52, 849-861 (1974)

DeMey, J., Janiau, M., De Brabander, M., Moens, W., Geuens, G.: Evidence for unaltered structure and in vivo assembly of microtubules in transformed cells. Proc. Natl. Acad. Sci. USA 75, 1339-1343 (1978)

Edelman, G.M.: Surface modulation in cell recognition and cell growth. Science 192, 218-226 (1976)

Edelman, G.M., Yahara, I.: Temperature-sensitive changes in surface modulating assemblies of fibroblasts transformed by mutants of Rous sarcoma virus. Proc. Natl. Acad. Sci. USA 73, 2047-2051 (1976)

Fonte, V., Porter, K.: Topographic changes associated with the viral transformation of normal cells to tumorigenicity. In: Eighth International Congress on Electron Microscopy, pp. 334-335. Australia: Australian Academy of Science 1974

Fuller, G.M., Brinkley, B.R., Boughter, J.M.: Immunofluorescence of mitotic spindles by using monospecific antibody against bovine brain tubulin. Science 187, 948-950 (1975)

Goldman, R.D., Follett, E.A.C.: Birefringent filamentous organelle in BHK-21 cells and its possible role in cell spreading and motility. Science 1969, 286-288 (1970)

Goldman, R.D., Chang, C., Williams, J.F.: Properties and behavior of hamster embryo cells transformed by human adenovirus type 5. Cold Spring Harbor Quant. Biol. 39, 601-614 (1974)

Goldman, R.D., Yerna, M.J., Schloss, J.A.: Localization and organization of microfilaments and related proteins in normal and virus-transformed cells. J. Supramol. Struct. 5, 155-183 (1976)

McNutt, N.S., Culp, L.A., Black, P.H.: Contract-inhibited revertant cell lines isolated from SV40-transformed cells. II. Ultrastructural study. J. Cell Biol. 50, 691-708 (1971)

Medina, D.: Preneoplastic lesions in mouse mammary tumorigenesis. Methods in Cancer Res. 7, 3-53 (1973)

Miller, C.L., Fuseler, J.W., Brinkley, B.R.: Cytoplasmic microtubules in transformed mouse X nontransformed cell hybrids: Correlation with in vivo growth. Cell 12, 319-331 (1977)

Nicolson, G.L.: Transmembrane control of the receptors on normal and tumor cells. II. Surface changes associated with transformation and malignancy. Biochim. Biophys. Acta 458, 1-72 (1976)

Osborn, M., Weber, K.: The display of microtubules in transformed cells. Cell 12, 561-571 (1977)

Osborn, M., Webster, R.E., Weber, K.: Individual microtubules viewed by immunofluorescence and electron microscopy in the same PtK$_2$ cell. J. Cell Biol. 77, R27-R34 (1978)

Pickett, P.B., Pitelka, D.R., Hamamoto, S.T., Misfeldt, D.S.: Occluding junctions and cell behavior in primary cultures of normal and neoplastic mammary gland cells. J. Cell Biol. 66, 316-332 (1975)

Pollack, R., Rifkin, D.: Actin-containing cables within anchorage-dependent rat embryo cells are dissociated by plasmin and trypsin. Cell 6, 495-506 (1975)

Pollack, R., Osborn, M., Weber, K.: Patterns of organization of actin and myosin in normal and transformed cultured cells. Proc. Natl. Acad. Sci. USA 72, 994-998 (1975)

Porter, K.R.: Cytoplasmic microtubules and Their Functions. In: Principles of Biomolecular Organization (G.E. Wolstenholme, M. O'Connor, eds.), pp. 308-356. Boston: Littel, Brown 1966

Puck, T.T.: Cyclic AMP, the microtubule-microfilament system, and cancer. Proc. Natl. Acad. Sci. USA 74, 4491-4495 (1977)

Sabatini, D.D., Bensch, K., Barrnett, R.J.: Cytochemistry and electron microscopy. The preservation of cellular ultrastructure and enzymatic activity by aldehyde fixation. J. Cell Biol. 17, 19-58 (1963)

Tucker, R.W., Sanford, K.K., Frankel, F.R.: Tubulin and actin in paired nonneoplastic and spontaneously transformed neoplastic cell lines in vitro: Fluorescent antibody studies. Cell 13, 629-642 (1978)

Voyles, B.A., McGrath, C.M.: Markers to distinguish normal and neoplastic mammary epithelial cells in vitro: Comparison of saturation density, morphology, and concanavalin A reactivity. Int. J. Cancer 18, 498-509 (1976)

Cytoplasmic Zone Analysis in the Study of the Polysomes of Differentiated Cells

J.-E. Edström and U. Lönn
Department of Histology, Karolinska Institutet
S-104 01 Stockholm, Sweden

I. Introduction

Homogenization-subfractionation techniques have been powerful in elucidating details of polysome assembly and function. Since whole tissue is used for such analysis they provide information averaged for a large number of cells and for the different parts of the cytoplasm of a given cell. Individuality of cells and regional differentiation of cytoplasm may, however, motivate analyses of polysome components related to a specific cytoplasmic location in morphologically identified cells. We have therefore elaborated micromanipulatory techniques permitting defined zones of cytoplasm to be analyzed for such constituents. In one of these techniques the analyses are related to the distance of the cytoplasm to the nuclear envelope (Edström and Lönn 1976), in another to the content of endoplasmic reticulum (ER) in defined parts of the cytoplasm (Edström and Thyberg, unpublished).

II. Methodological Principles

Analyses are performed on cell components isolated by micromanipulation from fixed tissue. Fixation is necessary for such isolations and has the advantage of giving improved preservation of high molecular weight RNA and minimizing postmortal redistribution of RNA. Cells from the larval salivary glands of the dipteran *Chironomus tentans* are used and in one of the techniques (Edström and Lönn 1976) dissected into nuclei and three concentric zones of cytoplasm at increasing distances from the nucleus (Fig. 1). These components, which are collected from a dozen cells under the phase contrast microscope, are extracted and analyzed for RNA at different times after RNA precursor administration. In these analyses 28 S and 5 S RNA serve as markers for the heavy ribosomal subunit (RSU), and 18 S RNA as marker for the light RSU. The 75 S RNA from the Balbiani rings (BR) (large tissue specific chromosome puffs) is a putative messenger (Daneholt et al. 1977) and 4 S RNA serves as a volume marker.

The polysome components can spread from the nucleus in different ways as indicated in Fig. 2, with or without the formation of measurable gradients. If a gradient forms, this is because the exported component interacts with a cytoplasmic

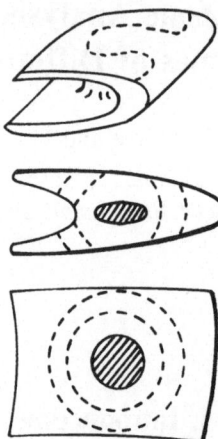

Fig. 1. Salivary gland cells and the dissection procedure for obtaining concentric zones of cytoplasm. Zones are defined on the basis of the distance from the nuclear envelope. For photomicrographs of the dissection procedure, see Edström and Lönn (1976)

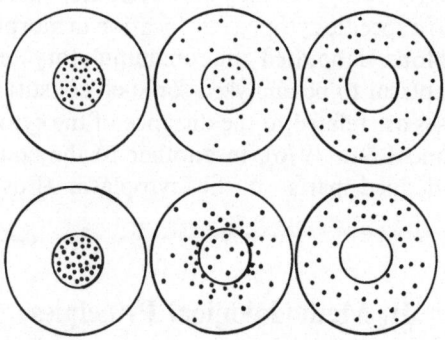

Fig. 2. Patterns of spread of RNA components in the cytoplasm after precursor administration. *Upper row* shows spread without formation of measurable gradients like for 4 S RNA, *lower row* shows temporary gradient formation like for 5 S, 18 S, and 75 S RNA

structure and/or because its half-life is short compared to its rate of peripheral spread. It is clear that gradients, when they form, are not an unspecific result because even the largest particle studied, the 75 S RNA (which in the cell is present in the form of RNP), can be induced to spread rapidly enough not to form any measurable gradients (Lönn 1978). The properties of the gradients can be studied in various ways. This permits conclusions regarding ·the state of the polysome components.

In a second technique, cytoplasm with a high and low content of ER can be compared (Edström and Thyberg, unpublished). The salivary gland cells contain a thin basal layer with a high content of mitochondria but with little endoplasmic reticulum (Kloetzel and Laufer 1969). This zone is too thin to interfere measurably

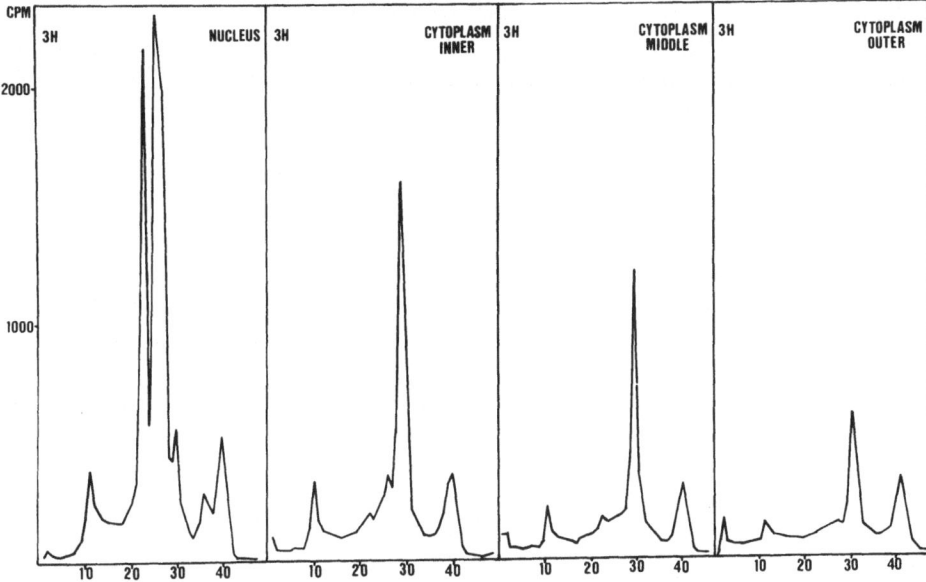

Fig. 3. Electrophoretic separations in 1% agarose gels of RNA extracted from 12 cells of an animal that was injected wtih 10 μCi of tritiated uridine (50 Ci/m-mole) and 10 μCi of cytidine (25 Ci/m-mole) 3 h before sacrifice. The cells were dissected into nuclei and three concentric cytoplasmic zones. RNA was denatured with 8 M urea in 0.02 M Tris buffer, 0.5% sodiumdodecylsulphate, pH 7.4 for 1 min at 50 °C immediately before application to the gels. For other experimental details, see Edström and Lönn (1976). The peak at slice 11-12 is 75 S RNA, at 29-30 18 S + denatured 28 S RNA in the cytoplasm and at slices 39-41 4 S RNA

with the previous type of analysis but it can be isolated and analyzed separately. We have only begun to exploit this technique which forms a valuable supplement to the previous one. Most of the work reviewed here will deal with results of the previous method.

III. Applications

A. RNA Species Exported by the Salivary Gland Cells

Figure 3 shows an analysis of RNA extracted from 12 salivary gland cells that were dissected 3 h after injection of tritiated nucleosides into the living larva. The four samples represent nucleus and three concentric zones of cytoplasm, inner, middle, and outer. The RNA was denatured before being applied to electrophoretic agarose gels. This converts the cytoplasmic 28 S RNA to components migrating like 18 S RNA and all rRNA will migrate in one peak. The nuclear counterpart is, however, resistant to denaturation. The nucleus shows the BR product, 75 S RNA, a series of preribosomal peaks, and 4 S RNA. In the cytoplasm 75 S RNA and ribosomal RNA appear in steep gradients but 4 S RNA is evenly distributed. The nuclear contamination of the inner zone is insignificant judged by the near absence in this analysis of nuclear 28 S, partially sap-localized ribosomal RNA precursor. Since the gradients extend over all three zones and nuclear contamination cannot always be checked, conclusions in the following have, as a rule, been based on results from only the two outer zones.

The 4 S RNA appears in roughly identical amounts in the three zones irrespective of labeling time (although what is apparently nuclear contamination may give an excess in the inner zone after short labeling times). It is quite likely to be present in proportion to cytoplasmic volume since it distributes between cytoplasmic zones like radioactively marked glycerol which is allowed to permeate the fixed tissue (Edström and Lönn 1976). The results therefore have been expressed as ratios of RNA components to 4 S RNA.

In the case of the RSU components gradients in radioactivity do not result from gradients of physical amounts of RNA but are due to decreasing specific radioactivities along the radius of the cell. This is clear because they have a half-life of several weeks (Edström et al. 1977) and because they distribute evenly between the three zones after a few hours (light RSU) or a few days (heavy RSU).

B. The 18 S RNA

The light RSU component as measured from 18 S/4 S RNA ratios is seen in steep gradients as soon as it appears in the cytoplasm. The gradients are not very long-lasting and in normal cases have disappeared 6 h after precursor injection (Edström and Lönn 1976). Administration of puromycin (PM) increases the rate of peripheral movement of the light RSU and the gradients are entirely extinguished within 45 min of drug exposure in vitro. Polysome formation therefore seems to be at least one factor underlying the formation of these gradients.

C. The 28 S RNA

The behavior of the 28 S RNA is more complex than that of 18 S RNA. Like the latter it forms steep gradients as soon as it appears (Edström and Lönn 1976). These decrease in steepness during the first few hours after precursor administration but are not entirely eliminated until after more than two days (Lönn and Edström 1976). During the early phase most of the gradient can be eliminated by PM, but later on it is entirely resistant to the drug. The gradient during its late PM-resistant phase does not form in starving, nongrowing animals (Lönn and Edström 1977a). On the other hand, drug-resistant gradients, allowed to form before starvation, become stabilized during a subsequent starvation period.

D. The 5 S RNA

The 5 S ribosomal RNA component mirrors the behavior of the 28 S RNA both in normal, growing (Edström and Lönn 1976; Lönn and Edström 1976) and starving, nongrowing animals (Lönn, unpublished). This is in agreement with expectations for a component which is added to the heavy RSU in the nucleus and which is then bound irreversibly to the RSU.

E. The 75 S RNA

Also the 75 S RNA forms steep gradients as soon as it appears in the cytoplasm. These gradients become shallower with time but can still be observed after two days (Lönn 1978). The gradients are resistant to PM and can form also in animals pretreated with the drug (Lönn, unpublished). In animals pretreated with cycloheximide (CH) the 75 S RNA leaves the nucleus but gradients do not form (Lönn 1978).

IV. Discussion

A. Significance of Puromycin-Sensitive RSU Gradients

The sensitivity of the early RSU gradients to PM suggests that polysome formation is at least one underlying factor. In cytoplasmic extracts most of the RSU are in fact present in structures sedimenting like polysomes 3 h after precursor injection with a characteristic sensitivity towards EDTA (Lönn, unpublished). These gradients thus suggest that the initial polysome formation occurs in the central part of the cytoplasm. We cannot decide whether polysome formation is preceded by a stage of free subunits in the cytoplasm as reported for other systems (Darnell 1968). If this is so, it is clear, however, that the free RSU are not able during this stage to spread over the whole cytoplasm.

The delay in peripheral spread of RSU in untreated cells could be caused by a low relative mobility of free polysomes, whether these are free in the cell sap or associated to a cytoskeleton (Lenk et al. 1977). The disappearance of the gradient with time could be explained by diffusion of free RSU between rounds of translation and/or diffusion of whole free polysomes.

The intitial central formation of polysomes is of interest against the finding that the RSU are inhibited in their export from the nucleus by PM and CH, an effect that is highly specific particularly for the RSU in *Chironomus* salivary gland cells (Lönn and Edström 1977b). Whereas this effect could possibly be explained by the elimination of a rapidly turning over protein necessary for export, an alternative possibility is that free RSU have to form polysomes in close physical association to the nuclear envelope in order for export to continue.

B. Significance of Puromycin-Resistant Heavy RSU Gradients

The heavy RSU can be observed in gradients resistant to PM, normally for at least two days (Lönn and Edström 1976), under conditions of starvation for six days (Lönn and Edström 1977a). The half-life of ribosomal RNA is of the order of several weeks (Edström et al. 1977) and the long-lasting gradients therefore cannot be a result of rapid degradation, nor can they be explained by persisting synthesis of RNA, since they are unaffected by synthesis inhibition with actinomycin D (Lönn and Edström 1976, 1977a). Therefore the gradients indicate that the heavy RSU bind to a structure strongly preventing free diffusion and that this binding occurs before they have reached distributional equilibrium.

About 50% of the heavy RSU are in a membrane fraction, resistant to PM 24 h after precursor injection, whereas the corresponding figure after six days is about 80% (Lönn 1977a). The heavy RSU therefore are only slowly becoming bound to the ER and appear at first to be present predominantly in free ribosomes judged by the PM sensitivity of the gradients also for the heavy RSU 3 h after precursor injection. A similar result has been obtained by Boshes (1970) who finds that free polysomes label more rapidly in ribosomal RNA (and heterogeneous RNA) than membrane-bound polysomes in *Drosophila* larvae. The polysomes first to form are thus not representative for the bulk of polysomes. The increasing percentage of heavy RSU attaching to membranes and the stability of the PM resistant heavy RSU gradients suggest that the membrane attachments are relatively stable.

We have also studied the distribution of RSU in microdissected cytoplasm with a high and a low content of ER as a function of time after RNA precursor injection. There is a significant increase in the relative content of RSU in the ER-rich cytoplasm with increasing age of the RSU, in good agreement with previous results (Edström and Thyberg, unpublished).

During starvation which stops the normal growth of the glands, drug resistant gradients do not form (Lönn and Edström 1977a). This could be understood if the bindings to the ER are irreversible in these cells, thus requiring formation of new ER membrane area for the attachment of heavy RSU.

C. Significance of 75 S RNA Gradient

The 75 S RNA gradient shows a superficial similarity to the heavy RSU gradient and it lasts for at least two days but not for six days (Lönn 1978). The 75 S RNA has a half-life of less than a day (Edström et al. 1978). The gradients are still apparent after 1–2 days. Therefore degradation may play a role in maintaining the gradients. Another consequence of the long duration of the gradients in relation to the half-life of the RNA is that the physical amounts of these gene products should be larger in the central cytoplasm than in the periphery. This agrees with the fact that the gradients do not reverse after longer labeling times. If most of the cytoplasmic 75 S RNA is present in polysomes as suggested by Daneholt et al. (1977), this would indicate a functional differentiation of the cytoplasm with the central parts preferentially translating 75 S RNA.

In animals pretreated with CH there is export of 75 S RNA (although the rate may be affected). In such annimals 75 S RNA may be exported without any evidence for the formation of measurable gradients (Lönn 1978) (Fig. 4). This finding is at present difficult to interpret. It could mean either that 75 S RNA is prevented from attaching to a structure which normally delays its peripheral spread and/or that the degradation is inhibited such that there is a higher chance of survival during the peripheral spread of the molecules. In animals sacrificed one day after

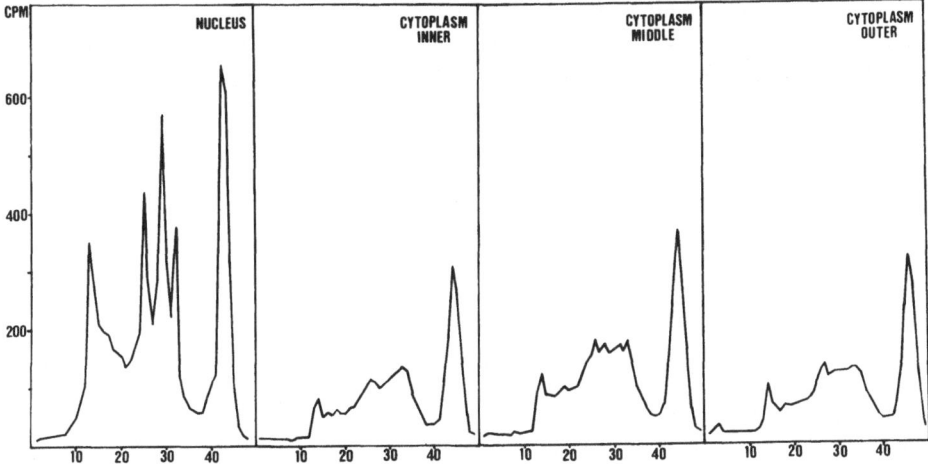

Fig. 4. Electrophoretic separations in 1% agarose gels of RNA extracted from 12 cells of an animal that was kept in normal culture medium with 10 μg/ml cycloheximide for 15 h, then injected with 10 μCi each of tritiated uridine and cytidine and sacrificed 3 h later. Dissection and analysis as for Fig. 3. The export of ribosomal RNA is completely inhibited and the relative labeling of its precursors in the nucleus is decreased. The export of 75 S RNA is decreased and/or delayed and the released 75 S RNA distributes in the cytoplasm without any measurable gradients

precursor injection the 75 S RNA is largely attached to ER membranes (Lönn 1977b), suggesting that such attachment may be one factor causing the gradients.

Irrespective of the action of CH, the results are of importance in showing that a large RNP particle like the one containing 75 S RNA is able to spread rapidly enough not to give rise to measurable gradients. This gives increased significance to the gradients that have been observed.

V. Conclusions

Polysome formation in the cytoplasm of *Chironomus* salivary gland cells is not only a temporally ordered process, it also bears specific gross topographic relations to the cytoplasm. The RSU liberated from the nucleus assemble into "free" polysomes in the central part of the cytoplasm and spread subsequently towards the periphery. During this spreading phase the heavy RSU will become more permanently attached to structures that are probably identical with the ER membranes. This gives rise to relatively stable gradients, suggesting a considerable gross level structural stability of this component of the cytoplasm. Also the BR transcripts, 75 S RNA, appear in gradients. Here an attachment to the ER as well as a short half-life are factors of importance for creating gradients. This RNA is present in larger physical amounts in the central cytoplasm, possibly indicating a differentiation of the cytoplasm along the radius of the cell with regard to translatory functions. A regional specialization of the cytoplasm may thus be dependent on factors such as spreading rates and half-lives of gene products. Further work will be required to understand the functional implications of the gradients described here. In particular it will be necessary to study other defined gene products.

Acknowledgments. The work described here was supported by the Swedish Cancer Society, Magnus Bergwall Foundation, and the Karolinska Institute.

References

Boshes, R.A.: Drosophila polyribosomes. The characterization of two populations by cell fractionation and isotopic labelling with nucleic acid and protein precursors. J. Cell Biol. 46, 477-490 (1970)
Daneholt, B., Andersson, K., Fagerlind, M.: Large-sized polysomes in *Chironomus tentans* salivary glands and their relation to Balbiani ring 75 S RNA. J. Cell Biol. 73, 149-160 (1977)
Darnell, J.E., Jr.: Ribonucleic acids from animal cells. Bacteriol. Rev. 32, 262-290 (1968)
Edström, J.-E., Lönn, U.: Cytoplasmic zone analysis: RNA flow studied by micromanipulation. J. Cell Biol. 70, 562-572 (1976)
Edström, J.-E., Ericson, E., Lindgren, S., Lönn, U., Rydlander, L.: Fate of Balbiani ring RNA in vivo. Cold Spring Harbor Symp. Quant. Biol. 42, 877-884 (1977)
Edström, J.-E., Lindgren, S., Lönn, U., Rydlander, L.: Balbiani ring RNA content and half-life in nucleus and cytoplasm of *Chironomus tentans* salivary gland cells. Chromosoma 66, 33-44 (1978)
Kloetzel, J.A., Laufer, H.: A fine-structural analysis of larval salivary gland function in *Chironomus thummi* (Diptera). J. Ultrastruct. Res. 29, 15-36 (1969)
Lenk, R., Ransom, L., Kaufman, Y., Penman, S.: A cytoskeletal structure with associated polyribosomes obtained from HeLa cells. Cell 10, 67-78 (1977)
Lönn, U.: Flow of heavy ribosomal subunits from cytosol to endoplasmic reticulum membranes. Med. Biol. 55, 292-295 (1977a)
Lönn, U.: Direct association of Balbiani ring 75 S RNA with membranes of the endoplasmic reticulum. Nature (London) 270, 630-631 (1977b)
Lönn, U.: Delayed flow through cytoplasm of newly synthesized Balbiani ring 75 S RNA. Cell 13, 727-733 (1978)
Lönn, U., Edström, J.-E.: Mobility restriction in vivo of the heavy ribosomal subunit in a secretory cell. J. Cell Biol. 70, 573-580 (1976)
Lönn, U., Edström, J.-E.: Movements and associations of ribosomal subunits in a secretory cell during growth inhibition by starvation. J. Cell Biol. 73, 696-704 (1977a)
Lönn, U., Edström, J.-E.: Protein synthesis inhibitors and export of ribosomal subunits. Biochim. Biophys. Acta 475, 677-679 (1977b)

Mechanism of Morphogenetic Tissue Interactions: The Message of Transfilter Experiments

L. Saxén

Department of Pathology, University of Helsinki
SF-00290 Helsinki 29, Finland

I. The Problem

From early embryogenesis on and throughout organogenesis, cells of different origin and destiny exchange morphogenetic messages to ensure synchronized development of the various components of tissues and organs. Such interactions are known to govern processes like proliferation, migration, cell orientation, tissue organization, and morphogenetic cell death. Ultimately they result in the expression of new phenotypic characteristics of cells in a temporally and spatially strictly controlled manner. Experimental disruption of the normal tissue architecture or genetic defects in the intercellular relations lead to impaired embryogenesis and dysmorphogenesis indicating the vital role of cell interactions in the implementation of the "building plan" of the entire organism (reviews: Saxén et al. 1976a; Wessells 1977). Many observations have also suggested that such disrupted or defective cell interactions are involved in the development of neoplasia (see Tarin 1972).

Although recognized as a major guiding principle in embryogenesis, and despite intense efforts since the beginning of this century, the basic mechanism of "inductive" or morphogenetic tissue interactions has remained obscure and their molecular basis open to speculation. Only three examples can be listed where the chemistry of the signal substances carrying a morphogenetic message is relatively well established: the "vegetalizing factor" in primary embryonic induction (Tiedemann 1976), the glycosaminoglycans stimulating somite chondrogenesis (Lash and Vasan 1977), and the "mesenchymal factor" acting upon differentiating pancreatic epithelium (Pictet and Rutter 1977). The mode and site of action of these and other hypothetical signal substances are virtually snown.

In the present situation we felt it important to reopen some of the old problems related to the basic mechanism of inductive cell interactions. Especially, we wanted to explore the *localization and mode of transmission of the signal substances* in the hope of developing a basis for future molecular analyses. Several modes of transmission should be considered (Table 1). Originally, Grobstein (1955) listed three alternatives: long-range diffusion, "matrix interaction", and induction mediated by actual cell contacts. Based on subsequent transfilter experiments and direct electron microscope observations, we found it convenient to distinguish between two major types of interaction (Saxén 1977): interactions transmitted by

Table 1. Alternative modes of transmission of inductive signals (Grobstein 1955; Saxén et al. 1976a; Saxén 1977; Weiss and Nir 1979)

Long-range transmission (50,000 nm)
1. Free diffusion
2. Matrix interaction
Short-range transmission (5 nm)
3. Short-range diffusion
4. Interactions of surface-associated molecules
5. Transfer of molecules through intercellular channels

compounds acting over distances of the order of 50,000 nm, and those requiring close apposition of cells with an interspace of only a few nm. The former would include the two first alternatives of Grobstein's schema whereas at least three possible routes of interaction can be suggested for the "short-range transmission": short-range diffusion (Weiss and Nir 1979), interaction of surface-associated molecules and actual transfer of signal substances through intercellular channels (Saxén 1977).

In this short review these alternative transmission mechanisms are examined by the filter technique in four developmental events serving as model systems for morphogenetic tissue interactions.

II. Transfilter Experiments

A. Methods

Since 1959, when I first became personally acquainted with the methods of Clifford Grobstein, his transfilter technique with some modifications has been employed (Grobstein 1953, 1956). The method involves mechanical separation of the interactive tissue components after pretreatment with EDTA or trypsin-pancreatin and their subsequent fixation on the opposite sides of different membrane filters. The filter assembly is cultured in an organotypic system for various periods of time or the interactants are reseparated after critical "induction period" and subcultured separately. The main criteria for a response to the inductive stimulus have remained morphological though supplemented by electron microscopy, histochemistry, and immunofluorescence (for details see Saxén 1961; Wartiovaara et al 1974; Thesleff et al. 1977; Karkinen-Jääkeläinen 1978).

Two types of commercial filters have been used, the Millipore membranes (Millipore Co., Bedford, MA, USA) and the Nuclepore filters (Nuclepore Co., Pleasanton, CA, USA). The former is a nitrocellulose membrane of spongy structure whereas the Nuclepore filters are made of polycarbonate tape exposed to charged particles in a nuclear reactor followed by chemical etching. Consequently, these filter pores are relatively straight and of a uniform pore diameter. Millipore filters are available in two thicknesses, 100 to 120 μm ("thick") and approx. 25 μm ("thin"), whereas Nuclepore membranes are somewhat thinner, 10 to 15 μm. The mean pore diameter of the two types of filters used by us has varied from 0.05 to 8 μm.

B. Model-Systems

The following four interactive situations have been selected for the transfilter experiments: induction of the CNS in Amphibian gastrulae ("primary embryonic induction"), induction of lens tissue from the belly epidermis of 8-somite chick embryos, the interaction between the epithelium and mesenchyme in the tooth bud of 17-day mouse embryos, and the induction between the Wolffian bud epithelium and the metanephrogenic mesenchyme leading to kidney tubule formation. The latter experiments were performed with tissues from 11-day mouse embryos, and according to the original method (Grobstein 1956) the inductor was replaced by a heterotypic inductor, a piece of spinal cord from the same embryos.

C. Summary of the Results

1. Primary Induction

In the original experiments (Saxén 1961), the presumptive neuroepithelium and the blastoporal lip mesoderm of young Urodelan gastrulae were combined through a "thin" Millipore filter with an average pore size of $0.8\,\mu$. The interactants were reseparated after 24 h and the ectoderm subcultured for 10 days. The ectoderm developed neural structures in a large proportion of these experiments, the formations being either unclassifiable neural vesicles or brain structures belonging to the forebrain region. Since electron microscopy of the filters failed to demonstrate cytoplasmic material in the pores, the conclusion was drawn that neural induction did not require direct cell contacts between ectoderm and mesoderm but was mediated by transmissable signal substances (Saxén 1961; Nyholm et al. 1962). The results have recently been confirmed in similar experiments using Nuclepore filters with pore sizes varying from 0.1 to $8.0\,\mu m$ (Toivonen et al. 1975).

As far as the second, mesodermalizing type of primary induction is concerned, somewhat contradictory results have been reported. While Toivonen and Wartiovaara (1976) did not obtain mesodermal inductions in their transfilter experiments and suggest that actual cytoplasmic contacts are required for this type of interaction, Minuth (1978) described mesodermal structures in ectoderms which had been separated from the dorsal lip mesoderm by different types of Nuclepore filters. His light microscopy seemed to exclude cytoplasmic processes penetrating the filters. The tentative conclusion was therefore drawn that both neural and mesodermal inductions are transmitted without actual cell contacts. The conclusion finds further, indirect support from the many previous experiments demonstrating both types of inducing capacity of various killed tissues and cell-free preparations (Saxén and Toivonen 1962).

2. Induction of Lens

The first transfilter studies of lens induction were reported in 1965 by Muthukkaruppan who obtained lens differentiation in the presumptive lens epidermis when separated from the optic bud by thin Millipore filters with a nominal pore size of 0.45 µm. Two subsequent observations led us, however, to re-examine the situation: the Millipore filters used by Muthukkaruppan (1965) do not certainly exclude penetration of cytoplasmic processes (Lehtonen et al. 1975), and the target tissue used has been shown to be predetermined towards lens differentiation and needs only a permissive trigger to express this bias (Karkinen-Jääskeläinen 1978a). Therefore, a new series of experiments were recently performed with different filters and dialyzer membranes separating the inductor (the optic vesicle) and the target epidermis. The latter was dissected this time from the trunk region of early chick embryos to avoid lens-forming bias (Karkinen-Jääskeläinen 1978b).

The results showed that the trunk ectoderm when separated from the inductor by any type of Millipore or Nuclepore filter developed lens-like structures with synthesis of crystallin demonstrated by immunofluorescence. The filters included "thick" Millipore filters of varying pore sizes and Nuclepore filters with pores as small as 0.1 µm. The verification of the existence of transmissable signal substances was ultimately made by the use of a dialyzer membrane allowing the passage of molecules of MW less than 12,000 daltons. Again a large proportion of the uncommitted trunk epidermis preparations developed lentoids (Karkinen-Jääskeläinen 1978b). Cell contamination was excluded by the use of chick/quail chimeric combinations.

3. Induction of Odontoblasts

Differentiation of the mesenchymal cells of the tooth bud into polarized odontoblasts secreting predentine requires an inductive stimulus from the overlying epithelial component which subsequently gives rise to the ameloblast layer. Interposed Nuclepore filters which an average pore size of 0.2 and 0.6 µm allowed the passage of this message and electron microscopy revealed cytoplasmic processes in the pores. Filters with pores of the diameter 0.1 µm prevented odontoblast dif-ferentiation, and no mesenchymal cell processes were seen within the filter. These observations suggested that the interaction between the dental mesenchyme and epithelium requires, in contrast to the two inductive events already discussed, close association of the interactants (Thesleff et al. 1977). However, even during the early stages of odontogenesis in vivo, the interacting cells are separated by a continuous basement membrane preventing actual contacts between them. This basement membrane is removed when the two tissue components are separated after enzyme treatment, but becomes rapidly restored in the transfilter cultures subsequently showing odontoblast differentiation. Further experiments have suggested that the actual "morphogenetic contact" in this interactive situation is established between the mesenchymal cell surface and the basement membrane (Thesleff et al. 1978).

The key molecules in the basement membrane apparently guiding odontoblast differentiation are not known. Collagenase treatment has yielded inconsistent results, and the role of collagen in odontoblast differentiation is still debatable (e.g., Ruch et al. 1972; Kollar 1978). Less attention has been focused on the other components of the basement membrane, the glycosaminoglycans (GAGs). Our recent experiments testing the effect of the glutamine analogue DON (diazo-oxo-norleucine) on odontoblast differentiation have suggested that these compounds are involved. This inhibitor of GAG synthesis completely blocked odontoblast differentiation and predentine secretion in concentrations which did not affect growth and survival of the tissues (Hurmerinta et al. 1979).

4. Induction of Kidney Tubules

The filter technique used in all the above studies was originally devised for an experimental analysis of the induction of secretory tubules in the metanephrogenic mesenchyme (Grobstein 1953, 1956). When spinal cord was used as an inductor, thin Millipore filters with average pore sizes from 0.8 to 0.1 μm allowed induction in these studies. Electron microscopy did not reveal cytoplasmic material deep in the filters with the smallest pores, and it was concluded that the filter had prevented actual cell contacts. The obvious conclusion was drawn that the induction was mediated by extracellular compounds (Grobstein and Dalton 1957).

Twenty years after these pioneering studies, new filter types and improved electron microscopic techniques allowed experiments yielding results not compatible with the original idea of "matrix interaction" without cell contacts. These studies revealed a good correlation between the establishment of close cell contacts (penetration of cytoplasmic processes into different filter types) and subsequent tubule formation (Table 2). Furthermore, analysis of the kinetics of the transmission of inductive signals in various experimental conditions excluded free diffusion of molecules (Nordling et al. 1971; Wartiovaara et al. 1974; Saxén et al. 1976b). Our postulate of contact-mediated induction was also consistent with direct electron microscopic observations in intact kidneys. A discontinuous basement membrane is seen between the epithelium and the mesenchyme, and, unlike the situation in the tooth germ, the cells approach each other very closely. In places the two plasma membranes are separated by an interspace of approx. 5 nm; in places there is a wider interspace filled with ruthenium red positive material (Lehtonen 1975).

Based on hypothesis of contact-mediated tubule induction, a series of experiments was recently made where the response of the mesenchyme was roughly

Table 2. Induction of kidney tubules through different Nuclepore filters and penetration of cytoplasmic material into their pores (Wartiovaara et al. 1974; Saxén et al. 1976b; Saxén and Lehtonen 1978, and unpublished)

	Mean pore diameter (μm)			
	0.05	0.1	0.2	0.6
Percent induction	0	50	90	100
Cytoplasmic penetration	–	+	+ + +	+ + +

quantified, so the kinetics of induction could be further analyzed. The results ehowed that the ingrowth of the cytoplasmic processes to the pores was rapid; in less than 2 h most pores were filled with membrane-coated processes allowing morphological contact between the interacting cells. This phase was followed by a long "induction period" during which the two tissues in transfilter position cannot be separated without interfering with the induction. The length of this "minimum induction time" was shown to be a function of the diameter and density of the pores and the thickness of the filter (Saxén and Lehtonen 1978). Furthermore, the intensity of the response (the number of tubules) was a function of these parameters and the contact time. The results thus support the concept (Meier and Hay 1975) that induction is a function of the total contact area between the interacting tissues.

The observations do not distinguish between the alternative short-range transmission models in Table 1. Various approaches to test these models are in progress: transmission of molecules from the inductor to the target cells can be studied by microinjections and electrophysiological means, the synthesis of the surface-associated key molecules can be inhibited and the interacting surfaces can be separated at molecular level by coating them with various compounds. All the three possibilities are under exploration. Direct physical studies are at the stage of methodological experimentation, whereas the biochemical approach has yielded some preliminary results. Inhibition of the synthesis of the surface-associated GAGs by the glutamine analogue DON prevents tubule induction when applicated during the first 24 h of transfilter culture (the "induction period"). A clear dose-response of tubule inhibition was shown. DON treatment had no effect on tubule morphogenesis after the critical period (Ekblom et al. 1979).

For the coating of the interacting surfaces, polyanions were used. Treatment with these molecules during the induction period blocked differentiation in a reversible way. If subsequently changed to normal medium with the inductor left in place, tubule formation was obtained. The inhibitory effect, furthermore, seemed to be a function of the charge density of the three polyanions used. We suggest that the compounds produce a molecular barrier between the interacting surfaces, thus preventing their morphogenetically significant interaction (Ekblom et al. 1978).

III. The Message

Technically similar filter experiments in four different model systems of inductive tissue interactions show that different mechanisms are involved in these events. Primary induction and induction of lens tissue by the optic vesicle seem to be implemented by transmissible molecules acting over considerable distances. Close apposition of the interactive cells is required for the induction of odontoblast differentiation and the formation of kidney tubules. The former seem to represent a cell-matrix type of interaction between the mesenchymal cell membrane and the basement membrane whereas direct cell-to-cell contact seems, at present, the most plausible mechanism for tubule induction.

The results of the transmission mechanism in the model systems studied thus suggest quite different modes of interaction apparently mediated by different types of signal substances.

References

Ekblom, P., Nordling, S., Saxén, L.: Inhibition of kidney tubule induction by charged polymers. Cell Differ. 7, 345-353 (1978)

Ekblom, P., Lash, J.W., Lehtonen, E., Nordling, S., Saxén, L.: Inhibition of morphogenetic cell interactions by 6-diazo-5-oxo-norleucine (DON). Exp. Cell Res. 121, 121-126 (1979)

Grobstein, C.: Morphogenetic interaction between embryonic mouse tissues separated by a membrane filter. Nature 172, 869-871 (1953)

Grobstein, C.: Tissue interactions in the morphogenesis of mouse embryonic rudiments in vitro. In: Aspects of synthesis and order of growth (ed. D. Rudnick), pp. 223-256. Princeton: Princeton University Press 1955

Grobstein, C.: Trans-filter induction of tubules in mouse metanephrogenic mesenchyme. Exp. Cell Res. 10, 424-440 (1956)

Grobstein, C., Dalton, A.J.: Kidney tubule induction in mouse metanephrogenic mesenchyme without cytoplasmic contact. J. Exp. Zool. 135, 57-73 (1957)

Hurmerinta, K., Thesleff, I., Saxén, L.: Inhibition of tooth germ differentiation in vitro by diazo-oxo-norleucine (DON). J. Embryol. Morphol. 50, 99-109 (1979)

Karkinen-Jääskeläinen, M.: Permissive and directive interactions in lens induction. J. Embryol. Exp. Morphol. 44, 167-179 (1978a)

Karkinen-Jääskeläinen, M.: Transfilter lens induction in avian embryos. Differentiation 12, 31-37 (1978b)

Kollar, E.J.: The role of collagen during tooth morphogenesis: Some genetic implications. In: Development, Function and Evolution of Teeth (eds. P.M. Buttler, K.A. Joysen), pp. 1-12. New York-London: Academic Press 1978

Lash, J.W., Vasan, N.S.: Tissue interactions and extracellular matrix components. In: Cell and Tissue Interactions (eds. J.W. Lash, M.M. Burger), pp. 101-113. New York: Raven 1977

Lehtonen, E.: Epithelio-mesenchymal interface during mouse kidney tubule induction in vivo. J. Embryol. Exp. Morphol. 34, 695-705 (1975)

Lehtonen, E., Wartiovaara, J., Nordling, S., Saxén, L.: Demonstration of cytoplasmic processes in Millipore filters permitting kidney tubule induction. J. Embryol. Exp. Morphol. 33, 187-203 (1975)

Meier, S., Hay, E.D.: Stimulation of corneal differentiation by interaction between cell surface and extracellular matrix. J. Cell Biol. 66, 275-291 (1975)

Minuth, W.: Transfilter mesodermalizing of amphibian gastrula ectoderm in transfilter experiments. Med. Biol. 56, 349-354 (1978)

Muthukkaruppan, V.: Inductive tissue interaction in the development of the mouse lens in vitro. J. Exp. Zool. 159, 269-288 (1965)

Nordling, S., Miettinen, H., Wartiovaara, J., Saxén, L.: Transmission and spread of embryonic induction. I. Temporal relationships in transfilter induction of kidney tubules in vitro. J. Embryol. Exp. Morphol. 26, 231-252 (1971)

Nyholm, M., Saxén, L., Toivonen, S., Vainio, T.: Electron microscopy of transfilter neural induction. Exp. Cell Res. 28, 209-212 (1962)

Pictet, R.L., Rutter, W.J.: The molecular basis of the mesenchymal-epithelial interactions in pancreatic development. In: Cell Interactions in Development (eds. M. Karkinen-Jääskeläinen, L. Saxén, L. Weiss), pp. 339-350. New York-London: Academic Press 1977

Ruch, J.V., Karcher-Djuricic, V., Gerber, R.: Quelques aspects du role de la prédentine dans la differenciation des adamantoblastes. Arch. Anat. Microsc. 127-138 (1972)

Saxén, L.: Transfilter neural induction of Amphibian ectoderm. Dev. Biol. 3, 140-152 (1961)

Saxén, L.: Morphogenetic tissue interactions: An introduction. In: Cell Interactions in Development (eds. M. Karkinen-Jääskeläinen, L. Saxén, L. Weiss), pp. 145-151. New York-London: Academic Press 1977

Saxén, L., Lehtonen, E.: Transfilter induction of kidney tubules as a function of the extent and duration of intercellular contacts. J. Embryol. Exp. Morphol. 47, 97-109 (1978)

Saxén, L., Toivonen, S.: Primary Embryonic Induction, pp. 1-270. New York-London: Academic Press 1962

Saxén, L., Karkinen-Jääskeläinen, M., Lehtonen, E., Nordling, S., Wartiovaara, J.: Inductive tissue interactions. In: The Cell Surface in Animal Embryogenesis and Development (eds. G. Poste, G.L. Nicolson), pp. 331-407. Amsterdam: North-Holland 1976a

Saxén, L., Lehtonen, E., Karkinen-Jääskeläinen, M., Nordling, S., Wartiovaara, J.: Are morphogenetic tissue interactions mediated by transmissable signal substances or through cell contacts? Nature 259, 622-663 (1976b)

Tarin, D. (ed.): Tissue Interactions in Carcinogenesis, pp. 1-483. New York-London: Academic Press 1972

Thesleff, I., Lehtonen, E., Wartiovaara, J., Saxén, L.: Interference of tooth differentiation with interposed filters. Dev. Biol. 58, 197-203 (1977)

Thesleff, I., Lehtonen, E., Saxén, L.: Basement membrane formation in transfilter tooth culture and its relation to odontoblast differentiation. Differentiation 10, 71-79 (1978)

Tiedemann, H.: Pattern formation in early developmental stages of amphibian embryos. J. Embryol. Exp. Morphol. 35, 437-444 (1976)

Toivonen, S., Wartiovaara, J.: Mechanism of cell interaction during primary embryonic induction studied in transfilter experiments. Differentiation 5, 61-66 (1976)

Toivonen, S., Tarin, D., Saxén, L., Tarin, P.J., Wartiovaara, J.: Transfilter studies on neural induction in the newt. Differentiation 4, 1-7 (1975)

Wartiovaara, J., Nordling, S., Lehtonen, E., Saxén, L.: Transfilter induction of kidney tubules: Correlation with cytoplasmic penetration into Nuclepore filters. J. Embryol. Exp. Morphol. 31, 667-682 (1974)

Weiss, L., Nir, S.: On the mechanism of transfilter induction of kidney tubules. J. Theoret. Biol. 78, 11-20 (1979)

Wessells, M.K.: Tissue Interactions and Development, pp. 1-276. Menlo Park, Calif.: Benjamin 1977

In Memoriam

Nelson Tracy Spratt, Jr.

On February 15, 1975, Professor Nelson Tracy Spratt, Jr. died in Atlanta, Georgia, at the age of 63 after a progressive illness of somewhat in excess of a year. He had retired early because of his failing health and, with his wife Gladys, returned to the place of his birth and his boyhood years.

With his death, developmental biology lost a productive researcher, teacher, author, and creative contributor of no small stature and merit as well as a gentleman and friend. He was a modest man, but yet was not without firm convictions about and complete confidence in what he had chosen as his life's work.

Nelson Spratt was awarded the doctorate by the University of Rochester in 1940. After three years in the U.S. Air Force and six years on the Biology staff of Johns Hopkins University, he spent the remainder of his professional career at the University of Minnesota. Occasional leaves of absence took him to Puerto Rico and Mexico, into geographical areas of which he was extremely fond. Another year was spent as an administrator with the National Science Foundation in Washington, D.C. At the request of his colleagues and the Dean, he served as Chairman of the University of Minnesota Zoology Department for eight years.

An exceptional teacher with undergraduates and graduate students alike, Nelson Spratt had the admirable ability to strip away the nonessential details so that the germane fundamental principles could be seen, comprehended and applied. His breadth of knowledge and experience led him to utilize richly varied examples in his teaching. Marvelously well organized and lucid as a lecturer, he consistently presented material in a complete and unhurried manner, most generally without notes, and miraculously—at least in the eyes of his students—finished the day's subject precisely as the bell signalled the end of the period. He is most certainly missed as a teacher of quality.

He devoted his many productive years as a researcher to exploring and explaining the nutritional requirements, the morphogenetic movements, the organ-forming areas and the integrative mechanisms characteristic of early avian development. The method of in vitro avian embryo culture which he developed was the major one widely utilized for many years. Patient and persistent, he was admired by those who associated with him for his extraordinarily keen powers of observation and his exemplary thoroughness as well as his precision and inventiveness.

Both his research and his teaching stimulated and encouraged many graduate students to prepare for and pursue careers in developmental biology. His collaboration with post-doctoral colleagues was fruitful and rewarding. With an impressive list of publications to his credit, including two books and contributed chapters to still others, it is obvious that Nelson Spratt lived a richly productive professional life. He was well respected for his contributions by his colleagues in the several professional societies of which he was a member.

His contribution to the Marine Biological Laboratory summer embryology course at Woods Hole from 1955 to 1961 is still remembered and appreciated by those who were, as I was, privileged to have enrolled in it during his tenure on the staff for the course.

Although he lived for many years in the north, Nelson Spratt never totally lost his warm Georgia accent although it was colored by his several years at eastern universities. Generally giving the appearance of being unhurried and unhassled, he implicitly encouraged us as students to knock on his office door. He was a gentleman—truly a gentle and concerned man who listened to our problems, who corrected us when and where necessary and who always was constructive and encouraging. In a somewhat lighter vein we might note that through most of his adult life he possessed a deceptively youthful appearance that mislead many unsuspecting students and visitors—much to the amusement of those of us who observed it—into assuming that he was a graduate student rather than a Professor of Zoology. Only the twinkle in his eye revealed his sense of humor in such instances.

Nelson Spratt delighted in the analysis of ideas and in speculative discussions evolving from the interpretation and application of research data—his as well as other's—particularly in the areas of morphogenesis and differentiation. For this reason, it is especially appropriate that this conference session be dedicated by us to his memory.

Ross L. Shoger
Professor of Biology
Carleton College
Northfield, MN, USA

Migration and Replication of the Germ Cell Line in Rana pipiens

S. SUBTELNY

Department of Biology, Rice University, Houston, TX 77001, USA

I. Introduction

Over the past several years we have undertaken a series of studies on germ cell migration in the anuran, *Rana pipiens*, during their endoderm phase, between the early tailbud stages 16, 17 (Shumway 1940) and the time when they emerge from the endoderm in the swimming larva (stage 22). They involve extirpation experiments and germ cell transfer experiments, utilizing the grafting techniques of Blackler (1962) and Gipouloux (1962). We have also made an analysis of gonocyte numbers in normal larvae at stage 25 (operculum complete), and their replication within the early gonad at later stages of development. These studies differ from earlier investigations in two important ways:

First, we discovered that one can conveniently dissociate gonocytes from the gonadal ridges with Ca^{2+} and Mg^{2+} free modified Niu-Twitty medium containing 0.0375 mg/ml EDTA in anesthetized larvae after careful removal of the ventral body wall and viscera. These large cells with abundant yolk inclusions stand out dramatically in sharp contrast to the small dark cells of the gonadal rudiments, and they can in no way be confused with the surrounding mesodermal elements. Their morphological appearance clearly indicates their origin from the subjacent endoderm cells of the digestive tract from which they arose. This has allowed gonadal germ cells to be counted at precise stages in larval development with ease and with an accuracy not easily achieved with sectioned material (Subtelny and Ladner 1976). It has further permitted quantitative assessments of gonocytes between experimentals and controls in given experiments. Wylie and Rose (1976) also independently reported the dissociation of gonocytes in the same year.

Second, in graft experiments, primordial germ cells within the endoderm of tailbud embryos (together with the surrounding ecto-mesoderm) have been transplanted into "sterile" host tailbud embryos which earlier had been exposed to UV radiation (wavelength, 253.7 nm; total dose, 18,000 ergs/mm²) during first cleavage. Over the past several years the procedure has routinely yielded 100% gonadal sterility in larvae at stage 25, when examined by autopsy or in sectioned material, confirming Smith's (1966) earlier findings. The rare occurrence of a larva with one or several gonocytes among the irradiated controls in a given experiment could be attributable to faulty technique in the irradiation procedure (see Bounoure et al. 1954), or to differences in sensitivity of eggs from different females to the radiation. In a like manner, the gonads of 61 examined irradiated animals are totally lacking in any recognizable gonocytes over the following 2 days of development. This argues strongly that all or virtually all gonocytes counted in irradiated graft larvae at stage 25 are indeed derived from the donor grafts. By 7 days after stage 25 (at 18 °C), 92% (43/47) of the irradiated larvae are lacking in germ cells while small numbers of gonocytes (1 to 6) make their appearance in the other four animals. However, among 23 larvae examined 2 or 3 weeks later, at Taylor and Kollros (1946) stages V-VII, only seven (30%) have been found lacking in germ cells; the remaining animals possess restricted numbers (1–36) of normal-looking or small abnormal-looking gonocytes (unpublished observations). Züst and Dixon (1975,1977) and Smith (pers. comm.) have also noted a similar delayed appearance of gonocytes in UV-irradiated animals. For our purposes here, when assessments of gonadal germ cells are made on irradiated stage 25 graft larvae, the gonadal ridges of the irradiated larvae are considered to be free of host germ cells. That graft germ line cells are transferred to irradiated host larvae is supported by experiments between *Rana* species. Utilizing a clearly identifiable LDH isozyme marker, graft *R. sylvatica* gonocytes have been identified in the gonads of irradiated host *R. pipiens* froglets (Subtelny and Carrethers 1975).

II. The Significance of Germ Cell Counts

In *R. pipiens*, gonocytes become distributed to the paired genital ridges in larvae during stages 23–24, and migration has normally terminated by the latter stage. Cytological studies reveal that the gonocytes are essentially quiescent mitotically at this time and in stage 25 larvae, and remain so for an additional 1 to 2 days or more, depending on the individual organism, after which time they initiate intense replication followed by a rapid increase in germ cell numbers and a corresponding decrease in size. By the end of the first week after stage 25 (at 18 °C), the average number of gonocytes has doubled and the germ cells may enter a second replication cycle in certain individuals (unpublished observations). In this respect *R. pipiens* differs from *Xenopus laevis* in which the gonocytes do not begin to replicate until about 1 to 2 weeks after they populate the gonadal ridges (Kalt and Gall 1974; Ijiri and Egami 1975; Züst and Dixon 1977).

A. Variability of Gonocyte Numbers in Progeny from Different Matings

Table 1 lists examples of gonadal germ cell counts in stage 25 larvae among the progeny from ten different representative crosses. First, the data disclose no apparent sex differences in germ cell numbers when they initially become segregated into the genital ridges. There are no obvious discontinuities in the gonocyte populations recorded from the progeny of any mating that can be assigned to two discrete groups. Second, the overall data reveal that gonocyte numbers in stage 25 larvae vary considerably (21–137). It is of interest that no more than a score of germ plasm containing cells have been cytologically identified in blastulae of various anurans (Bounoure 1934; Blackler 1958; Whitington and Dixon 1975). If germ

Table 1. Direct counts of dissociated germ cells in *Rana pipiens* (stage 25)

Cross	No. of progeny	Av. number germ cells	Germ cell number in individual larvae
1	15	42	21, 24, 27, 29, 36, 40, 43, 43, 46, 48, 49, 52, 54, 56, 57
2	15	50	23, 27, 27, 30, 37, 38, 39, 46, 50, 55, 63, 67, 75, 80, 88
3	15	54	23, 31, 35, 36, 38, 50, 53, 57, 57, 58, 62, 70, 74, 76, 92
4	15	53	43, 43, 44, 45, 45, 49, 52, 53, 57, 60, 60, 61, 61, 61, 65
5	21	47	30, 30, 30, 34, 34, 35, 39, 41, 45, 49, 50, 51, 52, 52, 53, 56, 58, 62, 63, 67, 68
6	15	54	21, 28, 40, 41, 46, 49, 49, 51, 53, 56, 61, 76, 76, 79, 83
7	15	37	21, 26, 27, 31, 32, 34, 35, 37, 38, 39, 41, 42, 46, 49, 53
8	20	61	34, 36, 41, 42, 43, 44, 52, 53, 54, 58, 59, 63, 64, 72, 78, 79, 84, 92, 95, 101
9	15	71	24, 32, 35, 52, 55, 63, 74, 76, 77, 80, 86, 94, 101, 102, 116
10	15	89	25, 53, 75, 75, 80, 82, 84, 92, 94, 96, 98, 100, 118, 127, 137

plasm bearing cells do represent the direct progenitors of gonadal germ cells, and there is convincing cytological and experimental evidence that they do (see reviews by Blackler 1966, 1970; Beams and Kessel 1974; Eddy 1975; Smith and Williams 1975; also see Tanabe and Kotani 1974; Ikenishi and Kotani 1975; Züst and Dixon 1975; Wakahara 1977, 1978), then the much increased gonocyte numbers encountered in swimming larvae is, in itself, indicative of their replication during post-blastula stages of development. Direct cytological and autoradiographic studies in *Xenopus* provide strong evidence that the primordial germ cells divide two to three times during their endoderm phase, between the blastula and swimming larval stages (Dziadek and Dixon 1975, 1977; Whitington and Dixon 1975; Kamimura et al. 1976). Interestingly, a fourfold difference in germ plasm bearing cells (4–15) exists among individual *R. pipiens* blastulae/early gastulae (DiBerardino 1961), and Table 1 reveals that the higher gonadal germ cell populations among the progeny within each cross falls within a fourfold range in the vast majority of cases. We infer that the variations in gonocyte numbers recorded in stage 25 *R. pipiens* animals within each cross can be explained, in the main, by the numbers of germ plasm bearing cells individual embryos possess at the blastula stage, and by the two to three replications these cells undergo in each individual between this stage and the time they emerge from the endoderm to colonize the gonadal ridges.

B. Variability in Size of Gonocytes

An additional point of significance is that the dissociated gonocytes in a given individual exhibit a range in size (15.0–55.0 μm) and yolk content (Figs. 1 and 2). These differences among the gonocytes in *R. pipiens* are present at the time they emerge from the underlying endoderm mass in stage 22 swimming larvae, and it is

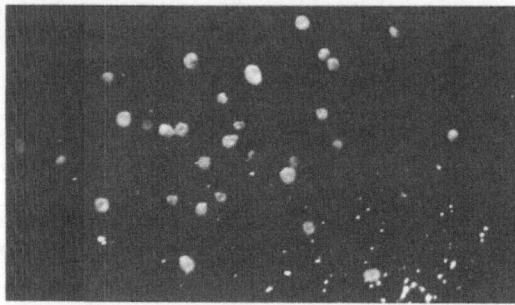

Fig. 1. Dissociated gonocytes from the gonadal ridges of an individual larva at stage 25. Note the considerable variation in their size. Most gonocytes are spherical or oval in shape; a few may possess a small satellite "bud", while others occasionally appear dumbbell in shape and may represent two incompletely separated cells. These morphological differences are present in the dissociated gonocytes before pipetting them into a culture dish

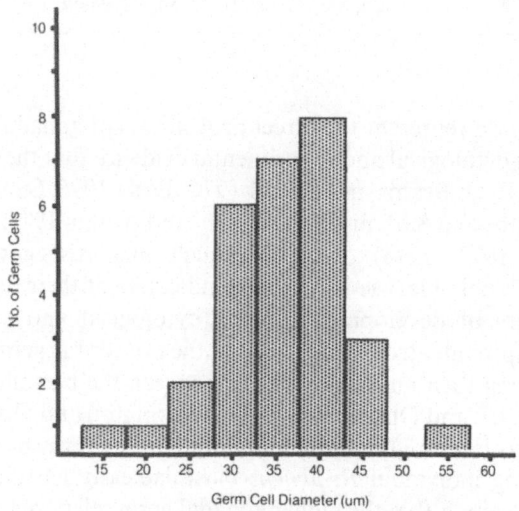

Fig. 2. Size distribution of gonadal germ cells from one larva at stage 25

not a consequence of their replication after they initially populate the gonadal ridges. At this stage the digestive tract is essentially an elongated yolk mass. Lateral constrictions in the anterior region delimit the stomach from the intestine while slight constrictions in the posterior region demarcate the midgut from the hindgut. The germ cells exit from the region confined to the posterior midgut and anterior hindgut. Our observations further indicate that the size and yolk inclusions of the gut endoderm cells within the latter two regions in stage 22 larvae vary and they are comparable to the range in size and yolk content of gonadal germ cells in stage 25 larvae. We infer that the observed differences in size and yolk contents of gonocytes in early gonadal ridges are primarily attributable to their location within the endoderm mass and the number of replication cycles they and their neighboring somatic endoderm cells undergo in a given individual at the time they exit from the endoderm mass.

C. The Question of Germ Cell Replication Rate Within the Endoderm

From their autoradiographic studies, Dziadek and Dixon (1975, 1977) suggest that primordial germ cells in *Xenopus* replicate at about the same rate as their surrounding somatic endoderm cells between the blastula stage and the time they arise from the endoderm. On the other hand, Kamimura et al. (1976) suggest that primordial germ cells have a higher rate of replication during their endodermal phase subject to their own mitotic control mechanisms. In our laboratory, Ladner (1978) has established that there is a significant difference in the average number of gonocytes between diploid and triploid siblings at the time the germ cells become localized in the gonadal ridges. Triploid larvae have about 63% the number of gonocytes present in diploid siblings at stage 25. It can be shown that this difference is not the result of pressure treatment used to produce triploids by suppression of second polar body formation. In addition, just as triploid somatic cells are 1.5 times the size of their diploid counterparts (Fankhauser 1945; Briggs 1947), germ cells from triploid larvae exhibit a corresponding difference in size relative to their diploid counterparts. Our measurement of endoderm cells in the posterior gut region of stage 22 triploid and diploid siblings indicate a range and average cell size comparable to gonocytes of triploid and diploid siblings, respectively. The implication of these findings is that the germ plasm does not seem to have a special role in regulating the replication rate of primordial germ cells. Mitotically the primordial germ cells behave just as the neighboring somatic cells during their residence within the endoderm of the developing embryo.

III. Localization of Germ Line Cells During Their Migration in the Endoderm

Cytological observations indicate that primordial germ cells are localized within the endoderm of the tailbud embryo and that they undergo considerable positional changes within the endoderm mass between this stage and the time they emerge from the endoderm in swimming larvae (Bounoure 1934; Blackler 1958; Whitington and Dixon 1975; Kamimura et al. 1976). In this laboratory, Duane Ringer has been carrying out a series of experiments involving extirpation of endoderm regions from embryos at successive stages during this period of embryogenesis. The numbers of gonocytes that colonize the gonadal ridges in the experimental larvae at stage 25 are then compared with the gonocyte populations in normal fertilized controls. The preliminary results indicate that at early tailbud stage 17, nearly all the germ cells are located in the ventral half of the endoderm mass distributed along the antero-posterior axis of the embryo; 30–40% are in the anterior half, while 60–70% of the germ line cells are located in the posterior half of the endoderm mass. Following elongation of the tailbud embryo (stage 19), the vast majority of the progenitors of the gonadal germ cells are still deep within the endoderm mass, but now localized in the posterior region. In swimming larvae (stage 21) more than 90% of the germ line cells are now localized in the dorsal ½ to ¼ of the posterior endoderm mass. These results are in accord with the descriptive account of the positional changes that germ plasm containing cells undergo within the endoderm at corresponding embryonic stages. The findings suggest that the experimental transfer of germ cells in graft experiments with neurula and tailbud embryos (Blackler 1962, 1966, 1970; Gipoulox 1962, 1970; also see Sect.IV) are the germ plasm bearing cells identified in histological material.

IV. Migration of Germ Cells Within the Endoderm

The displacements of primordial germ cells within the endoderm between the gastrula and early tailbud embryo have been attributed to passive movements accompanying normal embryological processes during this period in development. However, the movements of the primordial germ cells from their deep position within the endoderm during elongation of the tailbud embryo to the posterior dorsal crest region of the endoderm in swimming larvae have been viewed to be an active movement on the part of the germ line cells themselves. The question whether this oriented dorsal movement of the germ line cells is determined and responds to factors resident within the endoderm itself or to other factors has been the subject of an extended experimental analysis by Gipouloux (1970). This investigator joined two normal neurulae in parabiosis at the ventral abdominal region, then removed the head and dorsal structures (neural tube, notochord, somites, nephrotomes) from one (donor) embryo, thereby essentially fusing a graft of the endoderm mass (covered with ecto-mesoderm) to the other (host) embryo. Cytological examination of the gonadal ridges in the few surviving graft larvae indicated that they contained, on the average, 1.5 times as many germ cells as the normal fertilized controls. This suggested to Gipouloux (1962) that the partial donor embryo probably contributed the excess numbers of germ cells which must have migrated within the endoderm in a direction opposite to their normal direction to reach the host gonadal ridges. In another sort of experiment rotation of the dorsal-ventral orientation of the endoderm mass in neurulae resulted in an average of 6 gonadal germ cells in the few surviving experimental animals (Gipouloux 1963). The results lend further support to Gipouloux's earlier interpretation that the direction of germ cell migration within the endoderm is not fixed. Additional extirpation and graft experiments provide evidence that the germ line cells are caused to migrate to the genital ridges under the influence of chemical attraction substances emanating from dorsal axial structures (Gipouloux 1970; also see Giorgi 1974).

We have recently carried out germ cell transfer experiments by uniting at the ventral abdominal region two normal early tailbud embryos (stages 16, 17), or a normal embryo with a previously UV-irradiated "sterile" embryo, using Gipouloux's parabiosis procedure cited above. The dorsal axial structures were removed from a normal (donor) embryo in each kind of parabiosis combination, and the host animals containing the endoderm graft were reared to larval stage 25.

Table 2. Direct counts of germ cells in host parabionts (stage 25)

Type of experiment [a]	Total number embryos	Range in germ cell number	Average germ cell number
Normal donor – UV-irrad. host	77	6– 97	29 (0.56) [b]
Normal donor – normal host	60	14–170	76 (1.46) [b]
Normal fert. controls	57	23–136	52
UV-irrad. controls	64	0	—

[a]   The dorsal axial structures were extirpated from the donor parabionts
[b]   Proportion of germ cells in experimentals relative to the normal fertilized controls

The success of germ cell transfer was assessed by counting the gonocytes dissociated from the gonadal ridges of the host larvae and comparing these gonocyte populations with that of the normal fertilized controls in each experiment. The results of these experiments are summarized in Table 2. They provide strong evidence that germ line cells in the endoderm grafts contribute to the gonadal germ cell populations in the normal and irradiated host larvae. That the graft primordial cells must have altered the normal direction of their migration within the endoderm to populate the host genital ridges, is supported by histological sections of the host larvae. The studies reveal that the dorsal-ventral orientation of the regionally determined endoderm graft is maintained when fused with the endoderm of the host animals. A single, enlarged and abnormally folded but differentiated digestive tract is present in the host larvae. However, the presence and positions of paired derivatives of the digestive tract in the anterior region (liver, pancreas, gall bladder) and, in certain animals, the presence of two rectal tubes in the posterior digestive tract region betray the contributions and maintenance of the original dorsal-ventral orientation of the grafted endoderm mass. Thus, Gipouloux's original findings are confirmed. The direction of primordial germ cell migration within the endoderm can be altered and is not fixed.

The induced gonadal sterility in embryos irradiated with UV light during first cleavage has been viewed as evidence for the direct action of the UV radiation on a cytoplasmic germ cell determinant, the germinal plasm, which is localized subcortically in the vegetal pole region of the fertilized egg (see reviews by Eddy 1975; Smith and Williams 1975). Since the radiation can also cause abnormal cleavage at the vegetal pole and a delay in migration of the germ line cells, it throws into question whether its action may be indirect, producing unfavorable conditions for germ cell differentiation and migration in the irradiated animals (Züst and Dixon 1975, 1977). Migration of the primordial germ cells from the deep endoderm of the graft to the posterior dorsal crest of the endoderm in host animals is accomplished when the fused endoderm exists as a solid elongated mass of yolky cells. That is, in the irradiated host animals, the germ line cells must migrate from the normal endoderm graft, through the irradiated host endoderm to attain the genital ridges of the irradiated host larvae. Under the conditions of these experiments, none of the irradiated controls or experimental embryos exhibited abnormal cleavage at the vegetal pole. Nevertheless, all of the irradiated controls were found to possess gonadal ridges free of any recognizable germ cells when examined at larval stage 25. Although the success of transfer of graft germ cells to host animals varied considerably within each experiment, the overall results in Table 2 indicate that graft germ cells can migrate equally well within the endoderm of UV-irradiated embryos and within the endoderm of normal host embryos. On the average, similar proportions of graft germ cells (about 50%) appear to be transferred to both kinds of host animals. This suggests that gonadal sterility and delayed germ cell migration is not the indirect effect of the UV light on cleavage patterns. Rather, the results support the earlier conclusions (Bounoure et al. 1954; Smith 1966; Tanabe and Kotani 1974) that UV radiation acts directly on a germ cell determinant in causing delayed migration and gonadal sterility of the irradiated animals.

Acknowledgments. I should like to thank the following persons for help with some of the experiments described here: Loris Carrethers, Joseph O'Donnell, Duane Ringer, and Joseph Penkala. The research conducted in the author's laboratory and reported here was supported by a grant HD-05289 from the USPHS, and more recently by a grant GM02221 from the USPHS.

References

Beams, H.W., Kessel, R.G.: The problem of germ cell determinants. In: Int. Rev. Cytol., Vol. 39 (eds. G.H. Bourne, J.F. Danielli), pp. 413-479. New York-London: Academic Press 1974

Blackler, A.W.: Contribution to the study of germ-cells in the Anura. J. Embryol. Exp. Morphol. 6, 491-503 (1958)

Blackler, A.W.: Transfer of primordial germ-cells between two subspecies of *Xenopus laevis.* J. Embryol. Exp. Morphol. 10, 641-651 (1962)

Blackler, A.W.: Embryonic sex cells of amphibia. Adv. Reprod. Physiol. 1, 9-28 (1966)

Blackler, A.W.: The integrity of the reproductive cell line in the amphibia. Curr. Top. Dev. Biol. 5, 71-87 (1970)

Bounoure, L.: Recherches sur la lignée germinale chez la grenouille rousse aux premiers stades du développement. Ann. Sci. Nat. 10ᵉ Ser. 17, 67-248 (1934)

Bounoure, L., Aubry, R., Huck, M.L.: Nouvelles recherches expérimentales sur les origines de la lignée reproductrice chez la grenouille rousse. J. Embryol. Exp. Morphol. 2, 245-263 (1954)

Briggs, R.: The experimental production and development of triploid frog embryos. J. Exp. Zool. 106, 237-266 (1947)

DiBerardino, M.A.: Investigations of the germ plasm in relation to nuclear transplantation. J. Embryol. Exp. Morphol. 9, 507-513 (1961)

Dziadek, M., Dixon, K.E.: Mitosis in presumptive primordial germ cells in post-blastula embryos of *Xenopus laevis.* J. Exp. Zool. 192, 285-291 (1975)

Dziadek, M., Dixon, K.E.: An autoradiographic analysis of nucleic acid synthesis in the presumptive primordial germ cell of *Xenopus laevis.* J. Embryol. Exp. Morphol. 37, 13-31 (1977)

Eddy, E.M.: Germ plasm and the differentiation of the germ cell line. In: Int. Rev. Cytol. Vol. 43 (eds. G.H. Bourne, J.F. Danielli), pp. 229-280. New York-London: Academic Press 1975

Fankhauser, G.: The effects of changes in chromosome number on amphibian development. Q. Rev. Biol. 20, 20-78 (1945)

Giorgi, P.P.: Germ cell migration in toad *(Bufo bufo):* effect of ventral grafting of embryonic dorsal regions. J. Embryol. Exp. Morphol. 31, 75-87 (1974)

Gipouloux, J.D.: Les Tissus mésodermiques dorssaux exercent-ils une action attractive sur les gonocytes primordiaux situés dans l'endoderm chez l'embryon du crapaud commun *Bufo bufo* L. (Amphibien Anoure)? C.R. Acad. Sci. 255, 2179-2181 (1962)

Gipouloux, J.D.: Les gonocytes primordiaux peuvent subis une migration intraendodermique en direction opposée à la direction normale; demonstration expérimentale chez le Discoglosse, *Discoglossus pictus*. C.R. Acad. Sci. 256, 2028-2030 (1963)

Gipouloux, J.D.: Recherches expérimentales sur l'origine, la migration des cellules germinales, et l'edification des crêtes genitales chez les amphibiens anoures. Bull. Biol. Fr. Belg. 104, 21-93 (1970)

Ijiri, K.-I., Egami, N.: Mitotic activity of germ cells during normal development of *Xenopus laevis* tadpoles. J. Embryol. Exp. Morphol. 34, 687-694 (1975)

Ikenishi, K., Kotani, M.: Ultrastructure of the "germinal plasm" in *Xenopus* embryos after cleavage. Dev. Growth Differ. 17, 101-110 (1975)

Kalt, M.R., Gall, J.G.: Observations on early germ cell development and premeiotic ribosomal DNA amplification in *Xenopus laevis*. J. Cell Biol. 62, 460-472 (1974)

Kamimura, M., Ikenishi, K., Kotani, M., Matsuno, T.: Observations on the migration and proliferation of gonocytes in *Xenopus laevis*. J. Embryol. Exp. Morphol. 36, 197-207 (1976)

Ladner, M.B.: Influence of triploid germ cells on gonadal differentiation on *Rana pipiens*. Ph.D. Thesis, Rice University (1978)

Shumway, W.: Stages in normal development of *Rana pipiens*. I. External form. Anat. Rec. 78, 139-147 (1940)

Smith, L.D.: The role of a "germinal plasm" in the formation of primordial germ cells in *Rana pipiens*. Dev. Biol. 14, 330-347 (1966)

Smith, L.D.: (personal communication)

Smith, L.D., Williams, M.A.: Germinal plasm and determination of the primordial germ cells. In: The Developmental Biology of Reproduction; 33rd Symp. Soc. Dev. Biol. (eds. C.L. Markert, J. Papaconstantinou), pp. 3-24. New York-London: Academic Press 1975

Subtelny, S., Carrethers, L.: Interspecific germ cell migration in anurans. J. Cell Biol. 67, 421a (1975)

Subtelny, S., Ladner, M.B.: Germ cell number and size in the gonadal ridges of the frog, *Rana pipiens*. Am. Zool. 16, 230 (1976)

Tanabe, K., Kotani, M.: Relationship between the amount of the "germinal plasm" and the number of primordial germ cells in *Xenopus laevis*. J. Embryol. Exp. Morphol. 31, 89-98 (1974)

Taylor, A.C., Kollros, J.J.: Stages in the normal development of *Rana pipiens* larva. Anat. Rec. 94, 7-23 (1946)

Wakahara, M.: Partial characterization of "primordial germ cell forming activity" localized in vegetal pole cytoplasm in anuran eggs. J. Embryol. Exp. Morphol. 39, 221-233 (1977)

Wakahara, M.: Induction of supernumerary primordial germ cells by injecting vegetal pole plasm in *Xenopus* eggs. J. Exp. Zool. 203, 159-164 (1978)

Whitington, P. McD., Dixon, K.E.: Quantitative studies of germ plasm and germ cells during early embryogenesis of *Xenopus laevis*. J. Embryol. Exp. Morphol. 33, 57-74 (1975)

Wylie, C.C., Roose, T.B.: The formation of the gonadal ridge in *Xenopus laevis*. III. The behavior of isolated primordial germ cells. J. Embryol. Exp. Morphol. 35, 149-157 (1976)

Züst, B., Dixon, K.E.: The effect of u.v. irradiation of the vegetal pole of *Xenopus laevis* eggs on the presumptive primordial germ cells. J. Embryol. Exp. Morphol. 34, 209-220 (1975)

Züst, B., Dixon, K.E.: Events in the germ cell lineage after entry of the primordial germ cells into the genital ridges in normal and u.v. irradiated *Xenopus laevis*. J. Embryol. Exp. Morphol. 41, 33-46 (1977)

The Effects of Temperature-Sensitive Rous Sarcoma Virus and Phorbol Diester Tumor Promoters on Cell Lineages[1]

H. Holtzer, J. Biehl, M. Pacifici,
D. Boettiger, R. Payette, and C. West

Departments of Anatomy and Microbiology, School of Medicine
University of Pennsylvania, Philadelphia, PA 19104, USA

I. Introduction

My colleagues and I, as well as many other investigators (Pierce 1967; Foulds 1969; Pierce and Wallace 1971; Berenblum 1974; Boutwell 1976; Potter 1978) have performed experiments based on the assumption that those intracellular controls which are involved in the transformation of normal cells into malignant cells are also involved in the generation of diversity among normal cells (Holtzer et al. 1972, 1974, 1975c). In addition, we have assumed that these intracellular controls are the very mechanisms which permit a "cloned" zygote, a "cloned" blastula cell, or a "cloned" limb bud precursor cell in early compartments of their respective lineages to generate diversified progeny (Abbott et al. 1974; Dienstman et al. 1974; Holtzer 1978). In all instances the fundamental question is: How does a single replicating "mother" cell, following a quantal cell cycle, yield daughter(s) with a predictably different, but equally limited, set of metabolic options? A replicating blastula cell, a replicating erythrogenic hematocytoblast or Friend erythroleukemic cell (Holtzer et al. 1972, 1975c; Lajtha and Schofield 1974; Weintraub 1975; Holtzer 1978), a replicating presumptive chondroblast (Holtzer 1978), or a replicating presumptive myoblast (Holtzer et al. 1975c; Holtzer 1978) has no option following a quantal cell cycle, but to yield specifically different kinds of gastrula cells, erythroblasts, chondroblasts, and myoblasts, respectively. Presumably, daughter cells, as a consequence of a quantal cell cycle, acquire the capacity to transcribe regions of the genome that could not be transcribed in their mother[2]. It is equally likely that the daughter cells lose the capacity to transcribe regions that were uniquely transcribed in their mother cells. Similarly, embryonal carcinoma cells (Martin 1975), chondrosarcoma cells, hepatoma cells, or lymphoma cells have each inherited a different set of metabolic options which (1) not only distinguish these different types of neoplastic cells one from the other, but (2) distinguish them from their normal mother cells. In brief, as cells move from one compartment to the next in

1   This research was supported by grants from the National Institutes of Health (CA-18194, HL-18708, HL-15835 of the Pennsylvania Muscle Institute, GM-20138 and the Muscular Dystrophy Association).

2   In this discussion "proliferative cell cycles", the cell cycles most commonly studied, those in which mother and daughters have identical metabolic options, will be ignored (Holtzer et al. 1975c; Holtzer 1978).

any lineage—as well as from a "normal" compartment into a "malignant" compartment in any lineage—sets of genes that were *unavailable* for transcription in the mother become available for transcription in the daughter(s) due to structural changes in the daughter chromosomes which occurred as a result of a particular quantal cell cycle. This is the central theme of normal cell diversification; this is the central theme in the genesis and progression of neoplastic cells.

Specifically, to the degree that cell diversification is concerned with the transcription and translation of "luxury" molecules (Holtzer 1968; Holtzer and Abbott 1968; Holtzer et al. 1973) rather than of "housekeeping" molecules, we have focused on the shift in the kinds of luxury molecules synthesized by cells as they pass through the antepenultimate, penultimate, and ultimate compartments in the normal chondrogenic, myogenic, and melanogenic lineages. Then we have asked: How do these same cells behave when transformed with a temperature-sensitive mutant of the Rous sarcoma virus (ts-RSV), or when treated for various periods with the tumor promoter, phorbol-12-myristate-13-acetate (PMA)? The transforming activity of the strain of Rous virus used (ts-LA24) is probably due to the product of the SRC gene, which has been shown to be a protein kinase with a mol. wt.of 60,000 (Collett and Erikson 1978). Though the tumor promoter PMA induces ornithine decarboxylase and acts as a mitogen on some cell types, virtually nothing is known of how it might block cell differentiation (Kreibich et al. 1974; Cohen et al. 1977; Pacifici et al. 1977; Lowe et al. 1978; O'Brien and Diamond 1978).

II. Effect of Temperature-Sensitive Rous Sarcoma Virus on Chick Chondroblasts, Retinal Pigment, and Myogenic Cells

Normal chick chondroblasts form characteristic metachromatic, epithelioid colonies when grown in culture in the appropriate medium (Abbott and Holtzer 1968; Chacko et al. 1969; Dienstman et al. 1974). These cells roughly double every 24 h, and synthesize a chondroblast-unique extracellular Type IV sulfated proteoglycan (Okayama et al. 1976; Lowe et al. 1978) and Type II collagen chains (Schiltz et al. 1973; Mayne et al. 1974, 1975). If such chondroblasts are infected with a ts-RSV mutant and passaged several times in soft agar, the following are noted (Roby et al. 1976; Pacifici 1977): At permissive temperature (37° C) the cells (1) roughly double every 24 h (2) grow in soft agar; (3) form loose attachments when collagen is the substrate; (4) are morphologically indistinguishable from transformed fibroblasts; (5) display a high deoxyglucose uptake; and (6) most importantly, cease synthesizing their chondroblast-unique Type IV sulfated proteoglycan. Recent experiments using a cell-free translation system have shown that mRNAs from ts-transformed chondroblasts that were grown at permissive temperature synthesize a form of Type I rather than Type II collagens (Adams et al. 1980).

In reciprocal experiments, ts-transformed chondroblasts were shifted from permissive to nonpermissive (41° C) temperature for two passages, and then dropped to permissive temperature. At the nonpermissive temperature the infected cells were indistinguishable from uninfected cells in rate of replication, morphology, and types of sulfated proteoglycans synthesized. When dropped, to permissive temperature, the infected cells again assumed a transformed morphology, and ceased to synthesize their characteristic Type IV sulfated proteoglycan. Nevertheless, the cells replicated at their normal rate and synthesized and secreted "constitutive" sulfated proteoglycans synthesized and secreted by nonchondrogenic cells.

These experiments demonstrate that the morphology and the synthesis of the luxury molecules of chondroblasts are promptly and differentially blocked by the transforming factor regulated by the SRC gene. Of theoretical interest is the observation that the SRC gene product, though it blocks the synthesis of the chondroblast-unique Type IV sulfated proteoglycan, does not block the synthesis of the numerous housekeeping molecules required for replication and growth. Presumably the protein kinase coded for by the SRC gene (Collett and Ericson 1978; Anderson et al. unpublished data) differentially suppresses the synthesis of the luxury molecules synthesized by terminal chondroblasts, but only modestly alters the synthesis of the housekeeping molecules (Roby et al. 1976; Pacifici et al. 1977). We stress that, contrary to many views regarding differences between normal cells and their malignant progeny, there is no evidence that integration of the SRC gene and its readout renders transformed cells "more embryonic". The reversibly blocked ts-trsansformed cells have not lost their unique status in the chondrogenic lineage. In terms of cell diversification, transformed chondroblasts have more in common with normal chondroblasts than with other types of transformed cells. Though transformed chondroblasts do not actively transcribe the limited and unique sequence of genes which distinguish them from all other cells in the body, such blocked chondroblasts are not "epigenetically undifferentiated" (Holtzer and Abbott 1968; Holtzer et al. 1974; Holtzer 1978). They are *cryptic* chondrogenic cells. Transformed chondrogenic cells are able to maintain their position in the chondrogenic lineage for many generations. Somehow this chondrogenic commitment, this option to transcribe just those genes unique to functional chondroblasts, must be contained in some structural component of the cell. It is this unique structural component, transmitted generation after generation irrespectively of whether it is or is not expressed, that differentiates a chondroblast from all of its normal precursors, as well as from its malignant descendents, chondrosarcoma cells. Elsewhere we have postulated that this phenotypic commitment, or fixation, is the consequence of a particular sequence of quantal cell cycles (Holtzer 1978). As the precursor cells in the chondrogenic lineage pass from compartment to compartment, the stepwise alteration of their chromosomes that occurs during each S period is inherited by the daughter cells. It will be interesting to learn if these postulated changes in gene and/or chromosome structure, which occur during a quantal cell cycle involve: (1) rearrangement of the DNA, bringing together separate genes, as in the ontogeny of immunocompetent cells, (2) changes in regions of the DNA associated with histones or nonhistone proteins, or (3) the synthesis of unique RNA-excising molecules.

Similar results have been obtained with retinal pigment cells transformed with the ts-RSV (Roby et al. 1976; Boettiger et al. 1977). Ts-transformed retinal pigment cells cloned in soft agar at permissive temperature replicate as unpigmented, nondiscript "fibroblastic" cells. Microscopically they display no characteristics that allow them to be identified as retinal pigment cells. Following the shift to nonpermissive temperature, the infected cells reacquire the characteristic epithelioid attachments of normal retinal pigment cells and display melanosomes and synthesize melanin within several days (cf. Fig. 3 in Boettiger et al. 1977).

Ts-RSV infected myogenic cells reared for 10 days at permissive temperature behave similarly to transformed chondroblasts and pigment cells (Holtzer et al. 1975a). They synthesize "constitutive" myosin heavy and light chains common to many nonmyogenic cells, do not assemble thick and thin interdigitating myosin and actin filaments, and do not display cell surface properties that permit myoblasts to fuse. Shifted to nonpermissive temperature, many transformed myogenic cells withdraw from the cell cycle and form definitive, postmitotic myoblasts. If they are prevented from fusing by Cytochalsin-B (Holtzer et al. 1975b; Croop and Holtzer 1975) such postmitotic, mononucleated myoblasts initiate the synthesis of the definitive skeletal myosin heavy and light chains, and organize interdigitating thick and thin filaments into normal striated myofibrils (Chi et al. 1975). If not prevented from fusing, these postmitotic myoblasts form the characteristic multinucleated myotubes rich in postmitotic nuclei. Fiszman (1978) reports that ts-transformed myogenic cells reared at permissive temperature fail to accumulate ACh-receptors and fail to accumulate the muscle form of creatine phosphokinase.

There is one intriguing difference in the way in which transformed chondrogenic and transformed myogenic cells respond when shifted from nonpermissive to permissive temperature. Chondroblasts adjust by assuming the typical morphology of transformed cells and by shutting off the synthesis of their luxury molecules. Nevertheless, they continue to replicate and to synthesize their housekeeping molecules, including their constitutive sulfated proteoglycans and even some form of Type I collagen. When infected postmitotic myoblasts or myotubes are shifted from nonpermissive to permissive temperature, however, 100% of these post-mitotic cells vacuolate and die within 48–72 h.

This dramatic difference between transformed chondroblasts and myoblasts to the shift-down may be the consequence of differences in the state of the control of DNA synthesis in these two cell types. Chondroblasts in the terminal compartment of the chondrogenic lineage do not permanently withdraw from the cell cycle (Holtzer and Abbott 1968). In contrast, one of the earliest biochemical events after the birth of the definitive myoblast is irreversible withdrawal from the cell cycle (Holtzer et al. 1957; Okazaki and Holtzer 1965; Dienstman and Holtzer 1975; Yeoh and Holtzer 1977). It is worth stressing that there is no evidence in the literature that any molecule, any "growth factor", or any virus will induce postmitotic, mononucleated myoblasts or the nuclei within a myotube to reenter the cell cycle. The claim (Yaffe and Gershon 1967) that polyoma virus induces nuclei in myotubes to enter S has not been confirmed. The micrographs in that report probably confuse nuclear pulverization (Johnson and Rao 1971) with metaphase nuclei, or the uptake of ^{3}H-TdR for DNA repair for the DNA synthesis occurring during S. Similarly, the claims of Buckley and Konigsberg (1974, 1977) that mononucleated cells withdraw from the cell cycle *after* fusion, rather than withdrawal being a precondition for fusion, has now been shown to be untenable by several different laboratories (Holtzer et al. 1957, 1975c; Emerson and Beckner 1975; Moss and Strohman 1976; Nameroff and Munar 1976; Turner et al. 1976; Yeoh and Holtzer

1977). Conceivably, the myoblasts that have become postmitotic at nonpermissive temperature as part of their normal program of differentiation become subject to two irreconcilable signals when shifted to permissive temperature: (1) as a viral-infected cell they are constantly being primed to synthesize DNA and replicate, whereas (2) as postmitotic myoblasts there are stringent controls irreversibly blocking DNA synthesis. The presence of such conflicting signals in the same cell may lead to cell death.

III. Effects of Tumor Promoter, Phorbol-12-Myristate-13-Acetate on Chondrogenic and Myogenic Cells

The effects of the tumor promoter, phorbol-12-myristate-13-acetate, or PMA, on cultured chondroblasts mimics the differential effects of ts-RSV on infected and transformed chondroblasts at permissive temperature (Pacifici and Holtzer 1977; Lowe et al. 1978). PMA acts rapidly on the cell surface of normal chondroblasts. Approximately 40% of freshly plated chondroblasts adhere to the substrate during the first 24 h of normal culture and assume the polygonal morphology of functional chondroblasts. The remaining 60% persist as "floaters". The floaters attach to the substrate over the next 2 or 3 days, when they too assume a polygonal morphology. In contrast, 100% of the PMA-treated chondroblasts adhere to the substrate within 15 h, form long pseudopodia, do not exhibit contact inhibition of migration, and at all densities display much overlapping of processes. PMA-treated cells become multilayered and achieve densities many times greater than those exhibited by normal chondroblasts. PMA is a mitogen for chondroblasts (cf. Figs. 1–4 in Lowe et al. 1978).

Concurrent with this rapid change in morphology, PMA suppresses the synthesis of the chondroblast-unique Type IV sulfated proteoglycan. In addition, the PMA-treated chondroblasts alter their synthesis of Type II collagen chains toward a Type I collagen chain (West, Rosenbloom and Holtzer, unpublished).

Normal floating chondroblasts synthesize little if any fibronectin (cf. Figs. 5–7 in Lowe et al. 1978). Chondroblasts that attach to the substrate synthesize modest amounts of fibronectin. Both floaters and attached chondroblasts synthesize considerable quantities of actin (Holtzer et al. 1974) and a glycosylated protein with a mol. wt. of approximately 180,000 daltons. PMA-treated chondroblasts synthesize considerable quantities of fibronectin and actin, but cease to synthesize their characteristic quantities of the intracellular 180,000 dalton protein. The 180,000 dalton protein that accumulates intracellularly in control chondroblasts is likely to be a precursor form of collagen for it is rich in proline and readily digested with collagenase. It is noteworthy that BrdU-suppressed chondroblasts synthesize fibronectin and actin, but do not synthesize the chondroblast-unique Type IV sulfated proteoglycan nor the proline-rich 180,000 dalton protein; they also switch from synthesizing a predominantly Type II to a Type I collagen. In brief, the transforming protein synthesized by the SRC gene, the tumor promoter PMA, and the thymidine analog BrDU (Abbott and Holtzer 1968; Holtzer and Abbott 1968; Abbott et al. 1972; Holtzer et al. 1972; Levitt and Dorfman 1972; Holtzer et al.

1975c), all differentially suppress the unique molecules synthesized by terminal chondroblasts.

PMA also reversibly blocks the terminal differentiation of myogenic cells (Cohen et al. 1977). PMA added to primary cultures of chick presumptive myoblasts permits cell replication, but promptly blocks fusion and the emergence of multinucleated myotubes. PMA may block the formation of myotubes: (1) by keeping the replicating presumptive myoblasts in the cell cycle, or (2) by acting directly on the cell surface of the postmitotic myoblasts that emerge in these cultures. Though there is some indirect evidence that PMA maintains the myogenic cells in the cell cycle, the evidence that PMA acts directly on the surface of postmitotic myoblasts is unequivocal. Cytochalsin-B (Croop and Holtzer 1975; Holtzer et al. 1975b) or EGTA (Paterson and Strohman 1972) inhibits fusion of competent postmitotic myoblasts into multinucleated myotubes. When either Cytochalsin-B or EGTA is removed from cultures containing fusion-competent postmitotic myoblasts, over 50% of the myoblasts fuse within several hours (Cohen et al. 1977). If, however, after removing the Cytochalsin-B or EGTA, PMA is added to the medium, the postmitotic myoblasts do not fuse. As fusion of myoblasts in control medium can be completed within 60 min, this rapid blockage of fusion by PMA demonstrates that at least one site of action of the cocarcinogen is directly on the cell surface.

In typical primary cultures, myoblasts begin to fuse on day 2 and fusion peaks by day 5 or 6 (Yeoh and Holtzer 1977). Often cultures in PMA that show no signs of fusion by day 4 do, however, "break through" and fuse on day 5 and continue until day 8. The cause of the erratic behavior of PMA on fusion is still unclear. We have performed experiments which demonstrate that the PMA recovered from muscle cultures after 24 h will not block fusion when added to myoblasts competent to fuse. Clearly the PMA in such conditioned medium has in some manner been inactivated by the growing cells. It will be interesting to determine whether many kinds of cells, besides myogenic cells and 3T3 cells (Diamond et al. 1977) develop intracellular mechanisms to inactivate PMA or in some manner neutralize some of its biological effects (Kreibich et al. 1974; O'Brien and Diamond 1978).

PMA not only inhibits myogenesis and chondrogenesis of relatively mature myogenic and chondrogenic cells but inhibits earlier cells as well. Intact and dissociated limb buds from chick embryos, of stages 18 and 22 respectively, form postmitotic myotubes and nodules of cartilage in control medium (Dienstman et al. 1974; Holtzer 1978). When grown in suitable concentrations of PMA these limb buds replicate, but fail to differentiate into recognizable myoblasts or myotubes or chondroblasts.

Recently we (Toyama et al. 1979) have observed that PMA interferes with the synthesis of some of the muscle-specific contractile proteins in the *postmitotic myotubes*. The content of the myofibrils and the muscle-specific 10 nm filaments can be reversibly manipulated with PMA. After being exposed to PMA for 72 h myotubes, (1) lose all of their previously assembled striated myofibrils, and (2) display an exceedingly dense network of 100 Å filaments. The synthesis of several proteins which comprise the myofibrils is concordantly inhibited. The synthesis of alpha-actin and the skeletal-specific myosin heavy and light chains is drastically reduced, whereas the synthesis of the beta- and gamma-actins is not altered. The muscle-specific myosins and alpha actin is degraded, whereas the muscle-specific 10 nm filament protein, the 55 K protein (Croop and Holtzer 1975) persists. The fate of the constitutive 10 nm filament protein, the 58 K protein (Croop and Holtzer 1975; Holtzer et al. 1975b; Toyama et al. 1979; Bennett et al. 1979) has yet to be determined. These changes are remarkably reversible. Twenty-four hours after removal of PMA these myotubes are rich in striated, interdigitating thick and thin filaments. These experiments treating postmitotic myotubes with PMA suggest that cells need not be in the cell cycle for PMA to exert at least some of its effects.

IV. The Differential Effects of the Tumor Promoter, Phorbol-12-Myristate-13-Acetate on Cells Early and Late in the Pigment Cell Lineage

To determine whether PMA would have one effect on cells early, and another effect on cells late in the pigment cell lineage, experiments have been performed with two different systems: (1) Trypsin-dissociated neural crest cells from 40 h chick enbryos were reared in the presence and absence of PMA. Many of these neural crest cells are the unpigmented precursors to the definitive, solitary, dendritic trunk melanoblasts; (2) Trypsin-dissociated fully pigmented retinal epithelial cells from 7–day chick embryos also were reared in the presence and absence of PMA. If allowed to achieve confluence, these retinal pigment cells form epithelial sheets of polygonal cells. Virtually 100% of the cells in such colonies become fully pigmented.

The effects of PMA on neural crest cells and its effects on some of their progeny that normally differentiate into trunk melanoblasts were tested: Intact neural tubes with attached neural crest were removed from 40 h chick embryos and placed dorsal surface down on collagen coated petri dishes. After 2 days in culture, and after large numbers of cells had migrated out of the explant onto the collagen substrate, the still relatively intact neural tube was removed from the dish. The cells that had migrated and attached to the collagen substrate were allowed to replicate for another 24–48 h, trypsinized, and subcultured at densities between 5×10^4, 1×10^5 and 5×10^5 cells/petri dish. The densities had no noticeable effects on the results. At this stage no pigment or nerve cells could be detected histologically. Within the next 5 days approximately 20–30% of all the cells in the dish differentiated into typical dendritic melanoblasts. These normal melanoblasts display three to six short dendritic processes, 10–20 μ in length and contain 30–300 melanin granules/cell. EM micrographs display varying numbers of melanosomes in varying stages of melanization. The cultures were then trypsinized and secondary and tertiary cultures prepared. It is not uncommon for over 30% of the dispersed cells in tertiary cultures to display typical melanoblasts. Whether the remaining, nonpigmented cells in these cultures are precursor Schwann cells, sympathetic neurons, or unpigmented melanoblasts has not yet been determined (Payette, Biehl and Holtzer, in press).

There is a striking, reversible, blocking effect of PMA on this sequence of the progeny of the neural crest cells differentiating into definitive melanoblasts. If PMA is added to the cultures after the neural tube is removed, and if PMA is present in all subsequent stages of culturing, though the cells replicate as do the controls, recognizable pigment cells never appear. Equally dramatic is the gross morphological appearance of these cells reared in PMA. They have small cell bodies with two or three long, neurite-like processes that may exceed 100 to 200 μ in length. These processes vary between 1–3 μ in diameter. They do not display melanosomes in EM micrographs. Judging from their distribution in the petri dish, many of these melanoblasts are the products of a clone established from a single neural crest cell. This suppression of pigmentation is remarkably reversible. If tertiary cultures that had been in PMA for 12 days are transferred to normal medium, within 36 h many cells withdraw their elongated, neurite-like processes, and over the next 3 days the proportion of typical pigmented melanoblasts to nonpigmented cells equals the untreated controls that had been subcultured according to the same schedule. Cells that had been in PMA for 16 days when transferred to normal medium, fail to pigment and maintain their odd morphology.

If PMA is added to cultures of untreated melanoblasts, or PMA-treated melanoblasts that had been allowed to become melanoblasts by removing the PMA, within 3 or 4 days the number of pigment cells in both sets of cultures drops to approximately 1% of the total number of cells in the dish. If such cultures are transferred back to normal medium the percentage of pigment cells in the dishes rises again to 20–30%.

The effects of PMA differ somewhat on retinal pigment cells that have already pigmented when first treated with PMA. Trypsin-dissociated retinal pigment cells plated at densities up to 1×10^5 cells/35 mm petri dish, replicate and lose their pigment. If allowed to achieve confluence by repeated replication, within 5 days they form epithelial islands and virtually 100% of the cells in such colonies are fully pigmented. If PMA is added to such cultures it has no discernible effect on either the degree of pigmentation or morphology of the polygonal, pigmented epithelial cells.

If, however, PMA is added in earlier stages of culture, or at a time before the retinal cells establish epithelial colonies, the cells continue to replicate, but fail to form epithelial colonies. Instead such cells form mounds of overlapping, fibroblastic cells. The cells in these multilayered mounds do not pigment. EM micrographs do not reveal any melanosomes in these cells. If the dense islands of unpigmented cells that had been in PMA for 12 days are trypsinized and plated into normal medium, many after 5 or 6 days form epithelial islands of polygonal, fully pigmented cells. If however they are maintained in PMA for over 16 days, though they will replicate when transferred to normal medium, they do not form epithelioid colonies nor do they form pigmented cells.

There are many reports in the literature (see Weston and Butler 1973 for review) on the importance of "exogenous factors" in determining when and how individual cells in the pigment lineages – either trunk or retinal pigment cells – are induced to differentiate terminally. Our experiments suggest that with a minimum of interaction with other dispersed neural crest cells, or interaction with mesenchymal morphological structures such as the notochord or somites and the putative glycosaminoglycans they synthesize, some early neural crest precursor cells are already *endogenously* programed so as to permit terminal differentiation into melanoblasts. It is worth stressing that the transition from precursor neural crest cell to progeny that become definitive melanoblasts occurs in the relatively "neutral microenvironment" of our tissue culture medium, and in the absence of mysterious "inducing molecules" or complex "epithelial-mesenchymal" tissue interactions.

When acting on cells early in the pigment lineage or on melanoblasts that have "lost" the intracellular machinery to assemble melanosomes or synthesize melanin due to treatment with PMA, the tumor promoter appears to "freeze" cells in that state. It does not shift the PMA-treated cell into another lineage; it does not render them more "embryonic". On the other hand, once pigment cells have the ongoing intracellular machinery to assemble melanosomes and synthesize melanin, then the effect of PMA on these cells is not readily detectable. In this latter context, it is worth noting that PMA has little effect on frank pigmented melanoma cells (Hubermann, personal communication).

Clearly, the reversible way PMA interferes with the synthetic activity associated with pigmentation depends on the epigenetic and physiological status of the treated cell. It is equally clear that PMA can exert this effect without canceling a given cell's commitment to a given program of differentiation, though it will block the expression of that program.

V. Summary

We would stress that at least two gents ssociated with tumorigenesis the SRC protein kinase and PMA – are also agents that may reversibly block some events associated with terminal differentiation. Such agents, along with BrdU, clearly do not block the synthesis of the many "housekeeping" molecules required for replication and growth, nor do they inhibit the synthesis of those as yet unknown molecules that keep a cell commited to a specific compartment in a given lineage. On the other hand, both the transforming protein and PMA, as well as BrdU, differentially block the synthesis of those sets of "luxury" molecules that characterize some terminally differentiated cells. Learning more about this inverse relationship should lead to a better understanding of the intracellular mechanisms responsible both for cell transformation and for cell differentiation.

References

Abbott, J., Holtzer, H.: The loss of phenotypic traits by differentiated cells, V. The effect of 5-bromodeoxyuridine on cloned chondrocytes. Proc. Natl. Acad. Sci. USA 59, 1144-1151 (1968)

Abbott, J., Mayne, R., Holtzer, H.: Inhibition of cartilage development in organ cultures of chick somites by the thymidine analog 5-bromo-2-deoxyuridine. Dev. Biol. 28, 430-442 (1972)

Abbott, J., Schiltz, J., Dienstman, S., Holtzer, H.: The phenotypic complexity of myogenic clones. Proc. Natl. Acad. Sci. USA 71, 1506-1510 (1974)

Adams, S., Pacifici, M., Boettiger, D., Holtzer, H.: Switch from type II to type I collagen chains in ts-transformed chondroblasts. J. Cell Biol., in press (1980)

Bennett, G., Fellini, S., Toyama, Y., Holtzer, H.: Redistribution of intermediate filament subunits during skeletal myogenesis and maturation in vitro. J. Cell Biol. 82, 577-584 (1979)

Berenblum, I.: Carcinogenesis as a Biological Problem. Amsterdam: North-Holland 1974

Boettiger, D., Roby, K., Brumbaugh, J., Biehl, J., Holtzer, H.: Ts-Rous sarcoma virus transformed melanogenic cells. Cell 11, 881-890 (1977)

Boutwell, R.: The biochemistry of preneoplasia in mouse skin. Cancer Res. 36, 2631-2635 (1976)

Buckley, P., Konigsberg, I.: Myogenic fusion and the duration of the post-mitotic gap. Dev. Biol. 37, 193-212 (1974)

Buckley, P., Konigsberg, I.: Do myoblasts in vitro withdraw from the cell cycle? A re-examination. Proc. Natl. Acad. Sci. USA 74, 2031-2035 (1977)

Chacko, S., Holtzer, S., Holtzer, H.: Suppression of chondrogenic expression in mixtures of normal chondrocytes and Budr-altered chondrocytes grown in vitro. Biochem. Biophys. Res. Commun. 34, 183-189 (1969)

Chi, J., Fellini, S., Holtzer, H.: Differences among myosins synthesized in non-myogenic cells, presumptive myoblasts, and myoblasts. Proc. Natl. Acad. Sci. USA 72, 4999-5003 (1975)

Cohen, R., Pacifici, M., Rubinstein, N., Biehl, J., Holtzer, H.: Effect of a tumor promoter on myogenesis. Nature (London) 266, 538-540 (1977)

Collett, M.S., Erikson, R.L.: Protein kinase activity associated with the avain sarcoma virus SRC gene products. Proc. Natl. Acad. Sci. USA 75, 2021-2024 (1978)

Croop, J., Holtzer, H.: Response of myogenic and fibrogenic cells to cytochalasin-B and to colcemid. J. Cell Biol. 65, 271-285 (1975)

Diamond, L., O'Brien, T., Rovera, G.: Inhibition of adipose conversion of 313 fibroblasts by tumor promoters. Nature (London) 269, 247-249 (1977)

Dienstman, S., Holtzer, H.: Myogenesis: A cell lineage interpretation. In: Results and Problems in Cell Differentiation, Vol. 7 (eds. J. Reinert, H. Holtzer), pp. 1-25. Berlin-Heidelberg-New York: Springer 1975

Dienstman, S., Biehl, J., Holtzer, S., Holtzer, H.: Myogenic and chondrogenic lineages in developing limb buds grown in vitro. Dev. Biol. 39, 83-95 (1974)

Emerson, C., Beckner, S.: Activation of myosin synthesis in fusing and mononucleated myoblasts. J. Mol. Biol. 93, 431-477 (1975)

Fiszman, M.: Morphological and biochemical differentiation in RSV transformed chick embryo myoblasts. Cell Differ. 7, 89-101 (1978)

Foulds, L.: Neoplastic Development. New York-London: Academic Press 1969

Holtzer, H.: Induction of chondrogenesis: a concept in quest of mechanisms. In: Epithelial-Mesenchymal Interactions (eds. R. Fleischmajer, R. Billingham), pp. 152-164. Baltimore: Williams and Wilkins 1968

Holtzer, H.: Cell lineages, stem cells and the quantal cell cycle concept. In: Stem Cells and Tissue Homeostatis (eds. B. Lord, C. Potten, R. Cole), pp. 1-28. Cambridge-London-New York-Melbourne: Cambridge University Press 1978

Holtzer, H., Abbott, J.: Oscillations of the chondrogenic phenotype in vitro. In: Results and Problems in Cell Differentiation, Vol. I (eds. H. Beerman, J. Reinert, H. Ursprung). pp. 1-16. Berlin-Heidelberg-New York: Springer 1968

Holtzer, H., Marshall, J., Finck, H.: An analysis of myogenesis by the use of fluorescent antimyosin. J. Biophys. Biochem. Cytol. 3, 705-723 (1957)

Holtzer, H., Weintraub, H., Mayne, R., Mochan, B.: Cell cycle, cell lineages and cell differentiation. In: Current Topics in Developmental Biology, Vol. I (eds. A. Monroy, A. Moscona), pp. 229-256. New York-London: Academic Press 1972

Holtzer, H., Weintraub, H., Biehl, J.: Cell Cycle-dependent events during myogenesis, neurogenesis, and erythrogenesis. In: Biochemistry of Cell Differentiation, Vol. 24 (eds. A. Monroy, R. Tsanev), Proceedings of the 7th FEBS Meeting, pp. 41-54. New York-London: Academic Press 1973

Holtzer, H., Dienstman, S., Holtzer, S., Biehl, J.: Quantal cell cycles, normal cell lineages and tumorgenesis. In: Differentiation and Control of Malignancy of Tumor Cells (eds. W. Nakahara, T. Ono, T. Sugimura, H. Sugana). Proceedings of the 4th International Symposium of the Princess Takamatsu Cancer Research Fund, Tokyo, 1973. Tokyo: University of Tokyo Press 1974

Holtzer, H., Biehl, J., Yeoh, G., Meganathan, R., Kaji, A.: Effect of oncogenic virus on muscle differentiation. Proc. Natl. Acad. Sci. USA 72, 4051-4055 (1975a)

Holtzer, H., Croop, J., Dienstman, S., Ishikawa, H., Somlyo, S.: Effects of cytochalasin-B and colcemid on myogenic cultures. Proc. Natl. Acad. Sci. USA 72, 513-517 (1975b)

Holtzer, H., Rubinstein, N., Fellini, S., Yeoh, G., Chi, J., Birnbaum, J., Okayama, M.: Lineages, quantal cell cycles, and the generation of diversity. Q. Rev. Biophys. 8, 1-34 (1975c)

Holtzer, S., Barany, M., Holtzer, H.: Protein-bound ADP in myogenic and chondrogenic cells. Differentiation 2, 39-42 (1974)

Johnson, R., Rao, P.: Nucleo-cytoplasmic interactions in the achievement of nuclear synchrony in DNA synthesis and mitosis in multinucleated cells. Biol. Rev. 46, 97-130 (1971)

Kreibich, G., Süss, R., Kinzel, V.: On the biochemical mechanism of tumorogenesis in mouse skin. V. Studies of the metabolism of tumor promoting and non-promoting phorbol derivatives in vivo and in vitro. Z. Krebsforsch. 81, 135-149 (1974)

Lajtha, L., Schofield, R.: On the problem of differentiation in haemopoiesis. Differentiation 2, 313-320 (1974)

Levitt, D., Dorfman, A.: The irreversible inhibition of differentiation of limb bud mesenchyme by bromodeoxyuridine. Proc. Natl. Acad. Sci. USA 69, 1253-1257 (1972)

Lowe, M., Pacifici, M., Holtzer, H.: Effects of phorbal-2-myristate-13 acetate on the phenotypic program of cultured chondroblasts and fibroblasts. Cancer Res. 38, 2350-2356 (1978)

Martin, G.: Teratocarcinoma as a model system for the study of embryogenesis and neoplasia. Cell 5, 229-243 (1975)

Mayne, R., Schiltz, J., Holtzer, H.: Some overt and covert properties of chondrogenic cells. In: The Biology and Biochemistry of the Fibroblast (ed. J. Pikkarainen), pp. 61-78. New York-London: Academic Press 1974

Mayne, R., Vail, M., Miller, E.: Analysis of the changes in biosynthesis that occur when chick chondroblasts are grown in 5-bromo-2-deoxyuridine. Proc. Natl. Acad. Sci. USA 72, 4511-4515 (1975)

Moss, P., Strohman, R.: Myosin synthesis by fusion-arrested chick embryo myoblasts in cell culture. Dev. Biol. 48, 431-437 (1976)

Nameroff, M., Munar, E.: Inhibition of cellular differentiation by phospholipase C. II. Separation of fusion and recognition among myogenic cells. Dev. Biol. 49, 284-293 (1976)

O'Brien, T., Diamond, L.: Metabolism of tritium-labeled 12-O-tetradecanoylphorbol-13-acetate by cells in culture. Cancer Res. 38, 2562-2566 (1978)

Okayama, M., Pacifici, M., Holtzer, H.: Differences among sulfated proteoglycans in non-chondrogenic cells, presumptive chondroblasts, and chondroblasts. Proc. Natl. Acad. Sci. USA 73, 3224-3228 (1976)

Okazaki, K., Holtzer, H.: An analysis of myogenesis in vitro, using fluorescein-labelled antimyosin. J. Histochem. Cytochem. 13, 726-741 (1965)

Pacifici, M., Holtzer, H.: Effects of tumor-promoting agent on chondrogenesis. Am. J. Anat. 150, 207-212 (1977)

Pacifici, M., Boettiger, D., Roby, K., Holtzer, H.: Transformation of chondroblasts by Rous sarcoma virus and synthesis of the sulfated proteoglycan matrix. Cell 11, 891-899 (1977)

Paterson, B., Strohman, R.: Myosin synthesis in cultures of differentiating chick embryonic skeletal muscles. Dev. Biol. 29, 113-138 (1972)

Pierce, B.: Teratocarcinoma: model for developmental concept of cancer. In: Current Topics in Developmental Biology, Vol. 2 (eds. A. Monroy, A. Moscona), pp. 223-246. New York-London: Academic Press 1967

Pierce, B., Wallace, K.: Differentiation of malignant to benign cells. Cancer Res. 34, 127-137 (1971)

Potter, V.: Phenotypic diversity in experimental hepatomas: the concept of partially blocked ontogeny. Br. J. Cancer 38, 1-21 (1978)

Roby, K., Boettiger, D., Pacifici, M., Holtzer, H.: Effects of Rous sarcoma virus in the synthetic programs of chondroblasts and retinal melanoblasts. Am. J. Anat. 147, 401-405 (1976)

Schiltz, J., Mayne, R., Holtzer, H.: The synthesis of collagen and glycosaminoglycans by dedifferentiated chondroblasts in culture. Differentiation 1, 98-108 (1973)

Toyama, Y., West, C., Holtzer, H.: Differential response of myofibrils and 100 Å filaments to a co-carcinogen. Am. J. Anat. 156, 132-138 (1979)

Turner, D., Gmur, R., Siegrist, M., Buckhardt, E., Eppenburger, H.: Differentiation in cultures derived from embryonic chicken muscle. I. Muscle-specific enzyme changes before fusion in EGTA synchronized cultures. Dev. Biol. 48, 237-239 (1976)

Weintraub, H.: The organization of red cell development. In: Results and Problems in Cell Differentiation, Vol. 7 (eds. J. Reinert, H. Holtzer), pp. 27-42. Berlin-Heidelberg-New York: Springer 1975

Weston, J., Butler, S.: Cell interaction in neural crest development. In: Cell Interactions (ed. L. Silvestri). Amsterdam-New York: Elsevier 1973

Yaffe, D., Gershon, D.: Multinucleated muscle fibers: Induction of DNA synthesis and mitosis by polyoma virus infection. Nature (London) 215, 421-424 (1967)

Yeoh, G., Holtzer, H.: The effect of cell density, conditioned medium and cytosine arabinoside on myogenesis in primary and secondary cultures. Exp. Cell Res. 104, 63-78 (1977)

Control of Genome Integrity in Terminally Differentiating and Postmitotic Aging Cells

S.P. MODAK[1] and C. UNGER-ULLMANN[2]

[1]Unit of Developmental Biology, Institute of Pharmacology, CHUV
1011 Lausanne, Switzerland
and Department of Zoology, University of Poona, Pune 411007, India
[2]Österreichische Akademie der Wissenschaften, Institut für Molekularbiologie,
Salzburg, Austria

I. Introduction

During early morphogenesis, organ- or tissue-specific "stem" cell lines are established. Initially multipotent, the progeny become committed to the expression of an increasingly restricted number of specialized phenotypes. The process leading to acquisition by cells of specialized structures and/or functions is called cell differentiation. Implicit in such a process is the progressive loss of the potential for tissue metaplasia, i.e., the ability to be reprogramed and converted into another phenotype. A rare exception to this rule in vertebrates is the Wolffian lens regeneration in lentectomized newts where fully differentiated melanocytes from the dorsal iris dedifferentiate, proliferate and then redifferentiate into lens fiber cells (Yamada 1977).

The majority of differentiated cell types never divide but some of these maintain the potential for cell proliferation, while others do not. For example, differentiated fibroblasts, lymphocytes, chondrocytes, etc., maintain the ability to reenter the cell cycle and their progeny redifferentiate into the same basic cell type. Differentiated hemopoietic and intestinal epithelial cells do not divide but the cell population is replenished through a pool of proliferating "blast" cells. In hepatocytes, the division potential is expressed upon injury. Terminally differentiated cells lose their ability to divide more or less irreversibly and can be further subdivided into those which (1) maintain a transcriptionally active genome (e.g., neurons), (2) those which are transcriptionally inactive but can be reactivated (e.g., avian erythrocytes), and (3) those which lose their genome completely while maintaining a stable differentiated state (e.g., lens fibers).

DNA is the primary carrier of genetic information and its structural integrity is a prerequisite for gene expression. The functional stability of the genome depends on the redundancy on one hand, and the cellular repair potential on the other. Impairment of genome integrity results in either a complete loss or a modification of phenotypic expression. Thus, the molecular mechanisms involved in repair of damaged portions of DNA and their restitution into functionally intact informational units are fundamental to the maintenance of the genome integrity (Modak 1972). In proliferating cell populations the damaged parental DNA is repaired via either excision (prereplication) repair, photoreactivation, strand break rejoining or postreplication repair systems. Defective repair could cause an accumulation of lesions or mutations which might be either lethal, lead to an altered phenotype, or to neoplastic transformation. In postmitotic cells such damage may stimulate cell division or impair normal cellular processes by deletion or through physical alterations in one or many structural genes and thus challenge the stability to the differentiated state.

In this review, we discuss the role of DNA repair machinery and the chromatin in controlling the integrity of the genome in relation to the stability of the differentiated state of embryonic and aging postmitotic cells.

II. DNA Damage and Repair

The double stranded α-helical structure of DNA is stabilized by the base pairing and the stacking of the aromatic surfaces of the base pairs in the center of the helix so that the anionic phosphate groups exposed on the outside interact with the cationic groups of the aqueous environment. The hydrophobic interactions between the stacked bases is stronger than the hydrogen bonding between the bases. Double-stranded DNA can be denatured into single-stranded form by changes in pH, ionic strength, temperature, as well as by the formation of DNA adducts, pyrimidine dimers, apurinic acids, etc. The nature of DNA damage induced by different physical and chemical agents and the mechanisms of their repair have been reviewed in detail elsewhere (Brash and Hart 1978; Hart et al. 1979) and only a cursory description in given below.

The excision repair involves removal of the damaged portion of DNA by a sequential action by damage-specific endonuclease and exonuclease, and a subsequent replacement by DNA polymerase catalyzed gap-filling and rejoining with ligase. This type of mechanism is involved in excision repair of UV-induced pyrimidine dimers and similar other DNA distortions and modifications. The strand break repair involves rejoining of single-strand breaks produced by agents such as ionizing radiation. Generally speaking, double-strand breaks are not repaired and constitute lethal events. Damage in the parental DNA, which is not repaired before the replication, results in inhibition of DNA synthesis at the damaged site with a subsequent decrease in the size of the newly synthesized DNA. The gaps in the daughter DNA are replicated and sealed using the damaged DNA as template by postreplication repair using one or many of a variety of different pathways (Brash and Hart 1978; Hart et al. 1979).

The excision repair seems to be error-free (Kakunaga, T. 1974; Kakunaga, P. 1975) while the postreplication repair is suggested (Witkin and George 1973; Radman 1975; Witkin 1976) to be error-prone and may be involved in mutations and carcinogenesis and, possibly, aging. The mutagenic and carcinogenic potential of any DNA damaging agent is probably initiated when these enter the replicating machinery in rapidly dividing cells defective in excision repair as in Xeroderma pigmentosum (Maher and McCormick 1976; Maher et al. 1976).

III. Integrity of the Genome in Vivo

A. Terminal Cell Differentiation

Earlier, we have examined the incorporation of radioactive deoxyribo-nucleosides catalyzed by exogenous DNA polymerase-α and terminal transferase to probe the structure of DNA in fixed cell nuclei in situ (Modak et al. 1969; Modak and Bollum 1970, 1972; Price et al. 1971; Modak 1972; Modak and Traurig 1972). During the differentiation, nuclear DNA from lens fibers and vaginal keratinizing epithelium accumulate free 3'-OH and indicative of single-strand breaks (Modak and Bollum 1970, 1972; Modak 1972; Modak and Traurig 1972). Sedimentation in alkaline sucrose gradients also revealed that in differentiating lens fibers (Piatigorsky et al. 1973; Counis et al. 1977) and erythrocytes from chick, and rat muscle (Karran and Ormerod 1973), the single strand mol.wt. of DNA decreases. In nondividing lens fibers, cell nuclei progressively lose DNA, degenerate, and disappear (Modak and Perdue 1970). The nuclear degeneration begins in centrally located lens fibers after 8 days of lens development and the wave of pycnosis spreads peripherally following a strict temporal and spatial pattern (Modak and Perdue 1970). In degenerating nuclei, DNA undergoes massive breakage (Modak et al. 1969; Modak and Bollum 1970, 1972; Modak 1972) and native DNA fragments with short denatured regions having free 3'-OH ends appear and persist for a long time after the loss of nuclear structure (Modak and Bollum 1970, 1972; Modak 1972). Electrophoresis in native gels revealed that low mol.wt. multimeric DNA fragments appear in 15-day embryonic central lens fibers containing only degenerating nuclei (Appleby and Modak 1977). Recently, we have characterized the lens fiber DNA by two-dimensional electrophoresis in gels containing Tris-acetate and NaOH-EDTA buffers, respectively (Modak 1978). We found thet the low mol.wt. multimeric DNA fragments in central fibers are unnicked native molecules while high mol.wt. DNA in both central and peripheral fibers is considerably nicked. Thus, DNA in differentiating lens fibers accumulates single strand breaks and these are later converted into double strand breaks (Modak and Beard, Nuclei Acids Res., 8 (12), in press, 1980). We now report the results from additional studies on the temporal and spatial sequence of the appearance of low mol. wt. DNA fragments in lens fibers and their biochemical nature.

1. Appearance of Low Mol.Wt. Multimeric DNA Fragments in Differentiating Lens Fiber Cells

A large scale procedure was developed for isolation of fiber cell populations at different stages of differentiation and described in Fig. 1. This way we obtain central (CF), middle (MF), and outer (OF) fiber cell populations, as well as epithelium and annular pad complex (EP). MF and OF samples were obtained from 15-, 17-, and 19-day embryonic chick lenses, homogenized and lysed in EDTA-SDS-proteinase-K as described before. Lysates were electrophoresed in 2.5% polyacrylamide and 0.5% agarose mixed gels and processed as before (Appleby and

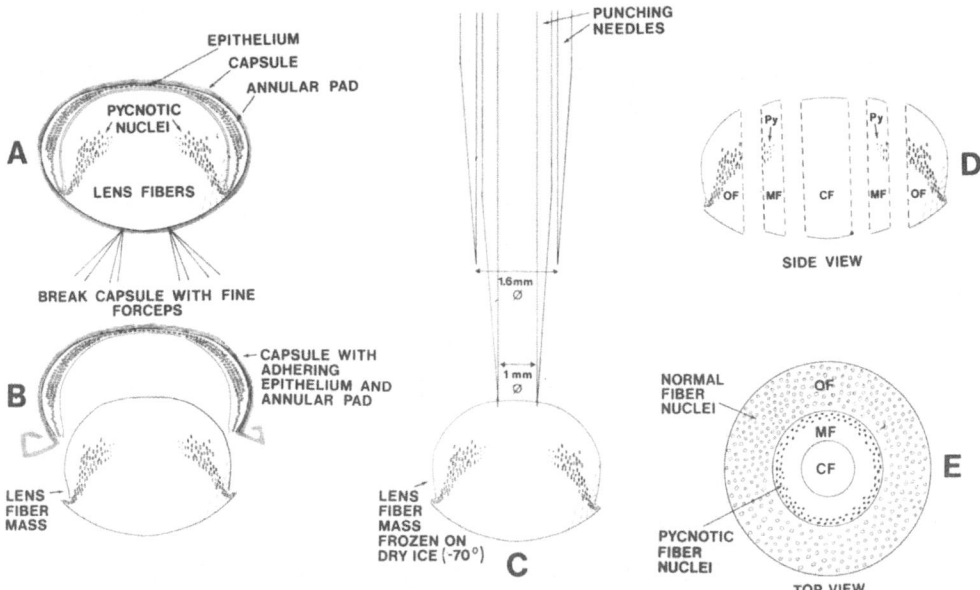

Fig. 1A-E. Dissection of chick embryo lens. **A** A schematic drawing of a section of a 19-day-old lens. The distribution and morphology of nuclei is shown, but cellular outlines are not shown. The lens capsule is punctured at the posterior pole with a pair of fine forceps. This results in a detachment of the capsule with the adhering epithelium and annular pad from the lens fiber mass. **B** Lens fiber mass is frozen in a petri dish placed on dry ice, and **C** punched successively with two hypodermic needles with inner diameters of 1 mm and 1.6 mm, respectively. We thus fractionate **(D, E)** fibers into CF-containing nonnucleated cells, MF-containing degenerating nuclei, and OF consisting of mostly transcriptionally active postmitotic nuclei

Modak 1977). We find that, while 15-day CF contain low mol.wt. multimeric DNA fragments with a unit monomeric length of 182 bp, these are absent in both MF and OF populations (Fig. 2). Low mol.wt. DNA fragments first appear in MF in 17-day lenses and persist in 19-day MF, but are absent in both 17- and 19-day OF. Since, at 17 and 19 days, the zone of nuclear degeneration corresponds to MF population and has not yet reached OF population (Modak and Perdue 1970), these results prove that at the final phases of nuclear degeneration massive DNA fragmentation occurs and this process follows the same spatial and temporal pattern as the nuclear degeneration itself.

2. Isolation and Characterization of Low Mol.Wt. Multimeric DNA-Protein Complexes

In order to understand further the molecular events involved in the production of multimeric low mol.wt. DNA fragments in degenerating lens fiber cell nuclei, we dissected 500 CF from 19-day lenses as shown in Fig. 1. Tissue was then homogenized in a buffer containing 10 mM Tris-HCl (ph 7.5), 1 mM EDTA,

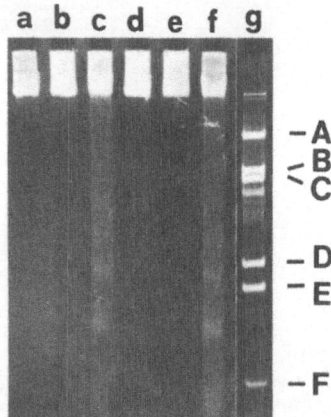

Fig. 2. Electrophoresis of DNA from 15-19-day MF and OF. 17- and 19-day MF contain monomer-tetramer size DNA fragments in addition to high mol.wt. DNA. 15-day MF and 15-19-day OF contain only high mol.wt. DNA. Gel is calibrated with Hind III restriction fragments of SV40 DNA. Unit length of monomeric DNA is found to be 170 bp for 17-day MF and 168 bp for 19-day MF. *a* 15-day MF; *b* 15-day OF; *c* 17-day MF; *d* 17-day OF; *e* 19-day OF; *f* 19-day MF; *g* SV40. Hind III restriction fragments: *A* 1154 bp; *B* 1160 bp; *C* 1105 bp; *D* 542 bp; *E* 439 bp; *F* 206 bp; *G* 158 bp (Appleby and Modak 1977)

0.5 mM EGTA, 5 mM β-mercaptoethanol, and 0.2 mM phenylmethylsulphonyl fluoride. The homogenate was first centrifuged at 1000 g (10 min, 2 °C) and the supernatant was centrifuged again at 10,000 g (15 min, 2 °C). Both pellets (P-I, P-II) and an aliquot of the supernatant (S-II) were kept aside and the remaining S-II was further centrifuged (216,000 g, 16 h, 2 °C) through a 15% sucrose (w/v) cushion. The resulting tightly packed pellet (P-III) was found to be covered by a fluffy layer (F). P-III, F, and the 15% sucrose interphase (IP) were collected separately, lysed in 1% sodium dodecyl sulfate and one half-part of each was extracted with phenol (Appleby and Modak 1977) and 0.4 N H_2SO_4 (Garrard and Bonner 1974) to obtain nucleic acids and basic proteins, respectively. In Fig. 3a we show the electrophoretic behavior of DNA in P-I, P-II, and S-II and conclude that most of high mol.wt. DNA in CF was removed in P-II fraction so that S-II contains only low mol.wt. multimeric DNA fragments. In the ultraspeed pellet P-III we find monomer-pentamer size DNA fragments of unit repeat of 160 bp (Appleby and Modak 1977) as well as 140 bp-long DNA migrating somewhat faster (Fig. 3b); neither F nor IP fractions contain DNA (not shown). The acid-soluble proteins electrophoresed on 15% SDS-polyacrylamide gels show the presence of histones H1, H3, H2B, H2A, and H4 in P-III only (Fig. 3c). There are, however, a large number of other protein bands common to P-III, F, and IP and these are lens proteins. From these experiments we conclude that the low mol.wt. multimeric DNA fragments appearing in lens fibers concomitant to the degeneration of nuclei are in fact chromatin subunits produced in vivo by breaks in the linker DNA. It should also be noted that, as compared to the distribution of histones from normal lens fiber nuclei which contain four to five times more histone H1 II than H1 I (Unger-Ullmann and

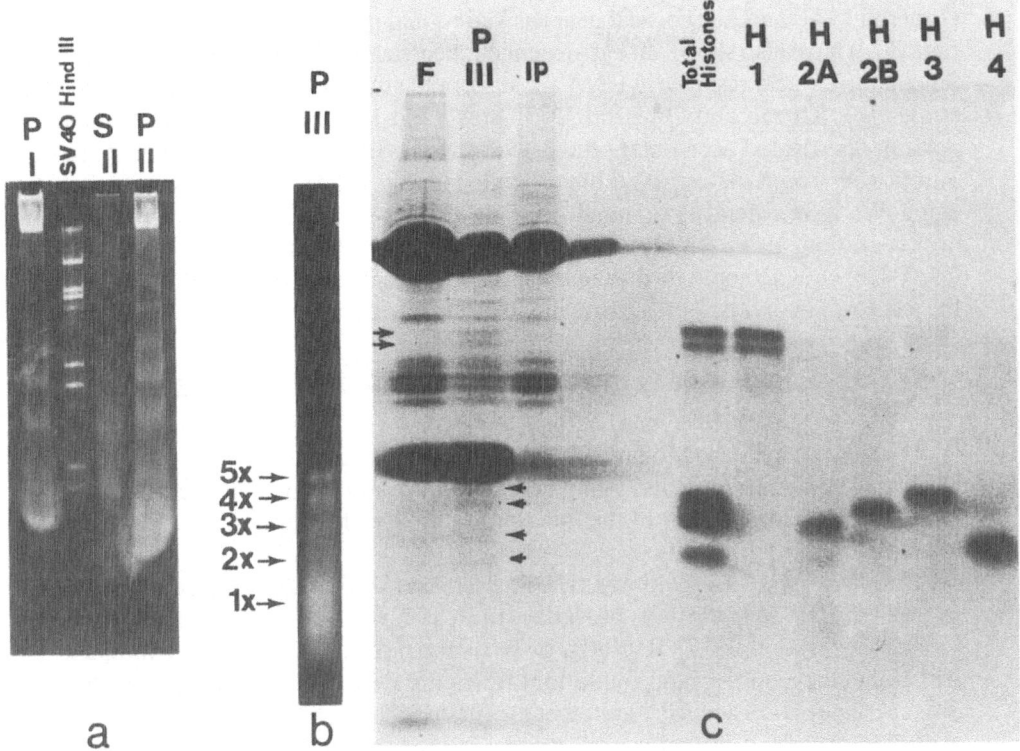

Fig. 3a-c. Gel analysis of CF DNA fractionated by differential centrifugation. CF homogenates are centrifuged at 1000 g, 10,000 g, and 216,000 g. **a** Fractions S-II, P-I and P-II are lysed, phenol extracted and electrophoresed in 1.6% native agarose gels and visualized with ethedium bromide staining. SV40 DNA Hind III restriction fragments are used to calibrate the gel. **b** Gel analysis of DNA in P-III. Note the absence of high mol. wt. DNA. **c** 0.4 N H_2SO_4-extracted protein was electrophoresed in 15% polyacrylamide gel using SDS-buffer (Unger-Ullmann and Modak 1979). Calf thymus histones were used as markers. *Arrows* show the location of histones H1 I, H1 II, H3, H2B, H2A, and H4, in the descending order found only in P-III fraction. Traces g H3, H2B, H2A and H4 are found in the fluffy layer *(F)*, while sucrose interphase *(IP)* completely lacks histones

Modak 1979), low mol.wt. chromatin fragments from CF contain both forms of H1 in roughly equal proportion. These results probably reflect significant differences in the overall conformation of transcriptionally active chromatin from OF and transcriptionally inactive and condensed chromatin from CF.

B. Postmitotic Aging Cells

Analysis of the template activity for DNA polymerase-α in cerebral cortical neurons, liver Kupffer cells, and cardiac muscle from 3- and 30-month-old mice revealed that DNA in these cells accumulated gaps and strand breaks with age (Modak and Price 1971; Price et al. 1971). Sedimentation in alkaline sucrose gradients of DNA from rat liver (Massie et al. 1972), rat muscle (Karran and Ormerod 1973) and dog cortical neurons and retinal photoreceptors (Wheeler and Lett 1974) have also shown an age-dependent decrease in the single-strand mol.wt. Since apurinic acid residues in DNA are susceptible to alkali, these studies (Massie et al. 1972; Karran and Ormerod 1973; Brash and Hart 1978) do not exclude a possible age-related accumulation of such residues. In conclusion, DNA in post-mitotic cells accumulates strand breaks and, possibly, other lesions during in vivo aging. We have recently examined (Modak, Gfeller-Leuba, Correa and Modak, in preparation, 1980) the DNA in cortical neurons of aging mice and find that free 3'-OH ends accumulate during the last quarter of the life span. In this context, the situation in dividing cells aging in vitro remains unexplored.

IV. DNA Repair in Terminally Differentiating and Aging Cells

We have postulated (Modak 1972) that the accumulation of lesions in DNA of terminally differentiating and aging cells is due to a defective repair machinery, resulting in the impairment of the functional stability of the genome by adversely affecting its transcribability.

Unscheduled DNA synthesis (UDS) decreases during the differentiation of muscle cells (Hahn et al. 1971; Stockdale 1971, 1972; Chan et al. 1976) and neural retina (Karran et al. 1977). Recently, we have measured the UDS in UV-irradiated lens from chick embryo and found that fibers are repair-inactive while epithelial cells carry out UDS actively (Tréton et al. 1979). DNA strand break rejoining has also been examined in various cell types. Chicken erythrocytes, rat muscle cells (Karran and Ormerod 1973), and chick lens fibers (Counis et al. 1977) do not rejoin X-ray-induced strand breaks, while rat spleen cells and thymocytes (Karran and Ormerod 1973), chick lens epithelial cells (Counis et al. 1977), and dog neurons (Wheeler and Lett 1974) do so effectively. On the other hand, rat neurons are deficient in excision repair of DNA adducts induced by ethylnitroso urea (Goth and Rajewsky 1974).

During the cellular aging in vitro excision repair or UDS decreases significantly in late passage cells (Goldstein 1971; Mattern and Cerutti 1975; Hart and Setlow 1976; Little 1976; Milo and Hart 1976).

Since a considerable number of enzymatic steps are involved in the excision and postreplication repair, a single defect in any one of these would result in a defective restitution of damaged DNA. In differentiating skeletal muscle (Stockdale 1970) and in aging mouse spleen (Barton and Yang 1975), DNA polymerase activity decreases and, in the latter, the DNA polymerase-β loses its fidelity of copying synthetic templates. A similar situation is found for DNA polymerases in diploid human fibroblasts (MRC-5) senescing in vitro (Linn et al. 1976).

V. The Role of Chromatin in the Conservation of Genome-Integrity

Eukaryotic DNA exists as a deoxyribonucleoprotein complex or chromatin (Ruiz-Carillo et al. 1975). The chromatin backbone is now well characterized and consists of a flexible string of closely packed beads or discs, also called "nu bodies" or "nucleosomes" (Olins and Olins 1974; Bradbury 1977; Thomas 1977). Each nucleosome contains a stretch of DNA (Hewish and Burgoyne 1973; Noll 1974) of a length varying between 165–212 bp among different species (Bradbury 1977; Thomas 1977) and a histone octamer composed of two molecules each of H2A, H2B, H3, and H4 (Kornberg 1974; Kornberg and Thomas 1974), to which is tightly bound a 140 bp long DNA giving rise to the "core particle" (Axel 1975; Sollner-Webb and Felsenfeld 1975), while the remaining 25–72 bp form the linker. Histone H1 is bound to the linker DNA (Shaw et al. 1976; Varshavsky et al. 1976; Whitlock and Simpson 1976; Todd and Garrard 1977). The size of DNA per nucleosome may vary among different tissues of the same organism (Morris 1976), different cell types in the same organ (Todd and Garrard 1977), and in cells at different stages of differentiation (Lohr et al. 1977; Weintraub 1978).

The linker DNA is preferentially hydrolyzed by endogenous (Hewish and Burgoyne 1973) or exogenous (Noll 1974) Ca^{++}/Mg^{++}-dependant nucleases giving rise to a discrete multimeric series of DNA fragments. Such fragments are also produced in vivo in radiation-induced cell death (Modak et al. 1976; Skalka et al. 1976). In differentiating lens fibers, strand breaks affect nuclear DNA at sites between nucleosomes (present data: Appleby and Modak 1977; Modak 1978).

In postmitotic cells 50% of chromatin-DNA is sensitive to staphylococcal nuclease (Clark and Felsenfeld 1971, 1972; Axel 1975; Sollner-Webb and Felsenfeld 1975; Modak et al. 1978), while in senescent mouse liver this fraction decreases to 38% (Modak et al. 1978). Conversely, as much as 80% of DNA in cells growing in vitro is sensitive to this enzyme (Compton et al. 1975; Bradbury 1977; Modak and D'Ambrosio, unpublished data). In a limit digest using this enzyme, DNA bound to the core particle has been shown to be resistant (Axel 1975; Sollner-Webb and Felsenfeld 1975) indicating that the differences in the nuclease-sensitive fraction among different cell types probably reflect differences in the number of nucleosomes per unit length of DNA (Modak et al. 1978). This suggests that compared with the postmitotic cells the chromatin in dividing cells either contains less nucleosomes or the DNA histone interactions in them are considerably weaker. Conversely, senescent postmitotic cell-chromatin may contain either a greater concentration of nucleosomes or it may be of a more rigid structure. It is proposed (Appleby and Modak 1977) that the histone octamer may either slide along the DNA or readily dissociate and render accessible to nuclease the previously octamer-bound DNA regions. Weintraub et al. (1976) have proposed that nucleosomes may transiently dissociate into half-nucleosomes and this may also explain the observed differences in the resistance of chromatin DNA to staphylococcal nuclease among dividing young postmitotic and senescent postmitotic cells (Modak et al. 1978).

Wilkins and Hart (1973) found that a significant portion of pyrimidine dimers induced by UV are masked in chromatin and can be made accessible to UV endonuclease by a pretreatment of cells with high salt, and suggested that the sites accessible in low salt are preferentially repaired. Indeed, newly repaired segments of DNA in UV-irradiated cells seem to be located in staphylococcal nuclease-sensitive regions of the chromatin (Cleaver 1977). Carcinogens, methylmethane sulfonate (MMS) (Bodell 1977), and Benzo(A)pyrene diol epoxide (Koostra et al. 1978) seem to bind preferentially to linker DNA, and MMS adducts seem to be excised preferentially from this region (Bodell 1977). While these studies definitely suggest that the chromatin structure controls the accessibility of the damaged site to DNA repair enzymes, in actual practice the situation is obviously far more complex. For example, in V79 cells 85–90% pyrimidine dimers remain unexcised after 6 h (D'Ambrosio and Setlow 1979) although only 20% of DNA is resistant to staphylococcal nuclease (Modak and D'Ambrosio, unpublished data). Thus, any model conferring upon the chromatin backbone the role of controlling the accessibility of damaged DNA to repair enzymes is naive unless the role of various interactions among DNA, histones, nonhistones, lipids, and the ionic microenvironment on one hand, and the repair enzyme activity on the other, are taken into account. Since UV also induce DNA-protein crosslinks, these will tend to immobilize the affected regions of chromatin. Finally, it is often forgotten that enzymes involved in DNA replication, transcription, and repair constitute large macromolecular complexes (at least as large as nucleosomes) and must require at least a temporary local destabilization of the DNA-histone complex. Indeed, the template-active regions of chromatin seem to be structurally distinct from the inactive regions (Axel et al. 1973; Garel and Axel 1973; Gilmour and Paul 1973; Weintraub and Groudine 1978).

VI. The Biological Significance of DNA Repair Potential

Since there exist numerous forms of DNA damage and DNA repair systems, a defect in any one of the steps in repair would result in accumulation of the lesion which would, however, be expressed only when it enters the replication or transcription complex, giving rise to either an altered genotype or phenotype, and may cause neoplasia or cell death. The data reviewed in the preceding sections clearly indicate that the nature of association between DNA and histones also control the accessibility of the damaged DNA site to repair enzymes as well as the accessibility of DNA to chemicals causing the damage.

Nevertheless, the cellular DNA repair potential seems to be one of the decisive elements controlling the restitution of functionally active genome and thus its life-span. In early mouse embryo the level of excision repair activity is high and it decreases to an apparent steady state towards the end of gestation (Peleg et al. 1977). This steady-state repair potential in young animals is found to be different among different placental mammals (Hart and Setlow 1974; Sacher and Hart 1978) and primates (Hall and Hart, unpublished 1978), and correlates to their life-span. At least for UDS induced by UV, it seems that the differences in the length of the life-span are related to the differences in the regulatory potential achieved toward the end of fetal life.

Limited information is available at present on the precise sequence of differentiation- and age-dependent decrease in the repair potential in postmitotic cells. Differentiated cells differ considerably in their cell proliferative potential as well as repair activity. Since the stability of the differentiated state depends not only on the repair capacity but also on the proliferation potential, we feel that these are closely related, and a loss of either would be tantamount to the onset of senescence, i.e., inability to maintain genome integrity, with consequent cell death. The proposed relationship, however, offers the paradox whereby the reentry into cell cycle would destabilize the differentiated state, while active error-free repair would do the inverse. In this context, it would be interesting to know whether in lower vertebrates showing higher incidence of tissue metaplasia the relative levels of repair activity show a much smaller difference among embryonic and adult cells.

It thus seems that the repair potential not only controls the stability of the differentiated state but also the maximum achivable life-span. Recent work by Smith-Sonneborn (1978) on *Paramecia* provides a spectacular demonstration that the effeciency of DNA repair mechanism is decisive in controlling the life-span. She found that UV irradiation of *Paramecia* and their subsequent growth in dark shortened life-span, while photoreactivation subsequent to UV irradiation resulted in the prolongation of the life-span.

VII. Summary

The postulate that terminally differentiating and aging postmitotic cells accumulate lesions in their DNA due to defective DNA repair enzyme machinery (Modak 1972) is supported by the new evidence gathered from a variety of cell and organ systems during the past 6 years.

New data are presented to show that the nuclear degeneration and the appearance of a repeating class of low mol.wt. DNA fragments in terminally differentiating lens fiber cells are both spatially and temporally related. Furthermore, it is shown that these fragments are produced by the digestion of lens fiber chromatin in vivo.

In view of recent knowledge on the structure of eukaryotic chromatin, we suggest that the chromatin structure controls the accessibility of the damaged DNA site for repair enzymes. Thus, the original postulate is modified to suggest that both the level of DNA repair enzymes and the chromatin structure control the maintainance of genome integrity.

The biological significance of DNA repair potential and cell proliferation potential is discussed in relation to the stability of the differentiated state.

DNA repair machinery holds the central place not only in the conservation of genome integrity, but also in defining the maximum achievable life-span for any species.

Acknowledgments. We are truly indebted to Dr. Tuneo Yamada, Swiss Institute for Experimental Cancer Research, and Dr. Ronald W. Hart, Ohio State University, Columbus, Ohio, USA, for critical discussions. We thank Mrs. Monique Reinhardt-Chappuis for expert technical assistance and Mr. Pierre Dubied for the art work.

This work was supported by grants 3.538.75 and 3.275.78 from the Swiss National Science Foundation.

References

Appleby, D.W., Modak, S.P.: DNA degradation in terminally differentiating lens fiber cells from chick embryos. Proc. Natl. Acad. Sci. USA 74, 5579-5583 (1977)
Axel, R.: Cleavage of DNA in nuclei and chromatin with staphylococcal nuclease. Biochemistry 14, 2921-2925 (1975)

Axel, R., Cedar, H., Felsenfeld, G.: Synthesis of globin ribonucleic acid from duck-reticulocyte chromatin in vitro. Proc. Natl. Acad. Sci. USA 70, 2029-2032 (1973)

Barton, R.W., Yang, W.K.: Low molecular weight DNA polymerase decreased activity in spleens of old Balb/c mice. Mech. Ageing Dev. 4, 123-136 (1975)

Bodell, W.J.: Non-uniform distribution of DNA repair in chromatin after treatment with methyl methanesulfonate. Nucleic Acids Res. 4, 2619-2628 (1977)

Bradbury, E.M.: Histone interactions and chromatin structure. In: The organization and expression of eukaryotic genome (eds. E.M. Bradbury, K. Javaherian), pp. 99-123. New York-London: Academic Press 1977

Brash, D.E., Hart, R.W.: DNA damage and repair in vivo. J. Environmental Pathol. Toxicol. 2, 79-114 (1978)

Chan, A.C., Ng, S.K.C., Walker, I.G.: Reduced DNA repair during differentiation of a myogenic cell line. J. Cell Biol. 70, 685-691 (1976)

Clark, R.J., Felsenfeld, G.: Structure of chromatin. Nature New Biol. 229, 101-106 (1971)

Clark, R.J., Felsenfeld, G.: Association of arginine-rich histones with GC-rich regions of DNA. Nature New Biol. 240, 226-229 (1972)

Cleaver, J.E.: Nucleosome structure controls rates of excision repair in DNA of human cells. Nature (London) 270, 451-453 (1977)

Compton, J.L., Hancock, R., Oudet, P., Chambon, P.: Biochemical and electron microscopic evidence that the subunit structure of chinese-hamster-ovary interphase chromatin is conserved in mitotic chromosomes. Eur. J. Biochem. 70, 555-568 (1975)

Counis, M.-F., Chadun, E., Courtois, Y.: DNA snythesis and repair in terminally differentiating embryonic lens cells. Dev. Biol. 57, 47-55 (1977)

D'Ambrosio, S.M., Setlow, R.B.: On the presence of UV-endonuclease sensitive site in mammalian DNA. In: DNA repair mechanisms (eds. P.C. Hanawelt, E. Freidberg, C.F. Fox), Vol. IX, pp. 499-503. New York: Academic Press 1978

Garel, A., Axel, R.: Selective digestion of transcriptionally active ovalbumin genes from oviduct nuclei. Proc. Natl. Acad. Sci. USA 73, 3966-3970 (1973)

Garrard, W.D., Bonner, J.: Changes in chromatin proteins during liver regeneration. J. Biol. Chem. 249, 3729-3736 (1974)

Gilmour, R.S., Paul, J.: Tissue-specific transcription of the globin gene in isolated chromatin. Proc. Natl. Acad. Sci. USA 70, 3440-3442 (1973)

Goldstein, S.: The biology of aging. New Engl. J. Med. 285, 1120-1129 (1971)

Goth, R., Rajewsky, M.F.: Persistence of 0^6-ethylguanine in rat brain DNA: correlation with nervous system-specific carcinogenesis by ethylnitrosourea. Proc. Natl. Acad. Sci. USA 71, 639-643 (1974)

Hahn, F.M., King, D., Yang, S.J.: Quantitative changes in unscheduled DNA synthesis in rat muscle cells after differentiation. Nature New Biol. 230, 242-244 (1971)

Hart, R.W., Setlow, R.B.: Correlation between deoxyribonucleic acid excision repair and life-span in a number of mammalian species. Proc. Natl. Acad. Sci. USA 71, 2169-2173 (1974)

Hart, R.W., Setlow, R.B.: DNA repair in late-passage human cells. Mech. Ageing Dev. 5, 67-7 1976)

Hart, R.W., D'Ambrosio, S.M., Ng, K.K., Modak, S.P.: Longevity, stability and DNA repair. Mech. Ageing Dev. 9, 203-223 (1979)

Hewish, D.R., Burgoyne, L.A.: Chromatin sub-structure. The digestion of chromatin DNA at regularly spaced sites by a nuclear deoxyribonuclease. Biochem. Biophys. Res. Commun. 52, 504-510 (1973)

Kakunaga, R.: The role of cell division in the malignant transformation of mouse cells treated with 3-methylcholanthrene. Cancer Res. 35, 1637-1642 (1975)

Kakunaga, T.: Requirement for cell replication in the fixation and expression of the transformed state in mouse cells treated with 4-nitroquinolin-1-oxide. Int. J. Cancer 14, 736-742 (1974)

Karran, P., Ormcrod, M.G.: Is the ability to repair damage to DNA related to the proliferative capacity of a cell? The rejoining of X-ray produced strand breaks. Biochim. Biophys. Acta 299, 54-64 (1973)

Karran, P., Moscona, A., Strauss, B.: Developmental decline in DNA repair in neural retina cells of chick embryos. J. Cell Biol. 74, 274-286 (1977)

Koostra, A., Slaga, T.J., Olins, D.E.: Binding of Benzo(A) pyrene diol epoxide to chromatin. Biophys. J. 21, 67a (1978)

Kornberg, R.D.: Chromatin structure: A repeating unit of histones and DNA. Science 184, 868-871 (1974)

Kornberg, R.D., Thomas, J.O.: Chromatin structure oligomers of the histones. Science 184, 865-868 (1974)

Linn, S., Kairis, M., Holliday, R.: Decreased fidelity of DNA polymerase activity isolated from aging human fibroblasts. Proc. Natl. Acad. Sci. USA 73, 2818-2822 (1976)

Little, J.B.: Relationship between DNA repair capacity and cellular aging. Gerontology 22, 28-55 (1976)

Lohr, D., Corden, J., Tatchell, K., Kovacic, R.T., van Holde, K.E.: Comparative subunit structure of HeLa, yeast and chicken erythrocyte chromatin. Proc. Natl. Acad. Sci. USA 74, 79-83 (1977)

Maher, V.M., McCormick, J.J.: Effect of DNA repair on the cytotoxicity and mutagenesis of UV radiation and chemical carcinogens in normal and Xeroderma pigmentosum cells. In: Biology of Radiation Carcinogenesis (eds. J.M. Yuhas, R.W. Tennant, J. Regan), pp. 129-145. New York: Raven Press 1976

Maher, V.M., Quellette, L.M., Curren, R.D., McCormick, J.J.: Frequency of UV-light-induced mutation is higher in Xeroderma Pigmentosum variant cells than in normal cells. Nature (London) 261, 593-595 (1976)

Massie, H.R., Baird, M.B., Nicolosi, R.J.: Changes in the structure of rat liver DNA in relation to age. Arch. Biochem. Biophys. 153, 736-741 (1972)

Mattern, M.R., Cerutti, P.A.: Selective excision of gammay ray damaged thymine from the DNA of cultured mammalian cells. Biochim. Biophys. Acta 395, 48-55 (1975)

Milo, G.E., Hart, R.W.: Age-related alterations in plasma membrane glycoprotein content and scheduled or unscheduled DNA synthesis. Arch. Biochem. Biophys. 176, 324-333 (1976)

Modak, S.P.: A model for transcriptional control in terminally differentiating lens fiber cells. In: Cell Differentiation (eds. R. Harris, P. Allin, D. Viza), pp. 339-342. Copenhagen: Munksgaard 1972

Modak, S.P.: Two-dimensional electrophoresis of native and denatured DNA from chromatin digests. Experiena 34, 57 (1978)

Modak, S.P., Bollum, F.J.: Terminal lens cell differentiation. III. Initiator activity of DNA during nuclear degeneration. Exp. Cell Res. 62, 421-432 (1970)

Modak, S.P., Bollum, F.J.: Detection and measurement of single-strand breaks in nuclear DNA in fixed lens sections. Exp. Cell Res. 75, 307-313 (1972)

Modak, S.P., Perdue, S.W.: Terminal lens cell differentiation. I. Histological and microspectrophotometric analyses of nuclear degeneration. Exp. Cell Res. 59, 43-56 (1970)

Modak, S.P., Price, G.B.: Exogenous DNA polymerase-catalyzed incorporation of deoxyribonucleotide monophosphates in nuclei of fixed mouse brain cells: Changes associated with age and X-irradiation. Exp. Cell Res. 65, 289-298 (1971)

Modak, S.P., Traurig, H.: Appearance of strand breaks in the nuclear DNA of terminally differentiating vaginal epithelium. Cell Differ. 2, 351-355 (1972)

Modak, S.P., von Borstel, R.C., Bollum, F.J.: Terminal lens cells differentiation. II. Template activity of DNA during nuclear degeneration. Exp. Cell Res. 56, 105-113 (1969)

Modak, S.P., Appleby, D.W., Chappuis, M.: Cytoplasmic informational DNA: Fact or Fantasy? J. Cell Biol. 70, 140a (1976)

Modak, S.P., Gonet, C., Unger-Ullmann, C., Chappuis, M.: Chromatin structure in aging mouse liver. Experiena 34, 57 (1978)

Morris, N.R.: A comparison of the structure of chick erythrocyte and chick liver chromatin. Cell 9, 627-632 (1976)

Noll, M.: Subunit structure of chromatin. Nature (London) 251, 249-251 (1974)

Olins, A.L., Olins, D.E.: Spheroid chromatin units (v Bodies). Science 183, 330-332 (1974)

Peleg, L., Raz, E., Ben-Ishai, R.: Changing capacity for DNA excision-repair in mouse embryonic cells in vitro. Exp. Cell Res. 104, 301-307 (1977)

Piatigorsky, J., Rothschild, S.S., Milstone, L.M.: Differentiation of lens fibers in explanted embryonic chick lens epithelia. Dev. Biol. 34, 334-345 (1973)

Price, G.B., Modak, S.P., Makinodan, T.: Age-associated changes in the DNA of mouse tissue. Science 171, 917-920 (1971)

Radman, M.: SOS repair hypothesis: Phenomenology of an inducible DNA repair which is accompanied by mutagenesis. In: Molecular Mechanisms for Repair of DNA (eds. P.C. Hanawalt, R.B. Setlow), pp. 355-368. New York: Plenum Press 1975

Ruiz-Carrillo, A., Wangh, L.J., Allfrey, V.G., Processing of newly snythesized histone molecules. Science 190, 117-128 (1975)

Sacher, G.A., Hart, R.W.: Longevity, aging and comparative cellular and molecular biology of the house mouse *Mus musculus* and the white footed mouse, *Peromyscus leucopus*. In: Birth Defects—Original Article Ser. 14, 71-98 (1978)

Shaw, B.R., Herman, T.M., Kovacic, R.T., Beaudreau, G.S., van Holde, K.E.: Analysis of subunit organization in chicken erythrocyte chromatin. Proc. Natl. Acad. Sci. USA 73, 505-509 (1976)

Skalka, M., Matyasova, J., Cejkova, M.: DNA in chromatin of irradiated lymphoid tissues degraded in vivo into regular fragments. FEBS Lett. 72, 271-274 (1976)

Sollner-Webb, B., Felsenfeld, G.: A comparison of the digestion of nuclei and chromatin by staphylococcal nuclease. Biochemistry 14, 2915-2920 (1975)

Smith-Sonneborn, J.: DNA repair and longevity assurance in *Paramecium tetraurelia*. Science 203, 1115-1117 (1978)

Stockdale, F.E.: Changing levels of DNA polymerase activity during the development of skeletal muscle tissue in vivo. Dev. Biol. 21, 462-474 (1970)

Stockdale, F.E.: DNA synthesis in differentiating skeletal muscle cells: initiation by ultraviolet light. Science 171, 1145-1147 (1971)

Stockdale, F.E., O'Neill, M.C.: Repair DNA synthesis in differentiated embryonic muscle cells. J. Cell Biol. 52, 589-597 (1972)

Thomas, J.O.: Aspects of the structure of chromatin. In: The Organization and expression of eukaryotic genome (eds. E.M. Bradbury, K. Javaherian), pp. 83-98. New York-London: Academic Press 1977

Todd, R.D., Garrard, W.T.: Two-dimensional electrophoretic analysis of polynucleosomes. J. Biol. Chem. 252, 4729-4738 (1977)

Tréton, J., Modak, S.P., Courtois, Y.: Analysis of thimidine incorporation in the DNA of chick embryo lens epithelium and lens fibers irradiated with ultraviolet light. Exp. Eye Res., in press (1980)

Unger-Ullmann, C., Modak, S.P.: Gel electrophoretic analysis of histones in late chick embryo lens epithelium, lens fiber, liver, brain and erythrocyte. Differentiation 12, 135-144 (1979)

Varshavsky, A.J., Bakayev, V.V., Gerogiev, G.P.: Heterogeneity of chromatin subunits in vitro and location of histone H1. Nucleic Acids Res. 3, 477-492 (1976)

Weintraub, H.: The nucleosome repeat length increases during erythropoiesis in the chick. Nucleic Acid Res. 5, 1179-1188 (1978)

Weintraub, H., Groudine, M.: Chromosomal subunits in active genes have an altered conformation. Science 193, 848-856 (1976)

Weintraub, H., Worcel, A., Alberts, B.: A model for chromatin based upon two symmetrically paired half-nucleosomes. Cell 9, 409-417 (1976)

Wheeler, K.T., Lett, J.T.: On the possibility that DNA repair is related to age in nondividing cells. Proc. Natl. Acad. Sci. USA 71, 1862-1865 (1974)

Whitlock, J.P., Simpson, R.T.: Removal of histone H1 exposes a fifty base pair DNA segment between nucleosomes. Biochemistry 15, 3307-3314 (1976)

Wilkins, R.J., Hart, R.W.: Preferential DNA repair in human cells. Nature New Biol. 247, 35-36 (1973)

Witkin, E.M.: Ultraviolet-induced mutation and inducible DNA repair in Escherichia coli. Bacteriol. Rev. 40, 869-907 (1976)

Witkin, E.M., George, D.L.: Ultraviolet mutagenesis in Pol A and Uvr A Pol h derivatives of Escherichia coli B/R: Evidence for an inducible error-prone repair system. Genetics 73 (Suppl.), 91-108 (1973)

Yamada, T.: Control mechanisms in cell-type conversion in Newt lens regeneration. In: Monographs in Developmental Biology, Vol. XIII (ed. A. Wolsky), pp. 1-126. Basel: Karger 1977

On RNA Action in Differentiation:
Induction and Differentiation of Somites in Chick Embryo

S. RANZI

Zoological Department of the University, Milano, Italy

An application of molecular biology to embryological problems is the study of the action of RNA and particularly of the messengers (mRNA's) in the developmental processes (Niu and Segal 1974).

In previous research work, we (Ranzi et al. 1961) extracted ribonucleoproteins from liver, heart, muscles, kidney and the glandular stomach of adult chickens. The ribonucleoproteins were extracted following the Niu method. An allantois explanted from a 4-day-old chick embryo was filled with these ribonucleoproteins and grafted on the chorioallantois membrane of an 8-day chick embryo (Fig. 1). Some cells of the chorioallantois were transformed in 30% of the treated embryos. Glandular cells synthesizing glycogen are induced with liver ribonucleoproteins in the chorioallantois. Multinucleate elements with fibrillar cytoplasm are induced by heart ribonucleoproteins. The skeletal-muscle ribonucleoprotein induces the transformation of mesenchyme cells into muscular structures, on which cross-striated elements may also appear. Hollow vesicles covered with epithelia, that remind us of the epithelia of the nephric ducts, are induced by kidney ribonucleoprotein. Secreting epithelial cells are induced by glandular stomach ribonucleoprotein.

More recently, we prepared a myosin 26S mRNA extract from muscles of a 14-day-old chick embryo following the procedure of Heywood and Rourke (1974). This messenger encodes the heaviest chain of the myosin molecule. We tested the action of this mRNA on the development of the young chicken embryo (Cigada Leonardi et al. 1977).

Fertilized chick eggs were incubated at 38.5° C for 24 h until they reached the primitive streak stage. Older embryos were discarded. The embryos at primitive streak stage were cultured in vitro following New's method (1955). Some embryos were cultured as one type of experimental control. In other embryos, the anterior 0.6 mm part was excised and the remaining part (postnodal piece) was cultured either in Pannet-Compton solution and albumen as controls (Fig. 2) or in Pannet-Compton or albumen solution with mRNA. The RNA (40–60 µg/ml in Pannet-Compton solution) was diluted four to six times in culture medium both under and upon the explant; consequently each explant received 10 µg of RNA.

The first series of explants was treated with myosin 26S mRNA extracted from a leg muscle of a 14-day-old chick embryo following the method of Heywood and

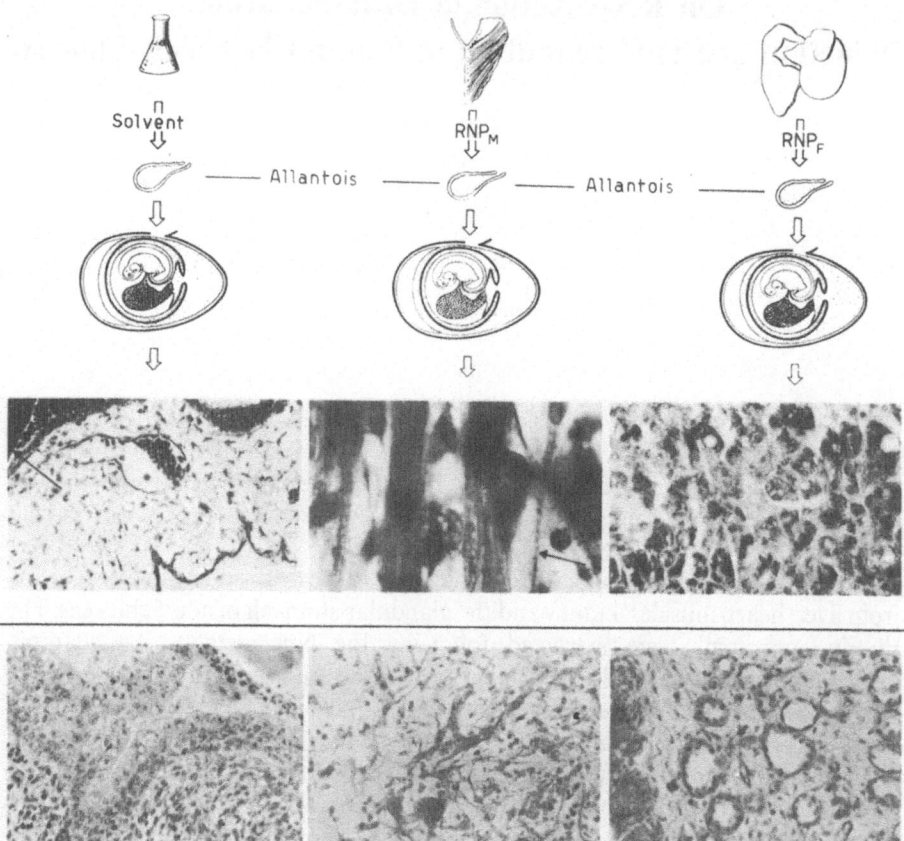

Fig. 1. Changes in chorioallantoic membrane induced by an allantois filled up with ribonucleoproteins of breast muscles RNP_M or with ribonucleoproteins of liver RNR_F. Under the
line from left: action of RNP of glandular stomach, heart, kidney

Rourke (1974). The explants were incubated at 38.5° C. After 20 h, 37.2% of treated
explants showed roundish mesodermic masses (Fig. 2). These masses were usually
found to be arranged in a single series, one near the other; the number of the masses
is different in the different cases and it varied from 2 or 3 to over 10, once it was 21.
These masses can be double-paired. There were never more than five pairs if the
explants were not stretched crosswise, but this seldom happened. These masses are
larger than the normal somites of the control embryos.

The area of the sixth somite in the control embryo is $3281.3 \pm 331.0 \, \mu^2$, and the
area of induced mesodermic masses is $5625.0 \pm 885.1 \, \mu^2$; the somite thickness of
control embryos is less than the thickness of the induced masses, consequently the
volume of the induced masses is greater than that of the somites. In the crosswise
stretched explants the induced mesodermic masses are laid out on the anterior edge
of the explant.

In section (Fig. 2) these masses seem to be a large somite. The cells appear
stretched, converging toward the center, where the mitoses are more frequent. This
arrangement corresponds to the arrangement which can be observed in the normal
somites. When examined under the electron microscope, the stretched mesodermal
cells appear to be young myoblasts like the normal myoblasts described by Shimada

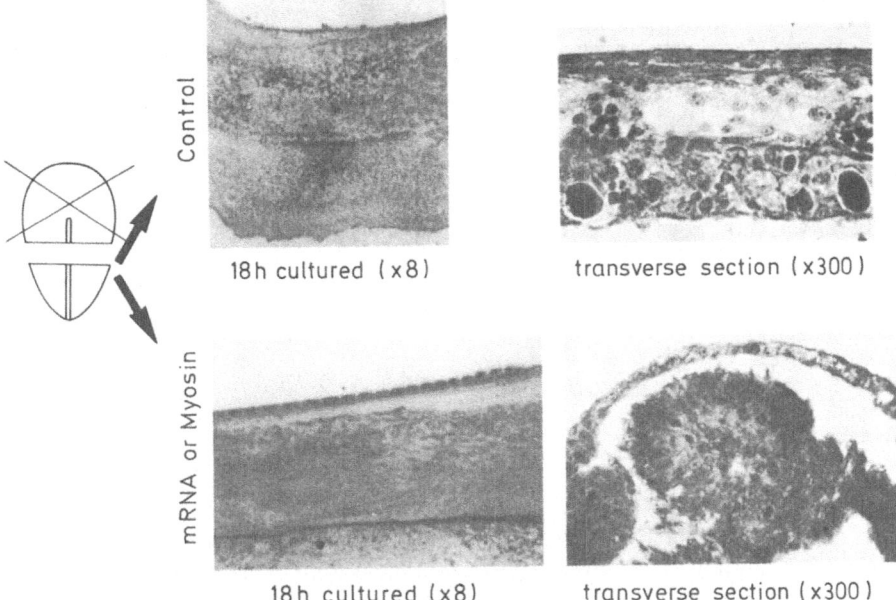

Fig. 2. Experiment on action of myosin RNA or of myosin on the development of postnodal explants of chick embryo

(1971) with numerous mitochondria and ribosomes in their basal region, distal in respect of the somite center.

In a second series of experiments, pooled mRNA's from embryonic muscle were used in the same concentration. This pool was separated on a Millipore filter following Brawerman (1974) or on Poly(U)-Sepharose column. In this pool, messengers of myosin and other muscular protein are present (Heywood and Rich 1968). The results were identical to those obtained in the first experiment. The frequency of somite induction (52.8%) seems to be greater.

Controls were performed in order to test the specificity of the reaction. For this purpose, postnodal explants were treated with mRNA pool extracted from chick liver or kidney. No mesodermic masses appeared. In contrast, mRNA extracted from rabbit muscles was able to induce somites in postnodal explants of blastoderm of chick embryo (Table 1). This observation is in complete accordance both with the cytodifferentiation of muscular elements in the chorioallantois of chicks obtained with ribonucleoprotein extracted from muscles of frogs (Ranzi et al. 1961) and rabbits (unpublished research), and with the observation of Gurdon et al. (1974) about the translation of globin mRNA of rabbit in *Xenopus* eggs.

Induced somites are stable for about 24 h. After this time they degenerate and disappear; in some cases they remain like undifferentiated vesicles.

We can conclude that myosin mRNA induces the formation of myoblasts. These myoblasts aggregate to form somites; the processes of embryonic cell aggregation are well known (Moscona and Moscona 1952). Notochord and the ventral part of the neural tube are necessary for the further development of somites (Packard and Jacobson 1976). The factor which comes from the rudiments is not present in our explants, consequently the somites induced in this way become small vesicles structured of dedifferentiated cells.

Table 1. Frequency of somite inductions in postnodal pieces of chick blastoderm treated with different extracts

	Explanted embryos	Surviving explants	Inductions	% Induction
Controls	83	56	0	0
Myosin mRNA	126	86	32	37.2
mRNA pool: muscle	204	144	76	52.8
mRNA pool: liver	12	8	0	0
mRNA pool: kidney	9	6	0	0
Myosin mRNA (rabbit)	8	6	4	66.7
Soluble myosin	23	20	8	40.0
Myosin mRNA + puromycin	30	9	0	0
Myosin + puromycin	9	6	3	50.0
F actin	5	4	0	0
G actin	18	14	0	0

This conclusion led us to test the action of soluble myosin. Chick muscle myosin was extracted following the procedure of Mommaerts and Parrish (1951) and put in culture medium of postnodal explants (15–10 µg/ml). Somites were induced in these explants.

We also wanted to test if the translation of myosin mRNA to myosin is necessary for the induction of somites. The translation can be blocked by puromycin. Postnodal explants were consequently treated with myosin mRNA (10 µg/ml) and puromycin (5 µg/ml). No mesodermic masses appeared.

In order to exclude that the puromycin blocks the induction of the somites, we put the explants in culture with myosin (15–10 µg/ml) and puromycin (5 µg/ml). We obtained 33% induction. We can conclude that the myosin induces somites without detectable new protein synthesis.

The induction obtained with the pool of muscle messengers raises the question of actin. The explants were cultivated in the presence of (monomeric or polymerized) muscle actin (30 µg/ml). No induction was observed.

Ebert (1955) discovered the presence of myosin in the chick blastoderm at the primitive streak stage. We prepared an antiserum against chick myosin. The antiserum reacted with the extract of chick embryo at the stage of late primitive streak. Additionally, the presence of myosin antigen at a comparable developmental stage in the frog embryo was demonstrated by Ranzi and Citterio (1956).

I do not know the meaning of the difference between our data and the findings of Holtzer et al. (1957) with fluorescent antibodies. They were not able to find myosin in somites before the formation of the myofibrils. Perhaps the reacting groups are not able to join with the fluorescent antibodies before the orientation and the complete formation of the myosin molecule. In our experiment only the heaviest chain of the molecule is present.

Our data indicate the myosin as the protein inductor of the somites. Consequently, we can conclude that, under the action of the organizer, DNA transcribes 26S mRNA of myosin, this mRNA is translated into myosin, and consequently somites appear. This hypothesis is presented in Fig. 3.

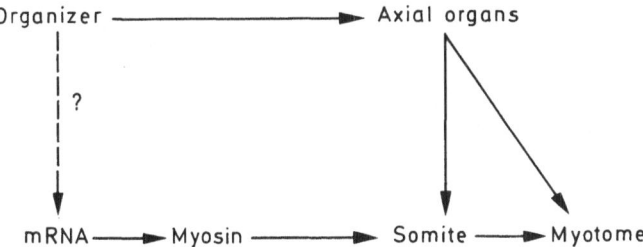

Fig. 3. Causal relations in the development of the somites

If we consider the question of the origin of the somites from an evolutionary point of view, we can assume that the sequence of actions for the differentiation of musculature was mRNA-myosin-muscular elements in the unsegmented ancestors. A second inductor appeared, coming from axial organs, after the appearance of the notochord and the segmentation. This inducer operated between the somites, induced by the myosin, and the muscular differentiation.

References

Brawerman, G.: Eukaryotic messenger RNA. Annu. Rev. Biochem. 43, 621-642 (1974)

Cigada Leonardi, M., De Bernardi, F., Maci, R., Ranzi, S.: Azione di mRNA di miosina sulla porzione post-nodale del blastoderma di pollo. Acta Embryol. Exp. 3, 369-370 (1977)

Ebert, J.D.: Some aspects of protein biosynthesis in development. In: Aspects of Synthesis and Order in Growth (ed. D. Rudnick), pp. 69-112. Princeton, NJ: Princeton Univ. Press 1954

Gurdon, J.H., Woodland, H.R., Lingrel, J.B.: The translation of mammalian globin mRNA injected into fertilized eggs of *Xenopus laevis*. I. Message stability in development. Dev. Biol. 39, 125-133 (1974)

Heywood, S.M., Rich, A.: In vitro synthesis of native myosin, actin and tropomyosin from embryonic chick polyribosomes. Proc. Natl. Acad. Sci. USA 59, 590-597 (1968)

Heywood, S.M., Rourke, A.W.: Cell-free synthesis of myosin. Methods Enzymol. 30, 669-674 (1974)

Holtzer, H., Marshall, J.M., Finck, H.: An analysis of myogenesis by the use of fluorescent antimyosin. J. Biophys. Biochem. Cytol. 3, 705-724 (1957)

Mommaerts, W.F.H.M., Parrish, G.A.: Studies on myosin. I. Preparation and criteria of purity. J. Biol. Chem. 188, 545-552 (1951)

Moscona, A., Moscona, H.: The dissociation and aggregation of cells from organ rudiments of the early chick embryo. J. Anat. 86, 278-286 (1952)

New, D.A.T.: A new technique for the cultivation of the chick embryo in vitro. J. Embryol. Exp. Morphol. 3, 310-331 (1955)

Niu, M.C., Segal, S.J. (eds.): The role of RNA in reproduction and development, Proc. of AAAS Symp. 1972, 358 p. Amsterdam: North-Holland 1973

Packard, D.S., Jacobson, A.G.: The influence of axial structures on chick somite formation. Dev. Biol. 53, 36-48 (1976)

Ranzi, S., Citterio, P.: Le comportement des differentes fractions protéiques au cours de developpement embryonnaire de *Rana esculenta*. Rev. Suisse Zool. 62, 275-281 (1956)

Ranzi, S., Gavarosi, G., Citterio, P.: Cytodifferentiation induced by ribonucleoproteins. Experimentia 17, 395 (1961)

Shimada, Y.: Electron microscope observations on the fusion of chick myoblasts in vitro. J. Cell Biol. 48, 128-142 (1971)

RNA Viruses, Cancer and Development

H.M. TEMIN

McArdle Laboratory, University of Wisconsin, Madison, WI 53706, USA

I. Introduction

RNA viruses that cause cancer and their relatives, members of the retrovirus family, have the special characteristic of existing either as a regular virus spread by transfer of virions or as a cellular gene spread by Mendelian inheritance. This characteristic gives these viruses a unique biology including a possible role in normal cellular processes.

In this article I shall briefly describe the retroviruses, present several hypotheses about their relationships to the cell genome (protovirus hypothesis), and describe some of the molecular stages in the formation and integration of retroviral DNA.

II. Retroviruses

Retroviruses form a large family of animal viruses whose virions are enveloped and contain an RNA genome and a DNA polymerase. Retroviruses replicate through a DNA intermediate, the provirus, and contain three or four genes.

Retroviruses can be classified by virion properties into subfamilies, genera and species (Vogt 1976). They can also be classified by their biological action – whether or not they cause neoplasia, whether or not they are defective, and whether or not they kill infected cells.

With this classification, retroviruses can be separated into strongly transforming (rapidly oncogenic), weakly transforming (slowly oncogenic), and not oncogenic. Examples of strongly transforming retroviruses are Rous sarcoma virus and murine sarcoma virus; of weakly transforming retroviruses, lymphoid leukosis virus and Gross murine leukemia virus; of not oncogenic retroviruses, Rous-associated virus-O and endogenous murine leukemia virus.

The only difference between otherwise homologous strongly transforming retroviruses and weakly or not oncogenic retroviruses is the existence of a specific gene, called *src*, in the RNA of strongly transforming retroviruses. The product of this gene in Rous sarcoma virus has recently been isolated and shown to have protein kinase activity (Collett and Erickson 1978).

III. Protovirus Hypotheses

The relationships of retroviruses to each other and to the cell genome have been described in a series of hypotheses called protovirus hypotheses.

Hypothesis 1: Strongly transforming retroviruses evolve from weakly transforming retroviruses and cellular DNA.

The relationship of strongly transforming retroviruses to other retroviruses and the cell genome has been studied by nucleic acid hybridization. The results are consistent with hypothesis 1. The evidence for this hypothesis is especially strong for the evolution of Rous sarcoma virus from lymphoid leukosis virus and chicken DNA (Stehelin et al. 1976) and for the evolution of Kirsten murine sarcoma virus from Kirsten murine leukemia virus and rat DNA (Scolnick et al. 1973).

Hypothesis 2: Weakly transforming retroviruses cause neoplasia by forming strongly transforming genes.

This hypothesis is supported by the work with AKR murine leukemia virus (Hartley et al. 1977) and with radiation murine leukemia virus (Haas 1978).

Hypothesis 3: Weakly transforming retroviruses evolve from cellular DNA.

This hypothesis is supported by the presence of complete endogenous viral genomes in normal cellular DNA, for example, avian leukosis virus genomes in chicken DNA and murine leukemia virus genomes in mouse DNA (Temin 1974; Cooper and Temin 1976). (See Note Added in Proof.)

Hypothesis 4: Some of the retrovirus-related proteins and nucleic acids that are present in many normal cells are precursors of retroviruses.

A continuous series of retrovirus-related particles and molecules exists ranging from exogenous virus, endogenous virus, endogenous virus with low specific infectivity, endogenous virus-related nucleotide sequences rescuable by phenotype mixing, endogenous virus-related nucleotide sequences rescuable by recombination, endogenous-virus-related nucleotide sequences not rescuable, and endogenous virus-related products and processes (Temin 1974). This series is consistent with this hypothesis. (See Note Added in Proof.)

Hypothesis 5: Retrovirus-related products and processes play a role in normal development.

The existence of different forms of endogenous virus-related glycoproteins in different organs (Elder et al. 1977) is consistent with this hypothesis. However, no direct evidence for this hypothesis has been published.

Hypothesis 6: Nonviral cancers are the result of the formation of strongly transforming genes by processes related to those involved in the evolution of retroviruses.

The recent reports of recovery by phenotypic mixing of strongly transforming genes from spoantaneous or chemically induced tumor cells (Rapp and Todaro 1978; Rasheed et al. 1978) are consistent with this hypothesis.

IV. Formation and Integration of Viral DNA

The ability of retroviruses to transfer information from RNA to DNA and integrate their DNA with cellular DNA is the feature that gives rise to their special

Formation of viral DNA

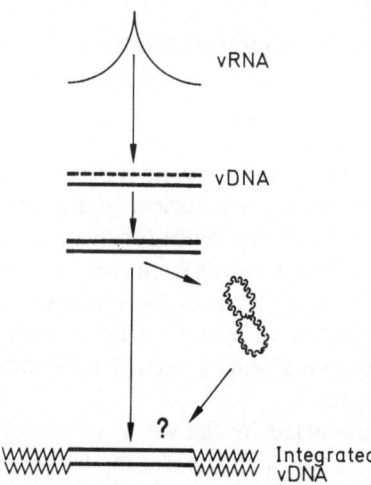

Fig. 1. Forms of retrovirus DNA in infected cells

Table 1. SNV DNA molecules in infected avian cells. [Based on Battula and Temin 1977, 1978; Fritsch and Temin 1977; Keshet and Temin 1978; see also Chen and Temin: J. Virol. 33, 1058-1073 (1980)]

	Acute phase (cell death)	Chronic phase (no cell death)
SNV DNA molecules per cell (approx.)		
not integrated	200	3
integrated	20	3
Sites of integration of viral DNA		
infectious	many	"one"
not infectious	many	many

characteristics. We have been studying these processes with an avian retrovirus—spleen necrosis virus (SNV).

Several forms of SNV DNA exist in infected cells (Fig. 1). The predominant ones are an unintegrated linear DNA of 6×10^6 daltons and an integrated one of greater than 50×10^6 daltons. These forms can easily by separated by the Hirt extraction procedure or by equilibrium cesium chloride density gradient centrifugation or gel electrophoresis. The forms of viral DNA can be quantified by transfection or nucleic acid hybridization assays (Table 1).

We have used a restriction enzyme, EcoRl, that does not cut SNV DNA to determine the number of integration sites of SNV DNA. SNV DNA is found in multiple size classes of EcoRl-cleaved DNA from SNV-infected cells (Fig. 2). Therefore, SNV DNA integrates at multiple sites in the DNA of infected avian cells.

When the infectivity of these different molecules of integrated DNA was assayed soon after infection (acute infection), we found that infectious molecules

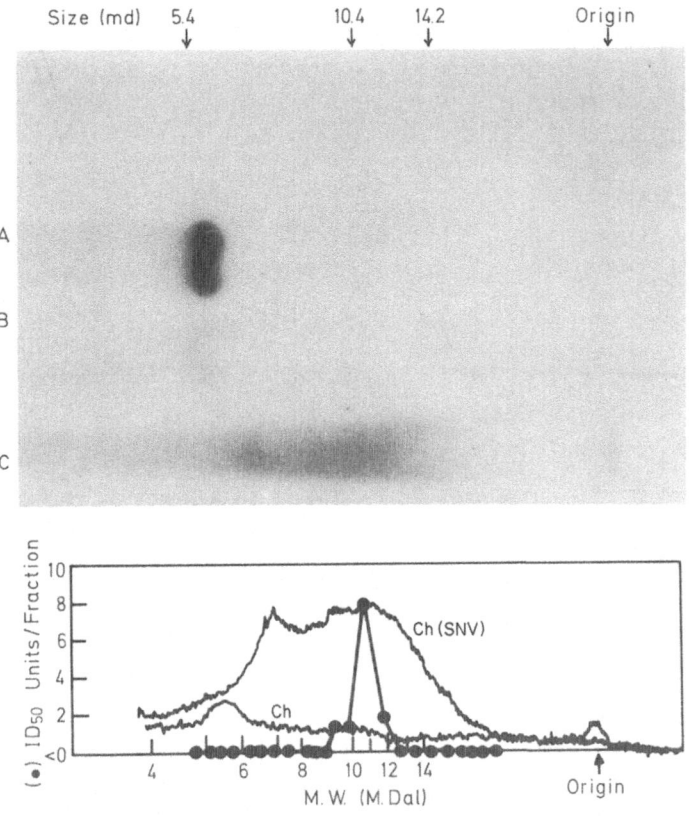

Fig. 2. Comparison of the size distribution of hybridizable fragments with the size distribution of infectious DNA fragments of EcoRl-digested DNA from chronically infected chicken cells. *Top:* Lane A: unintegrated viral DNA. Lane B: DNA from uninfected chicken cells. Lane C: High molecular weight DNA from chronically infected chicken cells. *Bottom:* Densitometry tracings of Lane B (Ch) and Lane C (Ch (SNV)) and infectivity of Lane C (ID_{50}). (Data from Keshet and Temin 1978)

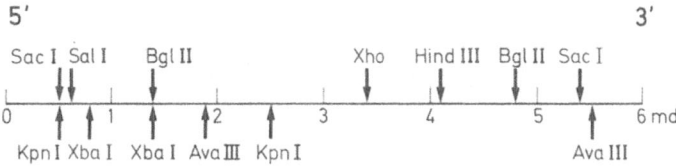

Fig. 3. Map of restriction enzyme cleavage sites in unintegrated SNV DNA. (Data from Keshet et al. 1979)

were present in all size classes. However, when the infectivity of these molecules of integrated DNA was assayed 3 or 4 weeks after infection (chronic infection), we found that only molecules of $10–11 \times 10^6$ daltons were infectious (Fig. 2).

We used a map of the restriction enzyme cleavage sites of eight enzymes in unintegrated SNV DNA (Fig. 3) to determine if the integrated DNA was co-linear

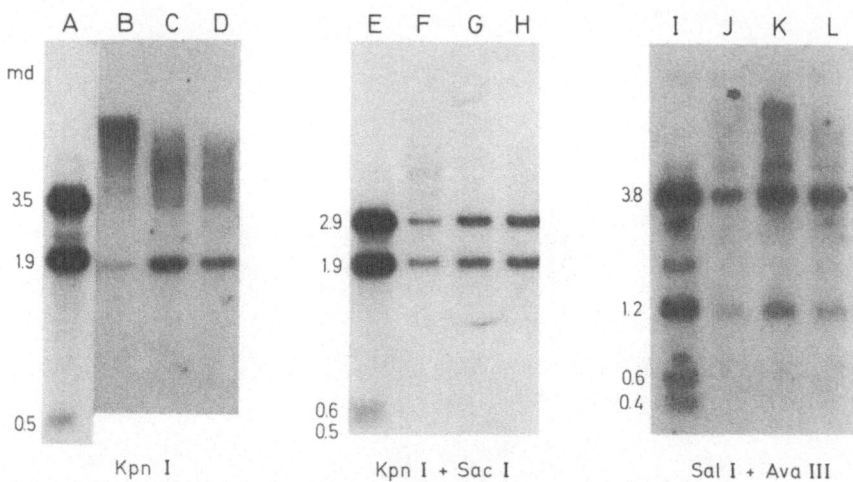

Fig. 4. Co-linearity with unintegrated DNA of infectious and not infectious fragments of EcoRl-digested high molecular DNA from chronically infected chicken cells. Lanes A, E, I: unintegrated DNA. Lanes B, F, J: greater than 11.5×10^6 daltons. Lanes C, G, K: 9.0 to 11.5×10^6 daltons. Lanes D, H, L: 7.0 to 9.0×10^6 daltons. (Data from Keshet et al. 1979)

with unintegrated DNA and if there were large insertions or deletions in the larger ($\geq 11.5 \times 10^6$ daltons) or smaller ($\leq 9 \times 10^6$ daltons) not infectious integrated molecules. Digestion of these molecules with these enzymes yielded the same sizes and amounts of internal bands as digestion of unintegrated SNV DNA. In addition, no fusion fragments were found.

Therefore, we can conclude that infectious and not infectious integrated SNV DNA are co-linear with unintegrated DNA, indicating a unique place at the termini of unintegrated viral DNA for integration, and that the not infectious molecules are not grossly abnormal (no large deletions or insertions and less than 3% nucleotide sequence changes), indicating that adjacent sequences can control the activity of the provirus (Fig. 4). [See also O'Rear et al.: Cell 20, 423-430 (1980).]

V. Summary and Conclusions

Retroviruses use a so far unique mode of information transfer from RNA to DNA and have mechanisms to integrate their DNA efficiently at multiple sites in the cell DNA.

Uninfected cells contain retrovirus-related molecules that may reflect the role of retrovirus-related processes in normal cellular functions.

Acknowledgements. The research in my laboratory is supported by grants CA-07175 and CA 22443 from the National Cancer Institute. I am an American Cancer Society Research Professor.

This paper is dedicated to Dr. Harold P. Rusch founder and longtime director of the McArdle Laboratory.

References

Battula, N., Temin, H.M.: Infectious DNA of spleen necrosis virus is integrated at a single site in the DNA of chronically infected chicken fibroblasts. Proc. Natl. Acad. Sci. USA 74, 281-285 (1977)

Battula, N., Temin, H.M.: Sites of integration of infectious DNA of avian reticuloendotheliosis viruses in different avian cellular DNAs. Cell 13, 387-398 (1978)

Collett, M.S., Erikson, R.L.: Protein kinase activity associated with the avian sarcoma virus *src* gene product. Proc. Natl. Acad. Sci. USA 75, 2021-2024 (1978)

Cooper, G.M., Temin, H.M.: Lack of infectivity of the endogenous avian leukosis virus-related genes in the DNA of uninfected chicken cells. J. Virol. 17, 422-430 (1976)

Elder, J.H., Jensen, F.C., Bryant, M.L., Lerner, R.A.: Polymorphism of the major envelope glycoprotein (gp70) of murine C-type viruses: virion associated and differentiation antigens encoded by a multi-gene family. Nature (London) 267, 23-28 (1977)

Fritsch, E., Temin, H.M.: Formation and structure of infectious DNA of spleen necrosis virus. J. Virol. 21, 119-130 (1977)

Haas, M.: Leukemogenic activity of thymotropic, ecotropic, and xenotropic radiation leukemia virus isolates. J. Virol. 25, 705-709 (1978)

Hartley, J.W., Wolford, N.K., Old, L.F., Rowe, W.P.: A new class of murine leukemia virus associated with development of spontaneous leukemia. Proc. Natl. Acad. Sci. USA 789-792 (1977)

Keshet, E., Temin, H.M.: Sites of integration of reticuloendotheliosis virus DNA in chicken DNA. Proc. Natl. Acad. Sci. USA 75, 3372-3376 (1978)

Keshet, E., O'Rear, J., Temin, H.M.: DNA of non-infectious and infectious integrated spleen necrosis virus (SNV) is colinear with unintegrated SNV DNA and not grossly abnormal. Cell 16, 51-61 (1979)

Rapp, U.R., Todaro, G.J.: Generation of oncogenic type C viruses: Rapidly leukemogenic viruses derived from C3H mouse cells *in vivo* and *in vitro*. Proc. Natl. Acad. Sci. USA 75, 2468-2472 (1978)

Rashed, S., Gardner, M.B., Huebner, R.J.: *In vitro* isolation of stable rat sarcoma viruses. Proc. Natl. Acad. Sci. USA 75, 2972-2976 (1978)

Scolnick, E.M., Rands, E., Williams, D., Parks, W.P.: Studies on the nucleic acid sequences of Kirsten sarcoma virus: a model for formation of a mammalian RNA-containing sarcoma virus. J. Virol. 12, 458-463 (1973)

Stehelin, D., Varmus, A.M., Bishop, J.M., Vogt, P.K.: DNA related to the transforming gene(s) of avian sarcoma viruses is present in normal avian DNA. Nature (London) 260, 170-173 (1976)

Temin, H.M.: On the origin of RNA tumor viruses. Annu. Rev. Genet. 8, 155-177 (1974)

Vogt, P.K.: The oncovirinae—a definition of the group. In: WHO Center for Collection and Evaluation of Data in Comparative Virology, Rep. No. 1 (ed. P. Thein), pp. 327-339. Munich 1976

Note Added in Proof

More recent work indicates that most endogenous retroviruses are the result of recent germ-line infection [Astrin, Buss, and Hayward: Nature 282, 339-341 (1979); Cohen and Varmus: Nature 278, 418-423 (1979); Frisby, Weiss, Roussel, and Stehelin: Cell 17, 623-634 (1979)]. However, even more recent work indicates that the nucleotide sequence of a provirus is similar in structure to that of bacterial movable genetic elements [Shimotohno, Mizutani, and Temin: Nature 285, 550-554 (1980)], thus supporting the protovirus hypothesis that retroviruses evolved from movable genetic elements in normal cellular DNA.

See also Chen and Temin: J. Virol. 33, 1058-1073 (1980) and O'Rear et al.: Cell 20, 423-430 (1980).

The Regulation of Differentiation
in Murine Virus-Induced Erythroleukemic Cells

CH. FRIEND

The Mollie B. Roth Laboratory, The Mount Sinai School of Medicine of the City University of New York, New York, NY, USA

Continuous lines of erythroleukemia cells, which originate from the spleens of mice with Friend virus-induced leukemia, provide a convenient experimental system in which the molecular control of erythrodifferentiation can be examined in some detail (Friend et al. 1966; Ikawa and Sugano 1966; Ostertag et al. 1972; Freedman and Lilly 1975). The cultures contain cells at various stages of differentiation. The majority are large, primitive erythroid cells with prominent nucleoli, but a few more mature cells recognizable as normoblasts are present. We had developed these lines with the idea of studying leukemia as a disease resulting from a block in maturation, and had hoped to find means of stimulating the cells to differentiate. To a certain extent, we achieved this goal.

When the medium in which the leukemia cells are grown is supplemented with dimethyl sulfoxide (DMSO), the mechanisms regulating normal development are switched on (Friend et al. 1971). The cells become responsive to some of the signals directing them to proceed to differentiate along the erythroid pathway, up to the normoblast stage. They recover the ability to express many of the specialized functions of red blood cells (Harrison 1976; Friend 1979). These are listed in Table 1. The treated cells synthesize adult hemoglobin characteristic of the DBA/2 mice from which they were derived and undergo many changes, both cellular and biochemical, which parallel the alterations which occur during normal

Table 1. Changes associated with erythroid differentiation in DMSO-treated Friend cells

---

Morphologically recognizable normoblasts
Synthesis of heme
Synthesis of hemoglobin
    Accumulation of globin mRNA
    Globin chain synthesis
Alterations in purine metabolism
Increase in carbonic anhydrase activity
Appearance of erythrocyte membrane antigens
Increase in spectrin and glycophorin synthesis
Decrease in cell volume
Limited capacity for self-renewal

Table 2. Classes of compounds inducing erythroleukemic cell differentiation

Polar organic solvents
Bisacetamides
Short chain fatty acids
Hemin
Purines and purine analogues
Cardiac glycosides
Metabolic inhibitors

erythropoiesis (Boyer et al. 1972; Ross et al. 1972; Ikawa et al. 1973; Kabat et al. 1975; Reem and Friend 1975, 1976; Arndt-Jovin et al. 1976; Aviv et al. 1976; Gusella et al. 1976; Eisen et al. 1977; Loritz et al. 1977).

Although the treated cells proceed to differentiate, they continue to synthesize virus. In fact, in the presence of DMSO, virus production is increased (Sato et al. 1971). Many viruses can be observed budding from the surface and in the cytoplasmic vacuoles of the treated cells, indicating that cells actively synthesizing type C viruses are not necessarily malignant. It appears that cell differentiation and virus proliferation are not mutually exclusive events. The fact that variant clones are inducible for hemoglobin but not for virus and others are stimulated to produce virus but not hemoglobin suggests that the mechanisms controlling differentiation and virus synthesis are most likely under separate control (Ikawa et al. 1976; Sherton et al. 1976; Tsuei et al. 1977).

The mechanisms by which DMSO acts in unfreezing the block in maturation are not as yet clear. The wide spectrum of effects that have been attributed to this compound have recently been reviewed in an effort to bring into focus those of its biological properties that might be implicated in modulating gene expression in the erythroleukemia cells (Friend and Freedman 1978). Its ability to alter membrane permeability and microviscosity have received the most attention. Whether these alterations are related to the single-stranded breaks detected in the DNA of the treated cells (Scher and Friend 1978; Terada et al. 1978b) remains to be determined.

In addition to DMSO, compounds of a variety of classes (Table 2) also have the ability to induce differentiation (Scher et al. 1973; Leder and Leder 1975; Tanaka et al. 1975; Bernstein et al. 1976b; Ebert et al. 1976; Gusella and Housman 1976; Lyman et al. 1976; Reuben et al. 1976; Ross and Sautner 1976; Terada et al. 1978). For some as yet unknown reason, erythropoietin, an inducer of normal erythroid cells, does not affect hemoglobin synthesis in these erythroleukemia cells. Most of the inducers also cause some alterations in cell membranes. The polar solvents, such as DMSO, produce a decrease in membrane fluidity and permeability (Arndt-Jovin 1976), as well as single-stranded breaks in the DNA of cells (Scher and Friend 1978; Terada et al. 1978b). The bisacetamides, such as HMBA, cause scissions in DNA (Terada et al. 1978b) and may alter membrane permeability, although they do not directly change membrane fluidity (Reuben et al. 1978). The fatty acids, hemin and the purines are naturally occurring products, but there is no evidence to indicate

Table 3. Classes of compounds
inhibiting induced erythroid
differentiation

Halogenated pyrimidines
Phorbol diesters
Local anesthetics
Interferon
Corticosteroids
Allylisopropylacetamide
3-Amino 1,3,4-triazole

---

that these compounds have a physiological function in the erythroleukemia cells. Butyric acid has been noted to affect the DNA structure (Scher and Friend 1978; Terada et al. 1978b) and may affect cell surface composition (Bernstein et al. 1976a). Hemin is not a particularly good inducer by itself, but it does act synergistically when added to the cultures together with DMSO (Ross and Sautner 1976). While the cell line in which it acts may be deficient in heme production and may only need an exogenous supply of heme to synthesize hemoglobin, there is a possibility that it, too, affects the membrane. The inducing purines, such as hypoxanthine, are thought to interact directly as the free base with some as yet unknown cellular target, since the incorporation of hypoxanthine into DNA or RNA does not appear to be necessary for differentiation to take place (Gusella and Housman 1976). However, it does cause DNA changes (Scher and Friend 1978). The cardiac glycoside, ouabain, which binds specifically to and inhibits the plasma membrane enzyme Na^+K^+ATPase, induces hemoglobin synthesis at concentrations which inhibit K-uptake (Bernstein et al. 1976b). It has been suggested that changes in the intracellular conentration of K^+ may be involved in the induction process, although K^+ dependence does not appear to be a property of the inducers of the other classes. A variety of metabolic inhibitors have inducing ability (Ebert et al. 1976). Actinomycin D has been found to cause single-stranded breaks in the DNA, but does not appear to affect the plasma membranes of the erythroleukemia cells (Terada et al. 1978a). It has, however, been reported to react with the membranes of other cells (Fico et al. 1977).

A direct approach which might shed some light on the regulatory processes that can be triggered in the erythroleukemia cells by so wide a variety of unrelated agents is to examine the compounds which inhibit differentiation. The inhibitory compounds come from as many different classes as do the inducers (Table 3). The inhibitory effect of the halogenated pyrimidine, 5-bromo-2'-deoxyuridine (BUdR), on DMSO (Scher et al. 1973) and butyric acid induction (Bick and Cullen 1976) is blocked by thymidine. It is thought to act at the level of transcription of globin mRNA.

The tumor-promoting phorbol diesters, but not their inactive analogues, reversibly block spontaneous and induced erythroid differentiation (Rovera et al. 1977; Yamasaki et al. 1977). In contrast to the inducers, which cause a decrease in macromolecular synthesis, these compounds stimulate DNA, RNA and protein synthesis and induce ornithine decarboxylase in a number of other cell culture systems (Baird et al. 1971; Yuspa et al. 1976; O'Brien and Diamond 1977; Peterson

Table 4. Inducers. Evidence for different mechanisms of action

1. Clones differ from each other in their characteristic pattern of alpha and beta globin RNA accumulation
2. Ratio of beta major to beta minor globin chains may vary, depending on:
   A) Inducer
   B) Culture conditions
   C) Origin of cell line
3. Variant clones resistant to one inducer may be responsive to treatment with another

et al. 1977). They have also been reported to cause a decrease in the major surface glycoprotein (Driedger and Blumberg 1977) and to affect lipid metabolism in membrane structures (Rohrschneider and Boutwell 1973).

Local anesthetics, such as procaine and tetracaine, are presumed to inhibit the effect of the inducers by increasing the membrane fluidity and permeability (Seeman 1972; Papahadjopoulos et al. 1975). Interferon may also interact with the cell membrane. In the presence of high concentrations of this agent, globin mRNA was decreased and little globin was synthesized, but heme synthesis was not appreciably affected. Therefore, both transcription and translation of globin mRNA were affected (Rossi et al. 1977).

Dexamethasone and hydrocortisone are among the most potent inhibitors in the corticosteroid class (Lo et al. 1978; Scher et al. 1978). The activity of the steroid hormones appears to depend on their ability to bind to a cytoplasmic receptor protein which then moves to the nucleus, where it interacts with the chromatin by intercalating between specific bases (Huggins and Yang 1962; Hendry et al. 1977). They are thought to act at the level of transcription, but the precise nature of receptor activation, transport and nuclear binding is not clear.

The last two agents specifically affect heme metabolism. Allylisopropyl-acetamide, which breaks down the heme moiety of heme proteins, reduced the amount of beta minor globin synthesized (Rovera et al. 1978). It is not known at what level it acts, nor whether exogenous heme would block the effect. Aminotriazole inhibits ALA dehydratase and heme synthesis. It completely blocks hemoglobin synthesis. If heme is added, the ability of the cells to synthesize globin mRNA is restored. This agent is presumed to act at the pretranslational level (Dabney and Beaudet 1977). Depending on the length of time the cells have been exposed, the inhibitory effects of these compounds may be reversible, allowing the cells to retain their phenotype.

Of particular interest is the fact that some of the inhibitors can also act as inducers in our system as well as others. BUdR may induce varying levels of globin mRNA in different clones of erythroleukemia cells (Preisler et al. 1973; Adesnik and Smitkin 1978) and induces differentiation in neuroblastoma cells (Schubert and Jacob 1970; Prasad et al. 1973). The phorbol diesters have been reported to induce erythroid differentiation in Rauscher virus erythroleukemia cell lines (Miao et al. 1978). Dexamethasone stimulates erythroid colony formation in normal mouse and human bone marrow and fetal liver cells in vitro (Golde et al. 1976a). Treatment with inhibitors such as BUdR, dexamethasone and the phorbol diesters also induce

Table 5. Inducers. Evidence for common mechanism of action

1. Most active at highest concentration tolerated by the cells
2. Cause depression in macromolecular synthesis
3. Affect membrane fluidity
4. Affect cell multiplication rate
5. Induction of differentiation in other cell types

some cell markers associated with normal macrophages and granulocytes in murine myeloid leukemia cells (Sachs 1978).

Several lines of evidence suggest that the inducers of differentiation in our model system may not all act via the same mechanisms and that a number of different sites may be involved (Table 4). Erythroleukemia cell clones may differ from each other in their characteristic pattern of alpha and beta globin RNA accumulation (Boyer et al. 1972; Ostertag et al. 1972; Orkin et al. 1975b; Alter and Goff 1977). The ratio of beta major to beta minor chains may vary, depending on the conditions of culture, the origin of the cell line, and on the inducer with which it is treated (Kabat et al. 1975; Nudel et al. 1977). Variant clones which are resistant to one inducer may differentiate in response to treatment with another (Orkin et al. 1975a; Gusella and Housman 1976).

On the other hand, there is also evidence to suggest the possibility that the inducers of differentiation may have mechanisms in common (Table 5). They are generally most active when used at the highest concentration tolerated by the cells, cause a depression in macromolecular synthesis, and affect the multiplication rate. The stimulation of differentiation by the inducers is not restricted to FL cells. DMSO, HMBA and heme also activate gene expression in neuroblastoma cells and in other developmental systems. DMSO induces differentiation in chicken yolk sac erythroid cells (Miura et al. 1976), rat erythroleukemia cells (Kluge et al. 1976), normal mouse fetal liver (Malpoix 1976), human lung cilia cells (Tralka and Rabson 1976), and mouse neuroblastoma cells (Kimhi et al. 1976) in culture. HMBA stimulates differentiation of mouse neuroblastoma cells (Palfrey et al. 1977) and of collagen synthesis in a cell line derived from a human glioblastoma multiforme (Rabson et al. 1977). Hemin has been reported to induce differentiation in neuroblastoma cells (Ishii and Maniatis 1978).

The blocking of differentiation by the inhibitory compounds also is not unique to the erythroleukemia cells. For example, BUdR inhibits normal erythropoiesis (Miura and Wilt 1971) and myogenesis (Stockdale et al. 1964) in chick embryo cells, and the ability of melanoma cells to produce melanin (Silagi and Bruce 1970). The phorbol diesters block adipose conversion of 3T3 cells (Diamond et al. 1977), and the differentiation of chick myoblasts (Cohen et al. 1977) and chondroblasts (Lowe et al. 1978). Dexamethasone inhibits granulopoiesis of normal mouse bone marrow cells in vitro (Golde et al. 1976b).

So far, the precise mechanism whereby some compounds induce differentiation and others inhibit their action remains elusive. The fact that the activity of each of the inducers can generally be blocked by a variety of inhibitors (Table 6) suggests that there may be steps at which their metabolic pathways converge. Further study

Table 6. The inhibition of induced differentiation in erythroleukemia cells

Inducer	Inhibitor			
	Halogenated pyrimidine (BUdR)	Phorbol diester (TPA)	Corticosteroid (dexamethasone)	Local anesthetics
DMSO	+	+	+	+
HMBA	+	+	+	+
Butyric acid	+	+	−	ND
Hypoxanthine	ND	+	+	+

ND: Not done to our knowledge

Table 7. Proposed model for inducer action

Malignant cell + inducer
↓
Change in membrane fluidity
↓
Activation of receptors at sensitive sites
↓
Induction of regulating signals
↓
Release of specific regulatory protein as end product
↓
Interaction of regulatory protein with chromatin
↓
Differentiation

of the interactions of a pair of closely related compounds, one a purine derivative which is a potent inducer, and the other a purine ribonucleoside, which acts as its inhibitor (Lacour et al. 1980), may allow us to obtain evidence to support this possibility.

Most of the compounds, as chemically and biologically diverse as they are, have been implicated to some extent in membrane function, particularly in the fluidity of the membrane. A model to explain their ability to influence gene expression can be constructed on the basis of the theories of Edelman in regard to surface modulating assemblies (Edelman 1976) and of Mueller et al. on the mechanism of action of the tumor promoters (Mueller et al. 1978). It is shown diagramatically in Table 7. The plasma membrane is visualized as a fluid structure with receptors having the ability to move on and in the cell surface. Microtubules, microfilaments and associated contractile and membrane proteins make up the surface modulating assemblies which may play a role in regulating growth control signals from the surface to the inside of the cells. DMSO, the most widely studied inducer in our system, affects at least one of these components. It promotes the formation of microtubules from microtubulin in the absence of microtubule-associated protein and may possibly have a similar function within the cells (Himes et al. 1976).

The inducers may affect the mobility and distribution of the cell surface receptors which perhaps, when situated in one region of the membrane, are inactive. In response to treatment, they may move to another region where they are activated. In order to influence gene expression, it may be necessary to have a specific combination of receptors translocated to the sensitive site. This part of the phenomenon would be analogous to what is considered to occur when divalent antibodies to receptors crosslink in a particular region of the cell surface to cause cap formation (Edelman 1976).

In our system, activation of the receptors may cause the release of an as yet unrecognized specific regulatory protein. The activity of such a component may be reflected in the DNA breaks detected in the induced cells. Its interaction with the chromatin may affect transcription or even cause gene translocation, thus dictating the direction the cell takes, whether towards continuous growth or towards differentiation. This may very well be a general mechanism governing growth and differentiation.

Acknowledgments. This work supported in part by NIC grants CA 10,000 and CA 13,047.

References

Adesnik, M., Smitkin, H.: Induction of erythroid differentiation in Friend leukemia cells by bromodeoxyuridine. J. Cell Physiol. 95, 307-318 (1978)

Alter, B.P., Goff, S.C.: Globin synthesis in mouse erythroleukemia cells in vitro: a switch in beta chains due to inducing agent. Blood 50, 867-876 (1977)

Arndt-Jovin, D.J., Ostertag, W., Eisen, H., Limek, F., Jovin, T.M.: Studies of cellular differentiation by automated cell separation. Two model systems: Friend virus-transformed cells and hydra attenuata. J. Histochem. Cytochem. 24, 332-347 (1976)

Aviv, H., Voloch, Z., Xastos, R., Levy, S.: Biosynthesis and stability of globin mRNA in cultured erythroleukemic Friend cells. Cell 8, 495-503 (1976)

Baird, W.M., Sedgwick, J.A., Boutwell, R.K.: Effects of phorbol and 4 diesters of phorbol on the incorporation of tritiated precursors into DNA, RNA and protein in mouse epidermis. Cancer Res. 31, 1434-1439 (1971)

Bernstein, A., Boyd, A.S., Crichley, V., Lamb, V.: Induction and inhibition of Friend leukemic cell differentiation: the role of membrane-active compounds. In: Biogenesis and Turnover of Membrane Macromolecules (ed. J.S. Cook), pp. 145-159. New York: Raven Press 1976a

Bernstein, A., Hunt, D.M., Crichley, V., Mak, T.W.: Induction by ouabain of hemoglobin synthesis in cultured Friend erythroleukemic cells. Cell 9, 375-381 (1976b)

Bick, M.D., Cullen, B.R.: Bromodeoxyuridine inhibition of Friend leukemia cell induction by butyric acid: Time course of inhibition, reversal and effect of other base analogs. Somat. Cell Genet. 2, 545-558 (1976)

Boyer, S.H., Wuu, K.D., Noyes, A.N., Young, R., Scher, W., Friend, C., Preisler, H.D., Bank, A.: Hemoglobin biosynthesis in murine virus-induced leukemic cells in vitro: structure and amounts of globin chains produced. Blood 40, 823-835 (1972)

Cohen, R., Pacifici, M., Rubenstein, N., Biehl, J., Holtzer, H.: Effect of a tumor promoter on myogenesis. Nature (London) 266, 538-540 (1977)

Dabney, B.J., Beaudet, A.L.: Increase in globin chains and globin mRNA in erythroleukemia cells in response to hemin. Arch. Biochem. Biophys. 179, 106-112 (1977)

Diamond, L., O'Brien, T.G., Rovera, G.: Inhibition of adipose conversion of 3T3 fibroblasts by tumor promoters. Nature (London) 269, 247-249 (1977)

Driedger, P.E., Blumberg, P.M.: The effect of phorbol diesters an chicken embryo fibroblasts. Cancer Res. 37, 3257-3265 (1977)

Ebert, P.S., Wars, I., Buell, D.N.: Erythroid differentiation in cultured Friend leukemia cells treated with metabolic inhibitors. Cancer Res. 36, 1809-1813 (1976)

Edelman, G.M.: Surface modulation in cell recognition and cell growth. Science 192, 218-226 (1976)

Eisen, H., Bach, R., Emery, R.: Induction of spectrin in erythroleukemic cells transformed by Friend virus. Proc. Natl. Acad. Sci. USA 74, 3898-3902 (1977)

Fico, R.M., Chen, T.K., Canellakis, E.S.: Bifunctional intercalators: relationship of antitumor activity of diacridine to the cell membrane. Science 198, 53-56 (1977)

Freedman, H., Lilly, F.: Properties of cell lines derived from tumors induced by Friend virus in Balb-C and Balb-C-H-2b mice. J. Exp. Med. 142, 212-223 (1975)

Friend, C.: The phenomenon of differentiation in murine erythroleukemic cells. In: The Harvey Lectures. Series 72, pp. 253-282. New York-London: Academic Press 1979

Friend, C., Freedman, H.A.: Effects and possible mechanism of action of dimethyl sulfoxide (DMSO) on Friend cell differentiation. Biochem. Pharmacol. 27, 1309-1314 (1978)

Friend, C., Patuleia, M.C., de Harven, E.: Erythrocytic maturation in vitro of murine (Friend) virus-induced leukemic cells. NCI Monogr. 22, 505-522 (1966)

Friend, C., Scher, W., Holland, J.G., Sato, T.: Hemoglobin synthesis in murine virus-induced leukemic cells in vitro. Proc. Natl. Acad. Sci. USA 68, 378-382 (1971)

Golde, D.W., Bersch, N., Cline, M.J.: Potentiation of erythropoiesis in vitro by dexamethasone. J. Clin. Invest. 57, 57-62 (1976a)

Golde, D.W., Bersch, N., Quan, S.G., Cline, M.J.: Inhibition of murine granulopoiesis in vitro by dexamethasone. Am. J. Hematol. 1, 369-373 (1976b)

Gusella, J.F., Housman, D.: Induction of erythroid differentiation in vitro by purines and purine analogues. Cell 8, 263-269 (1976)

Gusella, J., Geller, R., Clarke, B., Weeks, V., Housman, D.: Commitment to erythroid differentiation by Friend erythroleukemia cells: a stochastic analysis. Cell 9, 221-229 (1976)

Harrison, P.R.: Analysis of erythropoiesis at the molecular level. Nature (London) 262, 353-365 (1976)

Hendry, L.B., Witham, F.H., Chapman, O.L.: Gene regulation: the involvement of sterochemical recognition in DNA small molecule interaction. Perspect. Biol. Med. 21, 120-130 (1977)

Himes, R.H., Burton, P.R., Kersey, R.N., Pierson, G.B.: Brain tubulin polymerization in the absence of "microtubule-associated proteins". Proc. Natl. Acad. Sci. USA 73, 4397-4399 (1976)

Huggins, C., Yang, N.C.,: Induction and extinction of mammary cancer. A striking effect of hydrocarbons permits analysis of mechanisms of causes and cure of breast cancer. Science 173, 257-260 (1962)

Ikawa, Y., Sugano, H.: An ascites tumor derived from early splenic lesion of Friend's disease. Gann 57, 641-643 (1966)

Ikawa, Y., Furusawa, M., Sugano, H.: Erythrocyte membrane-specific antigens in Friend virus-induced leukemia cells. In: Unifying Concepts of Leukemia (eds. R.M. Dutcher, L. Chieco-Bianchi), pp. 955-967. Basel: Karger 1973

Ikawa, Y., Inoue, Y., Aida, M., Kameji, C., Shibeta, D., Sugano, H.: Phenotypic variants of differentiation-inducible Friend leukemia lines: isolation and correlation between inducibility and virus release. In: Comparative Leukemia Research (eds. J. Clemmesen, D.S. Yohn), pp. 37-47. Basel: Karger 1976

Ishii, D., Maniatis, G.M.: Hemin induces neuroblastoma differentiation. Nature (London) 274, 372-343 (1978)

Kabat, D., Sherton, C.C., Evans, L.M., Bigley, R., Koler, R.D.: Synthesis of erythrocyte-specific proteins in cultured Friend leukemic cells. Cell 5, 331-338 (1975)

Kimhi, Y., Palfrey, C., Spector, I., Barak, Y., Littauer, U.Z.: Maturation of neuroblastoma cells in the presence of DMSO. Proc. Natl. Acad. Sci. USA 73, 462-466 (1976)

Kluge, N., Ostertag, W., Sugiyama, T., Arndt-Jovin, D., Steinheider, G., Furusawa, M., Dube, S.K.: Dimethylsulfoxide-induced differentiation and hemoglobin synthesis in tissue cultures of rat erythroleukemia cells transformed by 7,12-dimethylbenz(a)anthracene. Proc. Natl. Acad. Sci. USA 73, 1237-1240 (1976)

Lacour, F., Harel, L., Friend, C., Huynh, T., Holland, J.G.: Induction of differentiation of murine erythroleukemia cells by aminonucleoside of puromycin and its inhibition by purines and purine derivatives. Proc. Natl. Acad. Sci. USA 77, 2740-2742 (1980)

Leder, A., Leder, P.: Butyric acid, a potent inducer of erythroid differentiation in cultured erythroleukemia cells. Cell 5, 319-323 (1975)

Lo, S.-C., Aft, R., Ross, J., Mueller, G.C.: Control of globin gene expression by steroid hormones in differentiating Friend leukemia cells. Cell 15, 447-453 (1978)

Loritz, F., Bernstein, A., Miller, R.G.: Early and late volume changes during erythroid differentiation of cultured Friend leukemic cells. J. Cell Physiol. 94, 275-286 (1977)

Lowe, M.E., Pacifici, M., Holtzer, H.: Effects of phorbol-12-myristate-13-acetate on the phenotypic program of cultured chondroblasts and fibroblasts. Cancer Res. 38, 2350-2356 (1978)

Lyman, G.H., Preisler, H.D., Papahadjopoulos, D.: Membrane action of DMSO and other chemical inducers of Friend leukaemic cell differentiation. Nature (London) 262, 360-363 (1976)

Malpoix, P.: Haemoglobin synthesis induced by Nu-methyl-2-pyrrolidinone and dimethyl sulfoxide in permanent cell lines derived from normal mouse foetal liver. Arch. Int. Physiol. Biochim. 84, 1090-1091 (1976)

Miao, R.M., Fieldsteel, A.H., Fodge, D.W.: Opposing effects of tumour promoters on erythroid differentiation. Nature (London) 274, 271-272 (1978)

Miura, Y., Wilt, F.H.: The effects of 5-bromodeoxyuridine on yolk sac erythropoiesis in the chick embryo. J. Cell Biol. 48, 523-531 (1971)

Miura, Y., Terasawa, T., Sawatani, S.: Dimethyl sulfoxide (DMSO) stimulates heme synthesis in quail embryonic yolk sac cells. Exp. Cell Res. 99, 197-200 (1976)

Mueller, G.C., Kensler, T.W., Kajiwara, K.: Mechanism of DNA and chromatin replication: possible targets of cocarcinogenesis. In: Carcinogenesis, Vol. 2, Mechanisms of Tumor Promotion and Cocarcinogenesis (eds. T.J. Slaga, A. Sivak, R.K. Boutwell), pp. 79-90. New York: Raven Press 1978

Nudel, U., Salmon, J.E., Terada, M., Bank, A., Rifkind, R.A., Marks, P.A.: Differential effects of chemical inducers on expression of β globin genes in murine erythroleukemia cells. Proc. Natl. Acad. Sci. USA 74, 1100-1104 (1977)

O'Brien, T.G., Diamond, L.: Ornithine decarboxylase induction and DNA synthesis in hamster embryo cell cultures treated with tumor promoting phorbol diesters. Cancer Res. 37, 3895-3900 (1977)

Orkin, S.H., Harosi, F.I., Leder, P.: Differentiation in erythroleukemic cells and their somatic hybrids. Proc. Natl. Acad. Sci. USA 72, 98-102 (1975a)

Orkin, S.H., Swan, D., Leder, P.: Differential expression of alpha and beta globin genes during differentiation of cultured erythroleukemic cells. J. Biol. Chem. 250, 8753-8760 (1975b)

Ostertag, W., Melderis, H., Steinheider, G., Kluge, N., Dube, S.: Synthesis of mouse haemoglobin and globin mRNA in leukemic cell cultures. Nature New Biol. 239, 231-232 (1972)

Palfrey, C., Kimhi, V., Littauer, U.Z., Reuben, R.C., Marks, P.A.: Induction of differentiation in mouse neuroblastoma cells by hexamethylene bisacetamide. Biochem. Biophys. Res. Commun. 76, 937-942 (1977)

Papahadjopoulos, D., Jacobson, K., Poste, G., Shepherd, C.: Effect of local anesthetics on membrane properties. I. Changes in the fluidity of phospholipid bilayers. Biochem. Biophys. Acta 394, 504-519 (1975)

Peterson, A.R., Mondal, S., Brankow, D.W., Thon, W., Heidelberger, C.: Effects of promoters on DNA synthesis in C3H/10T1/2 mouse fibroblasts. Cancer Res. 37, 3223-3227 (1977)

Prasad, K.N., Mandel, B., Kumar, S.: Human neuroblastoma cell culture: Effect of 5-bromodeoxyuridine on morphological differentiation and levels of neural enzymes. Proc. Soc. Exp. Biol. Med. 144, 38-42 (1973)

Preisler, H.D., Housman, D., Scher, W., Friend, C.: Effects of 5-bromo-2'-deoxyuridine on the production of globin messenger RNA in dimethyl sulfoxide-stimulated Friend leukemia cells. Proc. Natl. Acad. Sci. USA 70, 2956-2959 (1973)

Rabson, A.S., Stern, R., Tralka, T.S., Costa, J., Wilczek, J.: Hexamethylene bisacetamide induces morphologic changes and increased synthesis of precollagen in cell line from glioblastoma multiform. Proc. Natl. Acad. Sci. USA 74, 5060-5064 (1977)

Reem, G.H., Friend, C.: Purine metabolism in murine virus-induced erythroleukemic cells during differentiation in vitro. Proc. Natl. Acad. Sci. USA 72, 1630-1634 (1975)

Reem, G.H., Friend, C.: Purine and phosphoribosylpyrophosphate synthesis in differentiating murine virus-induced erythroleukemic cells in vitro. J. Cell Physiol. 88, 193-196 (1976)

Reuben, R.C., Wife, R.L., Breslow, R., Rifkind, R.A., Marks, P.A.: A new group of potent inducers of differentiation in murine erythroleukemia cells. Proc. Natl. Acad. Sci. USA 73, 862-866 (1976)

Reuben, R.C., Khanna, P., Gazitt, Y., Breslow, R., Rifkind, R.A., Marks, P.A.: Inducers of erythroleukemic differentiation. J. Biol. Chem. 253, 4214-4218 (1978)

Rohrschneider, L.R., Boutwell, R.K.: The early stimulation of phospholipid metabolism by 12-0-tetradecanoyl-phorbol-13-acetate and its specificity for tumor promotion. Cancer Res. 33, 1945-1952 (1973)

Ross, J., Sautner, D.: Induction of globin mRNA accumulation by hemin in cultured erythroleukemic cells. Cell 8, 512-520 (1976)

Ross, J., Ikawa, Y., Leder, P.: Globin messenger-RNA induction during erythroid differentiation of cultured leukemia cells. Proc. Natl. Acad. Sci. USA 69, 3520-3623 (1972)

Rossi, G.B., Dolei, A., Cioe, L., Benedetto, A., Malarese, G.P., Belardelli, F.: Inhibition of transcription and translation of globin messenger RNA in dimethyl sulfoxide-stimulated Friend erythroleukemia cells treated with interferon. Proc. Natl. Acad. Sci. USA 74, 2036-2040 (1977)

Rovera, G., O'Brien, T., Diamond, L.: Tumor promoters inhibit spontaneous differentiation of Friend erythroleukemic cells in culture. Proc. Natl. Acad. Sci. USA 74, 2894-2898 (1977)

Rovera, G., Aden, D., Surrey, S.: Allylisopropylacetamide restricts expression of beta minor globin gene in Friend cells. Nature (London) 272, 172-175 (1978)

Sachs, L.: Control of normal cell differentiation and the phenotypic reversion of malignancy in myeloid leukemia. Nature (London) 274, 535-539 (1978)

Sato, T., de Harven, E., Friend, C.: Ultrastructural changes in Friend erythroleukemia cells treated with DMSO. Cancer Res. 31, 1402-1407 (1971)

Scher, W., Friend, C.: Breakage of DNA and alterations in folded genomes by inducers of differentiation in Friend erythroleukemic cells. Cancer Res. 38, 841-849 (1978)

Scher, W., Preisler, H.D., Friend, C.: Hemoglobin synthesis in murine virus-induced leukemic cells in vitro. III. Effects of 5-bromo-2'-deoxyuridine, dimethylformamide and dimethyl sulfoxide. J. cell Physiol. 81, 63-70 (1973)

Scher, W., Tsuei, D., Sassa, S., Price, P., Gabelman, N., Friend, C.: Inhibition of DMSO-stimulated Friend cell erythrodifferentiation by hydrocortisone and other steroids. Proc. Natl. Acad. Sci. USA 75, 3851-3855 (1978)

Schubert, D., Jacob, F.: 5-Bromodeoxyuridine-induced differentiation of a neuroblastoma. Proc. Natl. Acad. Sci. USA 67, 247-254 (1970)

Seeman, P.: The membrane action of anesthetics and tranquilizers. Pharmacol. Rev. 24, 583-655 (1972)

Sherton, C.C., Evans, L.H., Polonoff, E., Kabat, D.: Relationship of Friend murine leukemia virus production to growth and hemoglobin synthesis in cultured erythroleukemia cells. J. Virol. 19, 118-125 (1976)

Silagi, S., Bruce, S.A.: Suppression of malignancy and differentiation in melanotic melanoma cells. Proc. Natl. Acad. Sci. USA 66, 72-78 (1970)

Stockdale, F., Okazaki, K., Nameroff, M., Holtzer, H.: 5-Bromodeoxyuridine. Effect on myogenesis in vitro. Science 146, 533-535 (1964)

Tanaka, M., Levy, J., Terada, M., Breslow, R., Rifkind, R.A., Marks, P.A.: Induction of erythroid differentiation in murine virus infected erythroleukemia cells by highly polar compounds. Proc. Natl. Acad. Sci. USA 72, 1003-1006 (1975)

Terada, M., Epner, E., Nudel, U., Salmon, J., Fibach, E., Rifkind, R.A., Marks, P.A.: Induction of murine erythroleukemia differentiation by actinomycin D. Proc. Natl. Acad. Sci. USA 75, 2795-2799 (1978a)

Terada, M., Nudel, U., Fibach, E., Rifkind, R.A., Marks, P.A.: Changes in DNA associated with induction of erythroid differentiation by dimethyl sulfoxide in murine erythroleukemia cells. Cancer Res. 38, 835-840 (1978b)

Tralka, T.S., Rabson, A.S.: Cilia formation in cultures of human lung cancer cells treated with dimethyl sulfoxide. J. Natl. Cancer Inst. 57, 1383-1388 (1976)

Tsuei, D., Haubenstock, H., Friend, C.: Virus production and erythroid differentiation in Friend erythroleukemia cells. In Vitro 13, 148 (1977)

Yamasaki, H., Fibach, E., Weinstein, I.B., Nudel, U., Rifkind, R.A., Marks, P.A.: Tumor promoters inhibit spontaneous and induced differentiation of murine erythroleukemia cells in culture. Proc. Natl. Acad. Sci. USA 74, 3451-3455 (1977)

Yuspa, S.H., Lichti, U., Ben, T., Patterson, E., Hennings, H., Slaga, T.J., Colburn, N., Kelsey, W.: Phorbol esters stimulate DNA synthesis and ornithine decarboxylase activity in mouse epidermal cell culture. Nature (London) 262, 402-404 (1976)

Activation of Normal Differentiation Genes and the Origin and Development of Myeloid Leukemia

L. SACHS

Department of Genetics, Weizmann Institute of Science, Rehovot, Israel

I. Introduction

I would like to discuss today the approach that I have been using to try and understand the control mechanisms that regulate the growth and differentiation of normal white blood cells and the origin and development of myeloid leukemia.

This approach was originally based on our in vitro studies on the differentiation of different types of white blood cells (Ginsburg and Sachs 1963, 1965; Pluznik and Sachs 1965, 1966; Ichikawa et al. 1966; Sachs 1964, 1974, 1978) including our identification (Ichikawa et al. 1966; Pluznik and Sachs 1965, 1966) of a normal regulatory protein that we now call MGI (macrophage and granulocyte inducer) (Landau and Sachs 1971), and our development of an in vitro colony forming assay for this protein with mouse (Ginsburg and Sachs 1963; Pluznik and Sachs 1965, 1966; Ichikawa et al. 1966) and human (Paran et al. 1970) cells. We have shown that this protein is required for the viability, growth and differentiation of normal macrophages and granulocytes (Sachs 1978). This discovery of MGI then made it possible to examine whether leukemic cells can still be induced to differentiate by this normal protein regulator. These experiments have shown that there is one type of myeloid leukemia cell, that we call MGI^+D^+, that can be induced by purified MGI (Fibach et al. 1972; Lotem and Sachs 1978) to differentiate normally to mature cells via the normal sequence of cell differentiation (Lotem and Sachs 1974, 1977; Sachs 1978). This type of leukemic cell has been identified in different strains of mice (Ichikawa 1969; Lotem and Sachs 1977) and humans (Paran et al. 1970), and normal differentiation in these cells can be induced in vitro and in vivo (Lotem and Sachs 1978). Differentiation in vivo can be enhanced by injecting MGI or MGI producing cells, and seems to be regulated by cells involved in the immune response (Lotem and Sachs 1978). Like normal mature macrophages and granulocytes, the mature cells induced from these leukemic cells are no longer malignant in vivo and no longer multiply in vitro (Sachs 1978). After we had identified (Ichikawa et al. 1966; Pluznik and Sachs 1965, 1966) the protein regulator that we now call MGI, we later called it mashran gm (Ichikawa et al. 1967) and it was then also referred to as colony stimulating factor (Metcalf 1969) and colony stimulating activity (Austin et al. 1971).

II. Origin and Further Development of Myeloid Leukemia

Our experiments have shown that these undifferentiated leukemic cells are malignant, not because they cannot be induced to differentiate by normal regulatory protein MGI, but because, unlike normal myeloid precursor cells, they no longer require MGI for cell viability and growth (Sachs 1978). The leukemic cells can, therefore, continue to multiply in the absence of MGI. These results have shown that leukemia can originate by the loss of a requirement of a normal regulatory protein for viability and growth in cells that can still be induced to differentiate normally by the normal protein regulator (Sachs 1978). This origin of leukemia is genetic and associated with a chromosome change (Azumi and Sachs 1977).

Experiments with different clones of myeloid leukemic cells have then shown that there can be further stages in the development of leukemia. The genetic change which allows the leukemic cells to grow in the absence of MGI can then be followed by other genetic changes that can produce different blocks in differentiation (Lotem and Sachs 1974; Hoffman-Lieberman and Sachs 1978; Lieberman and Sachs 1978; Sachs 1978). The isolation and study of such cell mutants have also made it possible to develop an experimental system that has been used genetically to dissect the controls that regulate induction of a variety of internal and external differentiation-associated markers that are switched on by the normal regulatory protein during differentiation to mature macrophages and granulocytes (Hoffman-Lieberman and Sachs 1978; Sachs 1978).

III. Cell Competence for Normal Differentiation

Experiments with leukemic cell clones with different degrees of competence for the induction of normal differentiation by MGI have shown, that differences in competence are associated with specific membrane changes including the mobility of certain surface receptors (Sachs 1974; Lotem and Sachs 1977), the ability for hormone desensitization (Simantov and Sachs 1978) and the production of type C RNA viruses (Lieberman and Sachs 1978). Genes for the expression and genes for the suppression of cell competence have been identified on two different chromosomes, numbers 2 and 12 in the mouse, and it was found that inducibility for differentiation by MGI is controlled by the balance between these genes (Azumi and Sachs 1977).

Studies with various compounds other that MGI, including those used in the present forms of cytotoxic cancer therapy, have shown that some of the stages of differentiation can be induced in appropriate clones of myeloid leukemic cells by various steroids, certain surface acting compounds and some compounds that interact with DNA (Sachs 1978). The use of appropriate cell mutants has shown that there can be different cellular sites for different compounds and that some compounds can act in mutant cells at differentiation sites that are no longer susceptible to the normal regulator MGI. In certain cases this activation of some stages of differentiation in the leukemic cells appear to be due to inhibition of the formation of repressors of the differentiation process (Sachs 1978). It was also found that some surface acting compounds can induce differentiation in clones with the appropriate genotype, by inducing in the cells that differentiate the production of the differentiation inducing protein MGI (Weiss and Sachs 1978).

IV. Autoregulation

This induction of MGI in cells that can be differentiated by MGI has shown that induction of differentiation by a normal regulatory protein may no necessarily be dependant upon interaction between different types of cells, but can be controlled by autoregulation. The induction, in the cells that differentiate, of regulatory proteins like MGI that can induce specific cell differentiation, may be a more general mechanism for the induction of differentiation by different inducers in various types of cells (Sachs 1978; Weiss and Sachs 1978). Our finding of another protein inducer (TCI) (Gerassi and Sachs 1979) that can induce the formation of normal T cell colonies (Gerassi and Sachs 1976, 1979) and can be produced by T cells, lends further support to the possibly important role of autoregulation in the control of normal cell growth and differentiation. Mouse erythroleukemic cells cannot be induced to differentiate by the normal erythroid inducing protein erythropoietin, but can be induced for some stages of differentiation by dimethylsulfoxide (Friend et al. 1971) and other surface acting compounds (Reuben et al. 1976). It will be of interest to determine if this also involves the induction of a specific normal differentiation inducing protein in the leukemic cells.

V. Therapeutic Possibilities

The results obtained also suggest novel forms of therapy for leukemia (Sachs 1978), which may be applicable to other diseases, based on the use of normal regulatory protein such as MGI to induce normal gene activation resulting in normal differentiation in malignant cells. MGI may also be useful in inducing a more rapid recovery of the normal cell population after the present forms of cytotoxic therapy. It may be further possible to use therapeutically other compounds that can induce MGI in vivo, or can affect mutant malignant cells at differentiation sites that are no longer susceptible to the normal regulator. Differences have been found in the competence of genotypically different malignant clones to be induced for some stages of differentiation by the chemicals and irradiation used in the commonly employed forms of cytotoxic therapy. This can also help to explain differences in response to treatments with cytotoxic therapy that have been found in different patients (Sachs 1978).

References

Austin, P.E., McCulloch, E.A., Till, J.E.: Characterization of the factor in L-cell conditioned medium capable of stimulating colony formation by mouse marrow cells in culture. J. Cell Physiol. 77, 121-134 (1971)
Azumi, J., Sachs, L.: Chromosome mapping of the genes that control differentiation and malignancy in myeloid leukemic cells. Proc. Natl. Acad. Sci. USA 74, 253-257 (1977)

Fibach, E., Landau, T., Sachs, L.: Normal differentiation of myeloid leukemic cells induced by a differentiation-inducing protein. Nature (New Biology) 237, 276-278 (1972)

Friend, C., Scher, W., Holland, J.G., Sato, T.: Hemoglobin synthesis in murine virus-infected leukemic cells in vitro: Stimulation of erythroid differentiation by dimethylsulfoxide. Proc. Natl. Acad. Sci. USA 68, 378-382 (1971)

Gerassi, E., Sachs, L.: Regulation of the induction of colonies in vitro by normal human lymphocytes. Proc. Natl. Acad. Sci. USA 73, 4546-4550 (1976)

Gerassi, E., Sachs, L.: Regulation of human T cell colonies by an inducing activity (TCI) produced by normal human and malignant mouse cells. J. Immunol. 121, 2547-2553 (1978)

Ginsburg, H., Sachs, L.: Formation of pure suspensions of mast cells in tissue culture by differentiation of lymphoid cells from the mouse thymus. J. Natl. Cancer Inst. 31, 1-40 (1963)

Ginsburg, H., Sachs, L.: Destruction of mouse and rat embryo cells in tissue culture by lymph node cells from unsensitized rats. J. Cell Comp. Physiol. 66, 319-324 (1965)

Hoffman-Lieberman, B., Sachs, L.: Regulation of actin and other proteins in the differentiation of myeloid leukemic cells. Cell 14, 825-834 (1978)

Ichikawa, Y.: Differentiation of a cell line of myeloid leukemia. J. Cell Physiol. 74, 223-234 (1969)

Ichikawa, Y., Pluznik, D.H., Sachs, L.: In vitro control of the development of macrophage and granulocyte colonies. Proc. Natl. Acad. Sci. USA 56, 488-495 (1966)

Ichikawa, Y., Pluznik, D.H., Sachs, L.: Feedback inhibition of the development of macrophage and granulocyte colonies. I. Inhibition by macrophages. Proc. Natl. Acad. Sci. USA 58, 1480-1486 (1967)

Landau, T., Sachs, L.: Characterization of the inducer required for the development of macrophage and granulocyte colonies. Proc. Natl. Acad. Sci. USA 68, 2540-2544 (1971)

Lieberman, D., Sachs, L.: Coregulation of type C RNA virus production and cell differentiation in myeloid leukemic cells. Cell 15, 823-835 (1978)

Lotem, J., Sachs, L.: Different blocks in the differentiation of myeloid leukemic cells. Proc. Natl. Acad. Sci. USA 71, 3507-3511 (1974)

Lotem, J., Sachs, L.: Genetic dissection of the control of normal differentiation in myeloid leukemic cells. Proc. Natl. Acad. Sci. USA 74, 5554-5558 (1977)

Lotem, J., Sachs, L.: In vivo induction of normal differentiation in myeloid leukemic cells. Proc. Natl. Acad. Sci. USA 75, 3781-3785 (1978)

Metcalf, D.: Studies on colony formation in vitro by mouse bone marrow cells. I. Continuous cluster formation and the relation of clusters to colonies. J. Cell Physiol. 74, 323-332 (1969)

Paran, M., Sachs, L., Barak, Y., Resnitzky, P.: In vitro induction of granulocyte differentiation in hematopoietic cells from leukemic and non-leukemic patients. Proc. Natl. Acad. Sci. USA 67, 1542-1549 (1970)

Pluznik, D.H., Sachs, L.: The cloning of normal "mast" cells in tissue culture. J. Cell Comp. Physiol. 66, 319-324 (1965)

Pluznik, D.H., Sachs, L.: The induction of clones of normal "mast" cells by a substance from conditioned medium. Exp. Cell Res. 43, 553-563 (1966)

Reuben, R.C., Wife, R.L., Breslow, R., Rifkind, R., Marks, P.A.: A new group of potent inducers of differentiation in murine erythroleukemic cells. Proc. Natl. Acad. Sci. USA 73, 862-866 (1976)

Sachs, L.: Analysis of regulatory mechanisms in cell differentiation. In: New Perspectives in Biology (ed. M. Sela), pp. 246-260. Amsterdam: Elsevier 1964

Sachs, L.: Regulation of membrane changes, differentiation, and malignancy in carcinogenesis. In: Harvey Lectures, Ser. 68, pp. 1-35. New York-London: Academic Press 1974

Sachs, L.: Control of normal cell differentiation and the phenotypic reversion of malignancy in myeloid leukemia. Nature (London) 274, 535-539 (1978)

Simantov, R., Sachs, L.: Differential desensitization of functional adrenergic receptors in normal and malignant myeloid cells. Relationship to receptor mediated hormone cytotoxicity. Proc. Natl. Acad. Sci. USA 75, 1805-1809 (1978)

Weiss, B., Sachs, L.: Indirect induction of differentiation in myeloid leukemic cells by lipid A. Proc. Natl. Acad. Sci. USA 75, 1374-1378 (1978)

Cellular Heterogeneities
in Acute Myeloblastic Leukemia

E.A. McCulloch, R.N. Buick, M.D. Minden, and C.A. Izaguirre

*Institute of Medical Science, University of Toronto
and The Ontario Cancer Institute, Toronto, Canada*

I. Introduction

The complexity of biological phenomena is well demonstrated by the heterogeneity of leukemic cell populations in acute myeloblastic leukemia (AML). This variation of the characteristics of the leukemic cells occurs within patients and between patients during the natural progression of the disease and following chemotherapy.

An analysis of this heterogeneity may accomplish several purposes. First, investigation may be directed towards understanding the generation of such heterogeneities in order to distinguish primary disturbances of growth and differentiation underlying AML from secondary phenomena of less importance. Second, for the oncologist, knowledge of the biological behavior of the leukemic cells may provide techniques to study malignancy during its progression; such data may provide objective criteria as a basis for staging AML that may prove useful in the design and interpretation of clinical trials.

This paper is planned as a description of some of the known heterogeneities in AML; where feasible, an opinion will be advanced concerning their generation.

II. Heterogeneity Within Individual Patients

Patients with AML suffer from two sets of leukemic manifestations: abnormal hemopoietic cellular growth and decreased marrow function. Microscopic examination of blood and bone marrow discloses the presence of a large percentage of abnormal cells. There is usually minimal evidence of production of red cells, platelets and granulocytes (myelopoiesis) whose functions are necessary for life. The abnormal cells, usually called blasts, are the defining population in AML.

Myelopoiesis including leukemic myelopoiesis, can now be studied in cell culture; using semisolid media and appropriate stimulators, granulopoietic (Pluznik and Sachs 1965; Bradley and Metcalf 1966), erythropoietic (Stephenson et al. 1971; Heath et al. 1976) and megakaryocytic (Metcalf et al. 1975; Nakeff and Daniels-McQueen 1976) colonies can be grown from mouse marrow. It is widely accepted (for a review see McCulloch and Till 1977) that the cells of origin of these

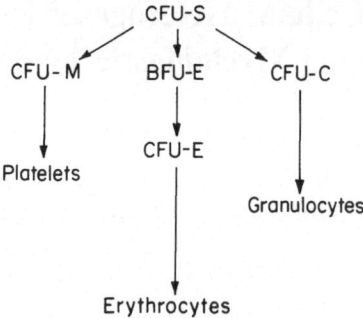

Fig. 1. Myelopoietic differentiation. CFU(colony forming unit)-S is a pluripotent hemopoietic stem cell detected by its capacity to form a spleen colony. CFU-M, BFU(burst forming unit)-E and CFU-C are early progenitors committed respectively to megakaryopoiesis, erythropoiesis and granulopoiesis. These cells have lost proliferative capacity in their transition from CFU-S, but gained new or increased sensitivity to regulators specific to their lineage. CFU-E is a precursor of erythroblasts. Reprinted from Cancer (McCulloch et al. 1978c) with permission of the publisher, J.B. Lippincott

colonies are very primitive precursors separated from pluripotent stem cells by relatively few differentiation events (Fig. 1). Until recently, the stem cells themselves eluded culture technology; they were usually detected and quantified by different techniques, based upon their capacity to form macroscopic clones in the spleens of suitable irradiated (Till and McCulloch 1961) or genetically anemic (McCulloch et al. 1964) recipient mice. In 1977, Johnson and Metcalf described, in cultures of CBA mouse fetal liver, colonies containing granulopoietic and erythropoietic cells; mixed colony formation depended upon stimulation by supernatant media from spleen cells cultured in the presence of pokeweed mitogen (Johnson and Metcalf 1977). They have provided evidence that the progenitors of these colonies may be a subpopulation of pluripotent stem cells, perhaps .characteristic of fetal hemopoiesis (Metcalf et al. 1979).

With the exception of the precursors of megakaryocytes, the mouse assays for committed progenitors have been adapted to human cells. Large erythropoietin dependent multilobulated colonies or bursts are considered to derive from the most primitive erythropoietic progenitors (BFU-E) (Iscove and Sieber 1975); small colonies of erythroblasts are the progeny of progenitors (CFU-E) closer to morphologically identified pro-erythroblasts (Tepperman et al. 1974). Granulo-poietic colonies depend upon the presence in the media of either feeder cells (usually derived from hemopoietic tissues) or products of these cells released into culture media. These colonies are considered to contain the progeny of a heterogeneous population of early granulopoietic progenitors (CFU-C) (for a review see McCulloch 1965). In 1978, Messner and Fauser reported colonies containing both granulocytes and erythroblasts in cultures of normal adult human marrow. Like the Johnson and Metcalf method, they found colony formation to be dependent upon supernatants of mitogen [phytohemagglutinin (PHA) rather than pokeweed] stimulated cells; unlike Johnson and Metcalf, erythropoietin was required for the growth of the erythropoietic components of the colonies. The place

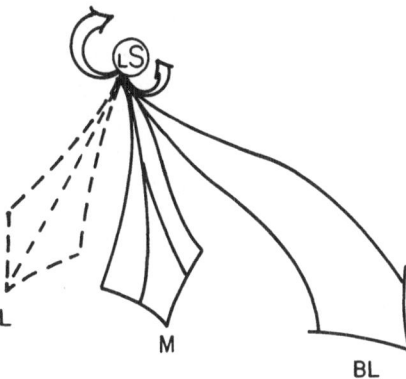

Fig. 2. A clone in AML. The clone is shown originating in transformed pluripotent stem cells (LS), which proliferates extensively and renews itself (Buick et al. 1979b). Myelopoietic (M) and lymphopoietic (L) lineages derive from LS; these are shown in a fashion symbolic of exponential expansion. Lymphopoiesis is shown in *dotted lines* because its relationship to LS in AML is not defined

of the cells of origin of these mixed colonies in hemopoietic differentiation remains to be determined.

Colonies of granulopoietic (Brown and Carbone 1971; Cowan et al. 1972; Moore et al. 1973, 1974; Spitzer et al. 1976) or erythropoietic differentiating (Lan et al. 1978) cells have been obtained from the marrow of many patients with AML. Chromosomal markers identified in fresh metaphase preparations from certain leukemic patients have been found in granulopoietic colonies (Duttera et al. 1972; Moore et al. 1972; Aye et al. 1973) and recognizable erythropoietic cells (Blackstock and Garson 1974). This genetic evidence indicates that leukemic transformation occurs at the level of pluripotent stem cells; further, the transformed cells continue to differentiate into at least two of the pathways of myelopoiesis. Differentiation may therefore be considered as a source of heterogeneity among leukemic populations in individual patients.

Blast cells of AML are homogeneous by morphological criteria. However, when studied in cell culture heterogeneity becomes evident. Colonies made up of cells morphologically similar to blast cells can be grown in cultures made viscid with methyl cellulose and containing, as stimulator, media conditioned by leukocytes in the presence of PHA (PHA-LCM) (Buick et al. 1979a; McCulloch et al. 1978b). The plating efficiency of the progenitors of blast colonies is small; from 1 in 10^{-2} to 1 in 10^{-4} cells of the blast population give rise to colonies. Colonies themselves, when examined by replating are found to contain both new blast progenitors (self renewal) and cells with blast morphology but lacking colony forming capacity (Buick et al. 1979b; McCulloch et al. 1979). The evidence has been interpreted to mean that cells belonging to a small population of blast progenitors can either renew themselves or give rise to descendents with reduced proliferative potential. The latter process may be considered analogous to differentiation in normal hemopoiesis, since differentiation to recognizable cells is associated with reduced proliferative potential. From this point of view, the blast population, like the myelopoietic population, is heterogeneous by reason of changes in cell properties occurring during growth.

In patients with chromosomal markers in blood or marrow, these have been identified in cells from blast colonies in culture (Buick et al. 1977; McCulloch et al. 1978b). Taken together with similar chromosomal evidence for myelopoiesis, leukemic colonies may be considered to originate in pluripotent stem cells. By differentiation or analogous processes, these give rise to the lineages of myelopoiesis and to blast cells; the latter populations have some of the characteristics of independent subclones because of their capacity for self renewal. A model of a leukemic clone is proposed schematically in Fig. 2. Regardless of the details of the model, heterogeneity within patients is a striking feature of leukemic hemopoiesis.

III. Heterogeneity Among Patients with AML

When the cell culture assays for hemopoietic progenitors were applied to marrow or peripheral blood from patients with acute myeloblastic leukemia (AML), marked patient-to-patient variation was observed (Moore et al. 1974; Curtis et al. 1975; Spitzer et al. 1976; Lan et al. 1978). In many patients the progenitors of granulopoiesis and erythropoiesis appeared to be reduced, while in some their numbers were increased. Moreover, numerical correlations were observed between the progenitors; that is, marrow with poor plating efficiency for granulopoietic progenitors usually also gave rise to few erythropoietic colonies when plated in the presence of erythropoietin (Lan et al. 1978). The question arose as to whether these patterns of growth were characteristic of subclasses of AML, not detected by conventional procedures.

An approach to this problem was suggested by the precedent of chronic myeloblastic leukemia (CML). This disease, like AML, originates in pluripotent stem cells (Whang et al. 1963; Fialkow et al. 1977); CML is closely related to AML and in its final blastic phase appears to transform into the acute disease. Genetic evidence is available, based on studies of the isoenzymes of glucose-6-phosphate dehydrogenase in heterozygous females with CML, that the disease is clonal, with all myelopoietic tissue being derived from a single progenitor (Fialkow et al. 1977). Such clonal dominance could arise readily if CML clones generally had growth advantages over normal clones; if transformation from normal to CML occurred in a single stem cell, dominance would be required for clonal survival. Even if more than one stem cell was transformed, sufficient time is available in the human life span for the observed clonal dominance to develop.

Equivalent genetic evidence for clonality has yet to be acquired for AML; nonetheless, it is reasonable to propose that the acute disease, like the chronic form, might be predominantly clonal at the time of presentation. If the hemopoietic population in each patient with AML is a clone, then techniques developed for assessing individual hemopoietic clones in mice (spleen colonies) may be applied appropriately. Spleen colonies are known to arise from a relatively homogeneous population of murine pluripotent stem cells. Yet, after 20 divisions, marked colony-to-colony variation is observed in cellular composition. This variation has been attributed to stochastic processes occurring during clonal expansion (Till et al. 1964; Korn et al. 1973); that is, alternative fates are available to stem cells and their early descendents; for example, each stem cell may either renew itself or differentiate and if differentiation occurs one of a number of alternative pathways may be initiated. If these alternative fates occur at random, governed only by definite probabilities, the observed variation is explained.

The patient-to-patient variation in AML is very similar to the colony-to-colony variation observed under experimental conditions in mice; considering each patient contains only a single hemopoietic clone, the heterogeneity observed between patients might, like that of spleen colonies, be based upon random events occurring during clonal expansion. A test of this hypothesis proved feasible because the usual treatment for AML involves reducing the size of the clone with chemotherapeutic drugs. If the observed patient-to-patient variation showed an inheritable pattern, then each clone regrowing after chemotherapy should reproduce its original pattern. Alternatively, a random model predicts that heterogeneity would be seen after reexpansion but that no correlation would be found in the rank of individual patients in the pre- and post-treatment distributions.

Granulopoietic progenitors were assessed in the marrows of a series of patients before and six weeks after chemotherapy (Till et al. 1978). At each time, the values for granulopoietic colony formation formed an asymmetrical distribution; however, as predicted in the stochastic model, patients were reassorted at random in the post-treatment distribution. These observations support the view that random events during clonal expansion are important in generating the observed heterogeneity between patients with AML.

Neither are blast populations homogeneous among patients with AML. Morphologically recognized blasts may be present in the marrow from 10% of nucleated cells (lowest value at which the diagnosis can be made) to nearly 100%. Similarly blasts may not be found in the peripheral blood at all or then may constitute 90–100% of a greatly increased peripheral leukocyte concentration. This variation provides an opportunity to relate blast cells to myelopoietic progenitors showing a similar range of numbers. The approach derives from the use of numerical correlation; for less closely linked classes many randomizing events will lineage relationships in normal hemopoiesis (Gregory and Henkelman 1977). The statistical method supposes that cells closely related in a differentiation lineage have few opportunities for randomization during growth and consequently maintain numerical correlatation; for less closely linked classes many randomizing events will reduce numerical correlations. When this approach was applied to myelopoietic progenitors in AML, the observed correlations among the progenitor classes were consistent with those expected from known differentiation patterns (Lan et al. 1978). In contrast, correlations were not found between blast cells and erythropoietic or granulopoietic progenitors. Specifically, the inverse correlation that might have been anticipated if blast cells represented immature forms blocked from differentiation was not observed. These data provide the basis for depicting the blast population independently from myelopoietic lineages in Fig. 2.

Heterogeneity among patients with AML may derive from random events during clonal expansion. If this interpretation of the data is correct, the variation is not a property of leukemia; rather, it is a manifestation of clonal, as opposed to polyclonal, hemopoiesis.

IV. Heterogeneity in Time

In general, transformation from normal to fully malignant cells is considered to be a multi-step process. Evidence is available that AML originates in a similar fashion. In many patients, overt AML is preceded by one of a number of recognizable preleukemic conditions (Pierre 1975; Linman and Bagby 1976). In some instances chromosomal abnormalities in preleukemia are consistent with the presence of one or more abnormal clones (Nowell 1977). The clonal nature of CML and its frequent progress into AML have already been mentioned, together with interpretation that CML clones have a growth advantage over normal clones. Abnormal clones in preleukemia may be considered to predominate since the distribution of differentiated cells within such clones provides the clinical picture upon which the diagnosis rests. A second general property shared by clones intermediate between normal and AML is genetic instability. This property, presumably varying from clone to clone, may underlie the observed clinical progression.

It is not excluded that genetically unstable clones with growth advantage may preceed AML in those cases without clinically recognized preleukemia. If such clones contained normal distributions of mature and functional cells their presence would not be detected. Precedence for this possibility comes from recent observations by Moore (Moore and Sheridan 1979); he has reported the isolation of hemopoietic clones from Friend virus infected long term cultures of mouse marrow. These clones were abnormal only in their capacity to survive multiple passages in irradiated animals, protecting their hosts from marrow failure at each passage. Such clones have the properties that might be anticipated of apparently normal populations preceding AML.

Clonal evolution continues following the appearance of obvious acute leukemia. Sometimes abnormal chromosomes are not recognized at onset but become apparent with progression of the disease and following treatment. These abnormalities together with those recognized at first presentation provide the basis for an important problem: how many of the clones persist at a point in time? Particularly, do normal clones survive the evolution of AML? Lacking genetic markers, clonal succession has not been mapped confidently. It has been observed, however, that abnormal chromosomes detected during relapse are reduced in number or disappear when remission is induced. This observation has been interpreted as evidence of reexpansion of normal clones when AML clones are suppressed. It is evident, however, from the considerations outlined above, that an apparently normal phenotype is not satisfactory evidence that expanded clones are not leukemic.

V. Heterogeneity in the Properties of Blast Cell Progenitors

The heterogeneities described in the previous sections are derived from measurements of progenitor number. Another approach is based on determining properties; three sets of data are available for blast cell progenitors. These are measurements of proliferative state (Minden et al. 1978), capacity for self renewal (Buick et al. 1979b), and sensitivity to chemotherapeutic agents (Buick et al. 1979a).

For minority populations not identified by morphological criteria, cell cycle measurements cannot be made using tritiated thymidine (^{3}HTdR) auto-radiography. Rather percentage of cells in the DNA synthetic (S) phase of the cycle is determined indirectly by measuring reduction in colony forming ability after brief exposure to S-phase specific inhibitors such as high specific activity ^{3}HTdR (Becker et al. 1965) or hydroxyurea (Byron 1972). Data is available for 15 patients with AML; uniformly, a high proportion of progenitors were found to be in the S phase. It may be concluded that blast progenitors are in a state of rapid cell cycle. A homogeneous high proliferative pattern is consistent with a role for blast cell progenitors in producing and maintaining a large population of morphologically identified blast cells, many with low proliferative potential.

Cells that maintain populations of descendents require the capacity to renew their own numbers. Replating in cell culture has been used to search for this property. Either cells collected from dishes containing many discrete colonies or individual colonies may be replated; secondary colonies obtained by this procedure may be characterized and compared with primaries. Using these methods, secondary colonies with morphological and cultural properties identical with primaries were obtained from cultures of blast cells from 11 of 15 patients (Buick et al. 1979b). However, marked heterogeneity was again observed in the group; the plating efficiency of pooled suspensions from primary colonies varied greatly and the distribution of values was asymmetrical, with blast progenitors from many patients showing little self-renewal capacity while in a few, self-renewal was extensive.

A similar variation among patients was observed in measurements of the sensitivity of blast cell progenitors to the anthracycline chemotherapeutic agent, adriamycin (Buick et al. 1979a). Dose response curves were obtained by exposing blast cells briefly to increasing concentration of drug, washing twice and then plating for colony formation. An exponential decrease in colony formation was observed with increasing drug concentration, permitting sensitivity to be expressed as the dose of adriamycin required to reduce colony formation to 10% of control (D_{10} adria). In a series of 16 patients with AML D_{10} adria values varied from 0.47 to 20.6 μg ml.

Earlier, when heterogeneity was discussed in respect to myelopoiesis, the question was posed whether or not a given growth pattern was heritable. The same question is clearly relevant to the observed variation in self-renewal capacity and adriamycin sensitivity of blast cell progenitors. At a technical level, consistent values have been demonstrated readily for the same cells using cryopreserved samples. Blasts can also be tested for self-renewal at the level of primary and secondary colonies; again, consistent results were obtained indicating that blast

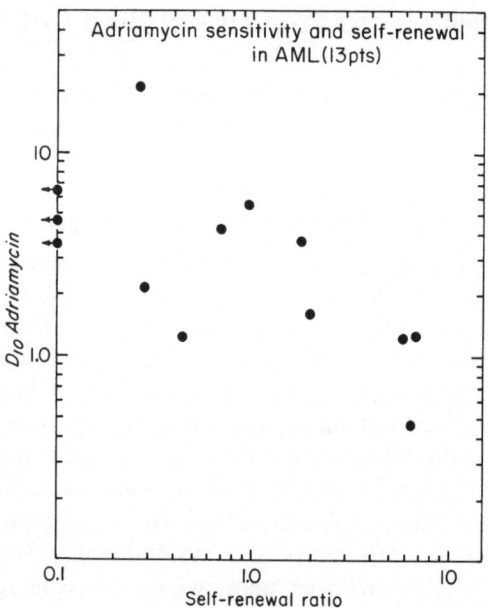

Fig. 3. A log-log plot of values for self-renewal and adriamycin sensitivity for 13 AML patients. The marked patient-to-patient variation for each parameter is apparent. A significant (Spearman rank correlation coefficient = 0.70, p < 0.5) correlation is seen between adriamycin sensitivity and self-renewal for this small series

progenitors that are progeny of cells with high self-renewal capacity can themselves be replated successfully. But, the crucial test is to examine the cells from the same patients before and after complete remission. Opportunities to make this comparison have not yet arisen. One unusual patient developing AML following ovarian carcinoma had blast cells resistant in culture to adriamycin and cytosine arabinoside; this patient failed to respond to treatment and cells could be obtained at different times during the course of her disease. These samples showed the same pattern of drug resistance.

In assembling the data on self-renewal and response to adriamycin an inverse correlation was observed between the two values (McCulloch et al. 1978a). This is shown in Fig. 3 for 13 patients where both measurements were made; (Spearmans rank correlation coefficient = 0.70 p < 0.05). At this level of significance, the association between the values might be fortuitous. If it is real, a link between such diverse functions as self-renewal and adriamycin sensitivity might implicate a common biological process, perhaps one affected at the time of leukemic transformation, and genetically conserved through subsequent cell generations.

VI. Conclusions

AML and its variants are known to be heterogeneous clinically. Features of the disease, such as percentage blasts in the marrow, and of the patient, such as age, are known to affect response to treatment. When cell culture methods are used to study hemopoietic progenitors in the disease, even more striking heterogeneity becomes apparent. Moreover, much of the heterogeneity cannot be attributed specifically to disease. For example, apparently striking abnormalities in numbers of myelopoietic progenitors may be attributed to random events occurring during the expansion of the single clone that dominates hemopoiesis in each patient.

The observed heterogeneity provides a significant determinant of research approaches. Model systems have limited value since even a model developed from human disease, cannot reflect the great patient-to-patient variation. Studies based on measurements of numbers of cells in AML are unlikely to detect fundamental lesions since variation in number may have a physiological rather than a pathological basis.

Even plans of management may be affected by heterogeneity; at present, patients with AML are usually treated using precisely defined protocols designed for clinical trials. Only approximately 50% of AML patients obtain remission with present regimes, and even this favorable response lasts for variable, but usually short times. Known prognostic factors account for less than 30% of the observed variance between patients. Perhaps the improvement in therapeutic outcome must await regimens that consider the biological characteristics of individual clones in each patient.

It remains to be determined if some of the patient-to-patient variation can be attributed to events at or close to leukemic transformation. A resolution of the question depends upon determining which characteristics of cellular populations are constant during the natural history of the disease, particularly when the course includes successful remission induction. If characteristics defining individual clones are identified, each patient may be perceived to contain a somatic mutant of hemopoiesis. Precedent from other systems indicates that mutants are powerful tools for fundamental studies of cellular processes. For clinical research, appreciation of the cellular basis for variation among patients may lead to meaningful individualization of treatment.

Acknowledgments. The work reported in this paper was supported by the Ontario Cancer Treatment and Research Foundation, the Medical Research Council of Canada (MT 1420) and the National Cancer Institute of Canada.

RNB was a Special Fellow of the Leukemia Society of America. His address is: Section of Hematology/Oncology, Dept. of Internal Medicine, Health Sciences Center, University of Arizona, Tucson, Arizona, 85724.

MDM is a Fellow of the Medical Research Council.

CAI is a Fellow of the National Cancer Institute of Canada.

References

Aye, M.T., Till, J.E., McCulloch, E.A.: Cytological studies of granulopoietic colonies from two patients with chronic myelogenous leukemia. Exp. Hematol. 1, 115-118 (1973)

Becker, A.J., McCulloch, E.A., Siminovitch, L., Till, J.E.: The effect of differing demands for blood cell production on DNA synthesis by hemopoietic colony-forming cells of mice. Blood 26, 296-308 (1965)

Blackstock, A.M., Garson, O.M.: Direct evidence for involvement of erythroid cells in actute myeloblastic leukemia. Lancet 2, 1178-1179 (1974)

Bradley, T.R., Metcalf, D.: The growth of mouse bone marrow cells in vitro. Aust. J. Exp. Biol. Med. Sci. 44, 287-300 (1966)

Brown, C.H., Carbone, P.P.: In vitro growth of normal and leukemic human bone marrow. J. Natl. Cancer Inst. 46, 989-1000 (1971)

Buick, R.N., Till, J.E., McCulloch, E.A.: Colony assay for proliferative blast cells circulating in myeloblastic leukaemia. Lancet 1, 862-863 (1977)

Buick, R.N., Minden, M.D., McCulloch, E.A.: Self renewal in culture of proliferative blast progenitor cells in acute myeloblastic leukemia. Blood 54, 95-104 (1979a)

Buick, R.N., Messner, H.A., Till, J.E., McCulloch, E.A.: Cytotoxicity of adriamycin and daunorubicin for normal and leukemic progenitor cells of man. J. Natl. Cancer Inst. 62, 249-255 (1979b)

Byron, J.W.: Comparison of the action of ^{3}H-thymidine and hydroxyurea on testosterone-treated haemopoietic stem cells. Blood 40, 198-215 (1972)

Cowan, D.H., Clarysse, A., Abu-Zahra, H., Senn, J.S., McCulloch, E.A.: The effect of remission induction in acute myeloblastic leukemia on colony formation in culture. Ser. Haematol. 5, 179-188 (1972)

Curtis, J.E., Cowan, D.H., Bergsagel, D.E., Hasselback, R., McCulloch, E.A.: Acute leukemia in adults: Assessment of remission induction with combination chemotherapy by clinical and cell culture criteria. Can. Med. Assoc. J. 113, 289-294 (1975)

Duttera, M.J., Whang-Peng, J., Bull, J.M.C., et al.: Cytogenetically abnormal cells in vitro in acute leukemia. Lancet 1, 715-717 (1972)

Fialkow, P.J., Jacobson, R.J., Papayannopoulou, T.: Chronic myelocytic leukemia: Clonal origin in a stem cell common to the granulocyte, erythrocyte, platelet and monocyte macrophage. Am. J. Med. 63, 125-130 (1977)

Gregory, C.J., Henkelman, R.M.: Relationships between early hemopoietic progenitor cells determined by correlation analysis of their numbers in individual spleen colonies. In: Experimental Hematology Today (eds. S.J. Baum, G.D. Ledney), pp. 93-101. Berlin-Heidelberg-New York: Springer 1977

Heath, D.S., Axelrad, A.A., McLeod, D.L., Shreeve, M.M.: Separation of the erythropoietin-responsive progenitors BFU-E and CFU-E in mouse bone marrow by unit gravity sedimentation. Blood 47, 777-792 (1976)

Iscove, N.N., Sieber, F.: Erythroid progenitors in mouse bone marrow detected by macroscopic colony formation in culture. Exp. Hematol. 3, 32-43 (1975)

Johnson, G.R., Metcalf, D.: Pure and mixed erythroid colony formation in vitro stimulated by spleen conditioned medium with no detectable erythropoietin. Proc. Natl. Acad. Sci. 74, 3879-3882 (1977)

Korn, A.P., Henkelman, R.M., Ottensmeyer, F.P., Till, J.E.: Investigations of a stochastic model of haemopoiesis. Exp. Hematol. 35, 1-14 (1973)

Lan, S., McCulloch, E.A., Till, J.E.: Cytodifferentiation in the acute myeloblastic leukemias of man. J. Natl. Cancer Inst. 60, 265-269 (1978)

Linman, J.S., Bagby, G.C., Jr.: The preleukemic syndrome: Clinical and laboratory features, natural course and management. Blood Cells 12, 11-31 (1976)

McCulloch, E.A.: Granulopoiesis in cultures of human hemopoietic cells. Clin. Haematol. 4, 509-533 (1975)

McCulloch, E.A., Till, J.E.: Stem cells in normal early haemopoiesis and certain clonal haemopathies. In: Recent Advances in Haematology, Vol. II (eds. A.V. Hoffbrand, M.C. Brain, J. Hirsch), pp. 85-110. Edinburgh-London-New York: Churchill-Livingstone 1977

McCulloch, E.A., Siminovitch, L., Till, J.E.: Spleen colony formation in anemic mice of genotype W/W^v. Science 144, 844-846 (1964)

McCulloch, E.A., Buick, R.N., Minden, M.D., Izaguirre, C.A.: Differentiation programmes underlying cellular heterogeneity in the myeloblastic leukemias of man. In: Hematopoietic Cell Differentiation (eds. D.W. Golde, M.J. Cline, D. Metcalf, C.F. Fox), pp. 317-333. New York: Academic Press 1978a

McCulloch, E.A., Buick, R.N., Till, J.E.: Cellular differentiation in the myeloblastic leukemias of man. In: Cell Differentiation and Neoplasia (ed. G.F. Saunders), pp. 211-221. New York: Raven Press 1978b

McCulloch, E.A., Buick, R.N., Till, J.E.: Normal and leukemic hemopoiesis compared. Cancer 42, 845-853 (1978c)

McCulloch, E.A., Howatson, A.F., Buick, R.N., Minden, M.D., Izaguirre, C.A.: Acute myeloblastic leukemia considered as a clonal hemopathy. Blood Cells 5, 261-282 (1979)

Messner, H.A., Fauser, A.: Distribution of fetal hemoglobin in individual colonies by RIA: A lead towards human stem cells. In: Cellular and Molecular Regulation of Hemoglobin Switching (eds. G. Stamatoyannopoulos, D. Nienhuis), p. 379. New York: Grune&Stratton 1978

Metcalf, D., MacDonald, H.R., Odarthchenko, N., Sordat, B.: Growth of mouse megakaryocyte colonies in vitro. Proc. Natl. Acad. Sci. USA 72, 1744-1748 (1975)

Metcalf, D., Johnson, G.R., Mandel, T.E.: Colony formation in agar by multipotential hemopoietic cells. J. Cell. Physiol. 98, 401-420 (1979)

Minden, M.D., Till, J.E., McCulloch, E.A.: Proliferative state of blast cell progenitors in acute myeloblastic leukemia. Blood 52, 592-600 (1978)

Moore, M.A.S., Sheridan, A.: Pluripotent stem cell replication in continuous human, prosimian and murine bone marrow culture. Blood Cells 5, 297-304 (1979)

Moore, M.A.S., Williams, N., Metcalf, D.: Characterization of in vitro colony forming cells in acute and chronic myeloid leukemia. In: The Nature of Leukemia (ed. P. Vincent), pp. 135-249. Proc. Int. Cancer Conf., Sidney, Australia 1972

Moore, M.A.S., Williams, N., Metcalf, D.: In vitro colony formation by normal and leukemic human hematopoietic cells: Interaction between colony forming and colony stimulating cells. J. Natl. Cancer Inst. 50, 591-602 (1973)

Moore, M.A.S., Spitzer, G., Williams, N., Metcalf, D., Buckley, J.: Agar culture studies in 127 cases of untreated acute leukemia: the prognostic value of reclassification of leukemia according to in vitro growth characteristics. Blood 44, 1-18 (1974)

Nakeff, A., Daniels-McQueen, S.: In vitro colony assay for a new class of megakaryocyte precursor: colony forming unit megakaryocyte (CFU-M). Proc. Soc. Exp. Biol. Med. 151, 587-590 (1976)

Nowell, P.C.: Preleukemia—Cytogenetic clues in some confusing disorders. Am. J. Pathol. 89, 459-473 (1977)

Pierre, R.V.: Cytogenetic studies in preleukemia: Studies before and after transition to acute leukemia in 17 subjects. Blood Cells 1, 163-170 (1975)

Pluznik, D.H., Sachs, L.: The cloning of normal "mast" cells in tissue culture. J. Cell. Comp. Physiol. 66, 319-324 (1965)

Spitzer, G., Dicke, K.A., Gehan, E.A., Smith, T., McCredie, K.B., Barlogie, B., Freireich, E.J.: A simplified in vitro classification for prognosis in adult acute leukemia: the application of in vitro results in remission predictive models. Blood 48, 795-807 (1976)

Stephenson, J.R., Axelrad, A.A., McLeod, D.C., Shreeve, M.M.: Induction of colonies of hemoglobin-synthesizing cells by erythropoietin in vitro. Proc. Natl. Acad. Sci. USA 68, 1542-1546 (1971)

Tepperman, A.D., Curtis, J.E., McCulloch, E.A.: Erythropoietic colonies in cultures of human marrow. Blood 44, 659-669 (1974)

Till, J.E., McCulloch, E.A.: A direct measurement of the radiation sensitivity of normal mouse bone marrow cells. Radiat. Res. 14, 213-222 (1961)

Till, J.E., McCulloch, E.A., Siminovitch, L.: A stochastic model of stem cell proliferation based on the growth of spleen colony forming cells. Proc. Natl. Acad. Sci. USA 51, 29-36 (1964)

Till, J.E., Lan, S., Buick, R.N., Sousan, T., Curtis, J.E., McCulloch, E.A.: Approaches to the evaluation of human hemopoietic stem cell function. In: Differentiation of Normal Neoplastic Hematopoietic Cells (Book B) (eds. B. Clarkson, P.A. Marks, J.E. Till), pp. 81-92. Cold Spring Harbor Symp. 1978

Whang, J., Frei, E.J., Tijo, J.H., Carbone, P.P., Brecher, G.: The distribution of the Philadelphia chromosome in patients with chronic myelogenous leukemia. Blood 22, 664-673 (1963)

Growth Regulations of Human Malignant Cells

A.M. Mauer, S.B. Murphy, and F.A. Hayes

*St. Jude Children's Research Hospital, 332 North Lauderdale
Memphis, TN 38101, USA*

I. Introduction

Until recently it was accepted that a hallmark of malignant cells was their lack of responsiveness to normal growth regulation. In fact, malignant cell populations were generally characterized as being rapidly growing and out of control.

Since that time, however, it has been appreciated that, in fact, there are three general areas of growth regulation for a malignant cell population. The first area of regulation is the duration of the cell cycle (generation time). Most malignant cell populations have cell cycle times longer than their normal counterpart. Some studies have indicated changes in cell cycle times during different growth phases of a tumor, but in general these changes contribute little to the rate of growth of the cell population. Thus, the cell cycle time does not appear to influence changes in growth characteristics of tumor cells.

Another important aspect of the growth rate of a tumor cell population is the rate of tumor cell loss. This rate can be measured in some experimental systems but is virtually impossible to measure in human tumors. Although it contributes substantially to the growth characteristics of tumors and is certainly an important area of modification during therapy, there do not appear to be specific regulatory mechanisms for determining cell loss rate.

The most important determinant for growth characteristics of tumor cells appears to be the proportion of the cell population in the cell cycle (growth fraction). Evidence has been presented that the tumor cell population is comprised of both dividing and resting cells. The resting cells, or at least a portion of them, return to active cell division both in the steady state (Saunders and Mauer 1969) and following chemotherapeutic perturbation (Gabutti et al. 1969). It is the size of the growth fraction in proportion to the total population of dividing and resting cells that determines the tumor cell doubling time, along with the important factor of cell loss. It is apparent from available studies that the main regulatory mechanism for tumor cell growth acts through adjusting the size of the growth fraction.

In this paper the evidence for three levels of growth regulation, systemic, regional, and cellular will be presented.

II. Methodology

The methods used for these studies are the determination of tumor cell labeling index by autoradiography after exposure to tritiated thymidine and mitotic index by determining the percent of mitotic figures within the tumor cell population. Samples have been obtained from blood, bone marrow and sites of solid tumor formation in patients with acute lymphocytic leukemia, non-Hodgkin's lymphoma and neuroblastoma.

III. Results

A. Evidence for Systemic Control

Early evidence for some systemic control mechanism was the indication that samples of bone marrow obtained from different sites after the injection of tritiated thymidine had similar values for labeling and mitotic indices (Mauer and Fisher 1966). This was subsequently confirmed by a study in which multiple marrow samples were obtained from the same patient over a 9 day period (Saunders and Mauer 1969). The size of the growth fraction varies with the phase of tumor cell growth and characteristically is least at the time of diagnosis and greatest at a time of subsequent relapse after a period of successful control with chemotherapy (Saunders et al. 1967). The results of these studies done in patients with leukemia have been confirmed at this institution for patients with non-Hodgkin's lymphoma and neuroblastoma.

B. Regional Influences

One of the earliest indications that tumor cell proliferation could be influenced by the site that the cell occupies developed from a study in which it was found that the labeling index of leukemic cells in blood was almost always less than that of cells in the bone marrow (Mauer and Fisher 1962). Recently, it has been possible to do simultaneously labeling indices at different sites in patients with neuroblastoma and non-Hodgkin's lymphoma. Representative results from these studies are shown in Table 1. Although there is a general tendency for cells growing freely in body fluids or at metastatic sites to have greater proliferative activity, that finding was not a consistent one.

C. Intrinsic Cellular Regulation

It has long been known that lines of cells grown in tissue culture have a characteristic density which they can achieve. From studies done to determine the labeling indices of different lymphoid malignancies, there is evidence that similar intrinsic mechanisms may also determine limits of cell growth.

Table 1. Labeling indices as a function of site of tumor cell growth

Tumor	Site		
Neuroblastoma	Primary	Lymph node	Bone marrow
Patient 1	16.1	21.3	
Patient 2	21.0	21.0	
Patient 3	0.4		14.8
Patient 4	4.0		19.2
Patient 5	10.8		8.0
Patient 6	18.0		23.4
Non-Hodgkin's lymphoma	Omentum	Acitic fluid	Pleural fluid
Patient 1	42	26	
Patient 2		28.5	39.4

Table 2. Bone marrow labeling indices in three lymphoid malignancies

Type of cell	Number of patients	Median labeling index (%)
Common ALL	44	4.8
T-cell ALL	11	13.1
Non-Hodgkin's lymphoma	16	36.5

Shown in Table 2 are the median labeling indices at diagnosis for the common form of acute lymphocytic leukemia, the E-rosette forming (T-cell) lympocytic leukemia and the marrow tumor cells of leukemic transformation of non-Hodgkin's lymphoma.

IV. Discussion

From the studies presented, there is clear evidence that systemic, regional, and intrinsic cellular mechanisms are involved in the regulation of tumor cell growth. All of these mechanisms appear to operate through the determination of the size of the tumor cell growth fraction.

The systemic regulation is most likely through the elaboration of a humoral factor which is in turn dependent upon tumor cell density, or total mass. It is unknown at this time whether the substance is stimulatory, resulting in an increased growth fraction, or inhibitory, leading to a greater proportion of the tumor cell population entering and staying in the resting cell fraction. Of greatest importance, from the standpoint of therapy, is the observation that the tumor cell growth characteristics can be perturbed by chemotherapy. The induction of a treatment regimen frequently results in an enhancement of the growth fraction which may in turn lead to a greater sensitivity of this cell population to drugs affecting primarily cells in the cell cycle. Not all cell populations can be so modified by chemotherapy, an observation which has been correlated with clinical response in neuroblastoma (Hayes et al. 1977). In that study, in those patients in whom a significant increase in labeling index was seen following chemotherapy, a clinical response of tumor cell regression was observed. In contrast, in patients in whom no alteration in growth fraction could be induced by chemotherapy, no clinical response was subsequently seen. These results need testing in other tumor systems, as well as further studies to determine reasons for both the kinetic and clinical unresponsiveness of the resistant tumor cells.

The reasons for regional influences may be found in differences in vascular supply with alterations in oxygen and nutrient delivery. However, there may also be other influences mediated through cell density and operating through the same mechanisms as the systemic growth regulation. These factors need further study because areas with increased numbers of resting cells may represent sanctuary areas for drug- and radiation-resistant tumors. With a better understanding of the regional effects on tumor cell growth, methods for enhancing sensitivity to treatment might be devised.

Of great interest is the observation that within a specific group of tumors derived from the same normal cell population, intrinsic differences in growth capacity may exist. Current evidence suggests that the standard or common ALL is the most primitive or least differentiated of the lymphoid malignancies. Acquisition of surface markers characteristic of T- or B-lymphocytes may correspond to the similar characteristics of differentiated normal lymphocytes of these cell lines. From the results of studies presented here, there is indication that with the acquisition of differentiation markers proliferative activity is increased.

The importance of the intrinsic cellular regulation of growth can be measured by the densities to which tumor cells grow in leukemia and in lymphoma. In general, the patient with standard or common ALL has a relatively small body burden of tumor at diagnosis and the smallest labeling index. Patients with T-cell leukemia and non-Hodgkin's lymphoma at leukemic transformation have progressively greater tumor masses and yet their tumor cell growth fractions are large. It would seem that the intrinsic mechanisms in these latter cell types allow growth to great densities and huge tumor masses and still maintain a large growth fraction.

It will be interesting to further test the hypothesis that with a greater degree of differentiation, tumors can also reach greater cell masses and maintain large growth potential. It will be important also to test this observation in the normal cell line for origin of the specific tumor type.

V. Summary

Evidence for growth regulation of tumor cell lines is growing. Three levels of regulation, systemic, regional, and cellular have been discussed in this paper. A hypothesis relating tumor cell proliferative activity with the degree of differentiation of the specific cell line has been presented. The characteristics of tumor cell growth have important implications for the clinical characteristics of the tumor and responsiveness to therapy.

Acknowledgments. This work was supported by USPHS research project grant CA15956 and CORE grant CA21765 from the National Cancer Institute and by ALSAC.

References

Gabutti, V., Pileri, A., Tarocco, R.P., Gavosto, F.: Proliferative potential of out-of-cycle leukemic cells. Nature 224, 375-376 (1969)

Hayes, F.A., Green, A.A., Mauer, A.M.: Correlation of cell kinetic and clinical response to chemotherapy in disseminated neuroblastoma. Cancer Res. 37, 3766-3770 (1977)

Mauer, A.M., Fisher, V.: Comparison of the proliferative capacity of acute leukemia cells in bone marrow and blood. Nature 193, 1085-1086 (1962)

Mauer, A.W., Fisher, V.: Characteristics of cell proliferation in four patients with untreated acute leukemia. Blood 28, 428-445 (1966)

Saunders, E.F., Mauer, A.M.: Reentry of nondividing leukemic cells into a proliferative phase in acute childhood leukemia. J. Clin. Invest. 48, 1299-1305 (1969)

Saunders, E.F., Lampkin, B.C., Mauer, A.M.: Variation of proliferative activity in leukemic cell populations of patients with acute leukemia. J. Clin. Invest. 46, 1356-1363

The Control of Tumor Metastasis

E. GORELIK, M. FOGEL, S. SEGAL, and M. FELDMAN

Department of Cell Biology, The Weizmann Institute of Science, Rehovot, Israel

I. Introduction

Most carcinomas and sarcomas in man are metastatic neoplasms. On the other hand, most solid tumors in mice, whether spontaneous or experimentally induced, are nonmetastatic. In the absence of mouse metastatic tumors, experimental models for the study of tumor metastasis have been rather limited, hence, the limitations imposed on the study of the mechanisms controlling metastatic spread and progression. In fact, some very basic questions have not been fully investigated, e.g., the question of whether a population of tumor metastasis cells is a random representative of the population of the local growth or whether it is a selected subpopulation possessing cell surface properties which might determine the metastatic properties. Recent studies have indicated that metastatic cells may differ from the primary cell population in a number of properties, such as susceptibility to drugs (Trope 1975), affinity to different organs (Fidler and Nicolson 1976; Nicolson et al. 1976) and chromosome number (Rabotti 1959; Chu and Ulmgren 1961).

We have carried out studies aimed at testing whether tumor populations from lung metastases possess cell surface antigens manifesting specificities different from those of the local tumor growth, and whether such differences, among other factors, may control their metastatic spread. Antigenic specificities, if they exist, may furnish the basis for interactions, via the host's immune system, between the primary tumor and its metastases. The analysis of such interactions was the second objective of our studies (Gorelik et al. 1978).

To approach these questions, we investigated the Lewis lung carcinoma (3LL) which arose spontaneously in a C57BL mouse and was maintained by subcutaneous transplantations in syngeneic mice. This tumor produces metastases in the lungs following subcutaneous, intramuscular or intra-footpad inoculation.

Table 1. Syngeneic C57BL/6 spleen cells sensitized in vitro against local (L-3LL) or metastatic derived (M-3LL) tumor cells manifest anti L-3LL and anti-M-3LL specificites

Sensitizing tumor cells	Target tumor cells	Net % lysis at lymphocyte-to-target cell ratio of:				
		Exp. 1		Exp. 2		Exp. 3
		25:1	12.5:1	25:1	12.5:1	25:1
L-3LL	L-3LL	−54.6	−57.6	−49.3	−31.1	−45.7
L-3LL	M-3LL	−13.1	+ 4.4	− 5.2	− 0.4	− 3.7
M-3LL	M-3LL	−21.5	−31.4	−47.1	−46.3	−37.1
M-3LL	L-3LL	− 9.2	− 4.2	− 7.0	− 9.5	− 8.2

II. Metastatic Cells Possess Cell Surface Antigens Different from Those of the Local Growth

We first tested whether tumor cell populations deriving from lung metastases (M-3LL) possess cell surface antigens of specificities different from those of the local growth (L-3LL). To answer this question, we sensitized C57BL spleen lymphocytes by culturing them for 5 days on monolayers of either M-3LL or L-3LL, in the absence of xenogeneic serum but in the presence of 1% syngeneic serum. We emphasize the application of syngeneic serum in our sensitization cultures, since we found that the hitherto used fetal calf serum elicits in vitro, in syngeneic cell systems, cell-mediated immunity directed against fetal calf specificities rather than against tumor specificities (Fogel et al. 1978b). On the other hand, in vitro sensitization in the presence of 1% syngeneic serum resulted in antitumor cytotoxic lymphocytes of specificities which could not be manifested when sensitization was carried out in xenogeneic serum. Anti-3LL cytotoxic lymphocytes generated in the presence of syngeneic serum did not cross-react with normal fibroblasts nor with other tumors of the same H-2 phenotype (Fogel et al. 1978b).

Thus, we sensitized C57BL spleen lymphocytes against L-3LL or M-3LL cells and then cross-tested them for cytotoxic activity, using the ^{51}Cr end-labeling technique (Moore et al. 1975). The results (Table 1) indicated that lymphocytes sensitized against L-3LL cells lysed L-3LL targets significantly more than M-3LL targets. Conversely, lymphocytes sensitized against M-3LL cells lysed M-3LL targets significantly more than L-3LL targets. It appears, therefore, that the cell populations of the tumor metastasis (M-3LL) and those of the local tumor (L-3LL) possess antigens of distinct specificities in addition to the antigens which are common to both.

We then tested whether the antigenic differences between M-3LL and L-3LL cells are recognized by the immune system of tumor-bearing mice (TBM). Syngeneic mice were, therefore, grafted subcutaneously with either 1×10^6 M-3LL cells or 1×10^6 L-3LL cells, and the spleens of the TBM were removed 14 days following tumor transplantation. The spleen cells were first incubated in vitro for 2 days, in the absence of tumor cells, to unmask their cytotoxic activity. The

Table 2. Mice bearing tumors derived from M–3LL or L–3LL cells generate cytotoxic spleen lymphocytes manifesting anti M–3LL and anti L–3LL specificities

Spleen cell donors	Tumor target cells	Exp. 1		Exp. 2	
		Ratio lymphocytes to targets	% net lysis	Ratio lymphocytes to targets	% net lysis
L–3LL TBM	L–3LL	25:1	−46	20:1	−30
		12.5:1	−25	10:1	− 6
M–3LL TBM	L–3LL	25:1	− 8	20:1	−17
		12.5:1	− 1.4	10:1	+ 1.0
L–3LL TBM	M–3LL	25:1	− 6.0	20:1	−12.0
		12.5:1	+ 7.0	10:1	+14.0
M–3LL TBM	M–3LL	25:1	−27.0	20:1	−21
		12.5:1	− 1.2	10:1	−25

"unmasked" lymphocytes were then cross-tested against M-3LL and L-3LL targets. We found (Table 2) that spleen lymphocytes from mice bearing tumors of L-3LL origin lysed L-3LL cells significantly more than they lysed M-3LL cells and vice versa, i.e., lymphocytes from M-3LL TBM lysed M-3LL targets significantly more than they lysed L-3LL cells. Thus, the antigenic specificities of L-3LL and M-3LL cells are recognized by the tumor-bearing animal. These cell populations elicit, both in vitro and in vivo, distinct immunogenic effects. These results suggest the possibility that immunoselection may be involved in the spread of metastatic cells. If the initial growth of L-3LL elicits, in vivo, an anti L-3LL response, then metastatic M-3LL cells, which are not susceptible to the anti L-3LL lymphocytes, could be selected out, while migrating via the circulation. Thus, a cell-mediated immune reaction directed against the antigenic specificities of the local tumor could furnish a selective means for metastatic spread.

III. Metastatic Cells Are More Resistant
to Natural Killer Cell Activity Than Cells of the Local Tumor

Recent studies in both mice and man focused attention on the possibility that naturally occurring killer cells (NK) may have a function in controlling tumor development (Herberman et al. 1975; Kiessling et al. 1975). Antigenic variants (M-3LL) escaping the anti L-3LL response could then be lysed by circulating NK cells. It seemed, therefore, of interest to test whether cells of the tumor metastasis differ from cells of the local tumor with regard to susceptibility to NK cells. To study this, normal spleen cells were maintained in culture for 2 days under conditions identical to those used to unmask the cytotoxic activity of spleen cells from TBM. They were then tested for cytotoxic activity against L-3LL and M3LL cells. We found (Table 3) that metastatic cells are significantly more resistant to NK cells than

Table 3. M–3LL cells are less susceptible than L–3LL cells
to the cytotoxic activity of normal syngeneic NK spleen
cells

Exp. no.	Ratio of cultivated spleen cells to tumor cells	% cytotoxicity against tumor target cells	
		L–3LL	M–3LL
1	50:1	− 35.8	− 16.0
	25:1	− 21.9	− 5.2
2	50:1	− 21.7	− 18.2
	25:1	− 22.8	− 13.8
	12:1	− 20.0	+ 14.6
	6:1	− 5.6	+ 13.5
3	50:1	− 24.1	+ 7.5
	25:1	− 20.1	+ 18.6
4	40:1	− 29.0	+ 3.0
	20:1	− 19.3	+ 5.9
5	50:1	− 33.6	− 14.9
	25:1	− 28.8	− 18.3

tumor cells from the local growth. These results suggest the possibility that natural killer lymphocytes may participate in the process resulting in selection of metastatic cells migrating out of the local growth.

IV. Lymphocytes Sensitized Against M-3LL Suppress the Development of Lung Metastases

Since tumor cells from lung metastases were found to manifest antigenic specificities different from those of the local tumor growth, we designed experiments to test whether anti M-3LL cytotoxic lymphocytes produced either in culture or in vivo will cause suppression of lung metastasis when admixed with tumor cells and injected to syngeneic animals. Spleen lymphocytes were sensitized against M-3LL or L-3LL monolayers. The sensitized lymphocytes were then mixed with either M-3LL or L-3LL tumor cells at ratios of 25:1 or 12.5:1 lymphocytes to tumor cells, and injected intra-footpad to syngeneic recipients. No significant differences were found in the growth of the local tumor. However, testing for the incidence of lung metastases 21 days following cell inoculation, we found (Table 4) that anti L-3LL cytotoxic lymphocytes did not reduce the incidence of metastases produced (rather they increased the number of lung metastases). On the other hand, anti M-3LL lymphocytes caused a significant reduction of metastasis when injected with L-3LL cells. A dramatic reduction was obtained when anti M-3LL lymphocytes were injected wtih L-3LL cells at ratios of 25:1 effector to tumor cells. When anti M-3LL lymphocytes were injected with M-3LL cells, such a reduction in incidence of metastasis was obtained at both 25:1 and 12.5:1 ratios. Thus, anti M-3LL cells generated in culture by sensitization against monolayers of metastatic cells conferred a significant protection against metastatic progression.

Table 4. Anti M–3LL lymphocytes generated in culture suppress
the development of lung metastases

Sensitizing tumor cells	Transplanted tumor cells	No. of pulmonary metastases at sensitized lymphocytes-to-tumor-cell ratios of:	
		25:1	12.5:1
Control, L–3LL cells alone		19.6	
None	L–3LL	21.7	27.3
L–3LL	L–3LL	28.6	35.2
M–3LL	L–3LL	3[a]	17.8
Control, M–3LL cells alone		21.1	
None	M–3LL	42.4	Not done
L–3LL	M–3LL	52[a]	42
M–3LL	M–3LL	7.7[a]	3.5[a]

[a] Differences significant according to the Mann-Whitney U test

A similar state of adoptive immunity against lung metastasis was obtained with spleen lymphocytes from mice bearing tumors deriving from M-3LL cells (Fogel et al. 1978a). These results suggest an approach towards adoptive immunotherapy directed specifically against metastatic tumor cells mediated via lymphocytes sensitized in vitro against metastatic cells.

V. Growth of Local Tumor Exerts a Suppressing Effect on Lung Metastasis

The observation that metastatic cells possess imunogenic determinants with specificities different from those of cells of the local tumor growth need not exclude the possibility that in vivo interactions may exist between these two cell populations. To test for such interactions, we studied the effects of excision of the local tumor growth on the development of lung metastasis (Gorelik et al. 1978). Mice were inoculated intra-footpad with 1×10^5 3LL cells; 21 days later, legs with tumors 8–10 mm in diameter were amputated. The result (Table 5) was that following excision of the local tumor the number of visible metastase and the cell mass of each metastasis increased significantly. If thus appears that the growth of the local tumor arrested the growth of metastasis. The removal of the local tumor was followed by progression in the lungs of otherwise arrested tumor cells, resulting in increased incidence of metastasis, and by increased growth of rate of metastases, resulting in increase in cell mass. One could argue that the procedure of surgical removal of the local tumor per se released tumor cells into the circulation and hence lung metastasis increased. Were this the case, then reinoculating the other leg with 3LL cells following tumor excision should still result in increased incidence of lung metastasis. However, grafting the nonamputated leg with a certain number of tumor cells prevented the increase in metastatic progression. This leads us to the

Table 5. Prevention of enhanced development of metastases by local reinoculation of tumor cells following tumor excision is tumor-specific[a]

Group no.	No. of mice	Treatment		No. of metastases in lungs	Volume of metastases (mm³)	Weight of lungs (mg)	Weight of spleen (mg)
		Ampu-tation	Rein-oculation of tumor cells				
1	8	−	−	26	0.2 ± 0.04	175 ± 9	416 ± 42
2	10	+	−	63[b]	5.4 ± 1.5[b]	481 ± 86[b]	175 ± 7[b]
2	10	+	1×10^6 3LL	20	0.8 ± 0.26	200 ± 15	184 ± 11[b]
4	11	+	1×10^6 B–16	48	3.2 ± 0.7[b]	245 ± 22[b]	167 ± 7[b]

[a] 1×10^5 3LL tumor cells were inoculated into the footpad; 15 days later, legs with developed tumors were amputated and simultaneously 1×10^6 3LL or syngeneic melanoma B–16 tumor cells were inoculated. Primary tumor diameter at time of excision was 9–11 mm. Lungs were examined 10 days following amputation

[b] Differs significantly from group 1 ($p < 0.05$)

conclusion that the growth of the local tumor elicits a controlling effect on metastatic progression. This effect seems to be specific to the tested tumor, since grafting another tumor of C57BL origin (e.g., the B-16 melanoma) after 3LL excision did not result in inhibition of the accelerated growth of lung metastasis (Table 5).

The fact that the grafting of antigenically unrelated C57BL tumors such as B-16 melanoma or EL4 lymphoma (Gorelik et al. 1978) did not prevent the accelerated progression of 3LL lung metastases suggests that an immunological mechanism may be involved in the interactions between the local and the metastatic populations of the 3LL tumor. The participation of the lymphoid system in the inhibiting effect which the local tumor exerts on lung metastasis was supported by observations of the response of the spleen cell population. Growth of the local tumor was associated with a progressive increase in spleen size (Table 5); excision of the local tumor, resulting in increased metastatic progress, was followed by a decrease in spleen cell population back to its normal size.

These observations were recently followed by a detailed analysis of the effects of splenectomy on the growth of both local 3LL tumors and their metastases (to be published). These studies indicated that interplay between distinct subpopulations of spleen lymphocytes may be associated with the control of lung metastasis exerted by the local tumor. Another observation indicating that lymphocytes participate in the control of metastasis concerns increase in lung metastasis incidence and cell mass following total body irradiation at a dose of 450 R (Gorelik et al. 1978). The excision of the local tumor from such x-irradiated tumor-bearing mice resulted in further increase in metastatic growth, showing that depletion of the lymphoid system does affect metastasis. The fact that the local tumor growth still exerted an arresting effect on lung metastasis in irradiated animals could be attributed either to the possibility that 450 R did not completely deplete the subpopulation responsible for the accelerated metastatic growth (suppressor cells?) or to the existence of mechanisms of control not mediated via the lymphoid system.

VI. Summary

Studies were made to test whether cells of tumor metastasis of the 3LL Lewis lung carcinoma (M-3LL) possess cell surface antigens different from those of cells of the local tumor growth (L-3LL). Sensitization of syngeneic lymphocytes against monolayers of M-3LL cells resulted in cytotoxic lymphocytes which lysed metastatic cells significantly more than they lysed local tumor cells (L-3LL). Conversely, lymphocytes sensitized against L-3LL monolayers lysed L-3LL cells more than they lysed M-3LL cells. This shows that there are antigenic differences between the local and the metastatic cell populations. Such differences are recognized by the tumor-bearing animal, as shown by the fact that similar results are obtained using spleen cells from animals grafted with M-3LL or with L-3LL cells. It appears, therefore that immunoselective processes may control metastasis development; metastatic cells could escape the anti L-3LL response which the initial local growth elicits and thus establish themselves in the lung prior to formation of an anti M-3LL reaction. Natural killer (NK) cells in the ciruclation or in the spleen could conceivably prevent invasion by M-3LL cells. We found, however, that M-3LL cells were significantly more resistant to the cytotoxic effect of NK cells and thus, again, they could confer a selective advantage on metastatic cells. Anti M-3LL but not anti L-3LL lymphocytes, when injected with tumor cells to syngeneic recipients, caused significant suppression of lung metastasis. This suggests an approach towards immunotherapy of metastasis, based on the introduction of lymphocytes sensitized in culture against metastatic cells.

Interactions between local and metastatic foci were demonstrated. Excision of the local tumor resulted in increased incidence and accelerated growth rate of lung metastases. Since reinoculation of tumor cells, after excision of a first tumor, prevented the accelerated development of lung metastases, we conclude that the local tumor growth exerts a suppressive effect on metastatic progress. This suppressive effect is specific for the given tumor, since other tumors of the same genotype did not arrest the progress of 3LL metastasis. The suppressive effect might be mediated, at least partly, via the lymphoid system and might in fact have an immunological basis.

Acknowledgment. This work was supported by NIH Contract No. N01-CB-74185.

References

Chu, E., Ulmgren, R.: Microspectophotometric determination of DNA in primary and metastatic mouse mammary tumors. J. Natl. Cancer Inst. 27, 217-220 (1961)

Fidler, J., Nicolson, G.: Organ selectivity for implantation survival and growth of B-16 melanoma variant tumor lines. J. Natl. Cancer Inst. 57, 1199-1202 (1976)

Fogel, M., Gorelik, E., Segal, S., Feldman, M.: In: Proceedings of the 12th Leukocyte Culture Conference. Beersheba, Israel (1978a)

Fogel, M., Segal, S., Gorelik, E., Feldman, M.: Specific cytotoxic lymphocytes against syngeneic tumors are generated in culture in the presence of syngeneic, but not xenogenic serum. Int. J. Cancer 22, 329-334 (1978b)

Gorelik, E., Segal, S., Feldman, M.: Growth of a local tumor exerts a specific inhibitory effect on progression of lung metastases. Int. J. Cancer 21, 617-625 (1978)

Herberman, R., Nunn, M.E., Lavrin, D.H.: Natural cytotoxic reactivity of mouse lymphoid cells against syngeneic and allogenic tumors. I. Distribution of reactivity and specificity. Int. J. Cancer 16, 216-229 (1975)

Kiessling, R., Klein, E., Pross, H., Wigzell, H.: "Natural" killer cells in the mouse. II. Cytotoxic cells with specificity for the mouse Moloney leukemia cells. Characteristics of the killer cells. Eur. J. Immunol. 5, 117-121 (1975)

More, R., Yron, I., Ben-Sasson, S., Weiss, D.: In vitro studies on cell-mediated cytotoxicity by means of a terminal tumor variants labeling technique. Cell Immunol. 15, 382-191 (1975)

Nicolson, G., Winkelhake, J., Nussey, A.: An approach to studying cellular properties iants selected in vivo for enhanced metastasis. In: Fundamental Aspects of Metastasis (ed. L. Weiss), pp. 291-303. Amsterdam: North-Holland 1976

Rabotti, G.: Ploidy of primary and metastatic human tumors. Nature (London) 183, 1276-1277 (1959)

Trope, C.: Different sensitivities to cytostatic drugs of primary tumor and metastasis of the Lewis carcinoma. Neoplasma 22, 171-180 (1975)

Hepatocarcinogenesis as a Problem in Developmental Biology

H.C. Pitot and A.E. Sirica

Departments of Oncology and Pathology
McArdle Laboratory for Cancer Research, The Medical School
University of Wisconsin, Madison, WI 53706, USA

I. Differentiation in Relation to Hepatocarcinogenesis

An association of differentiation with carcinogenesis and cancer has been recognized for more than a hundred years by pathologists and oncologists alike. The "embryonal rest theory" of Cohnheim (1889) was one of the first concepts of neoplasia that depended entirely upon a relationship of the neoplastic process to embryonic development. For most of the next century pathologists recognized Cohnheim's concepts by relating the nomenclature of malignant neoplasms to their embryonic bases (e.g., sarcoma and carcinoma) and utilized terms such as dedifferentiation, embryonal, blastoma, etc. in the description of both benign and malignant neoplasms. However, it has only been within the last two decades that some understanding of the cellular and biochemical mechanisms of the developmental biology of neoplasia has become apparent.

A. Fetal Antigens in Hepatocarcinogenesis

One of the original demonstrations of the biochemical bases for the "fetalization" of hepatomas was reported by Abelev (1963), who demonstrated that hepatomas obtained by the feeding of carcinogenic azo dyes to rats possessed antigen(s) which completely cross-reacted with antigens normally present only in fetal liver. Since this investigation numerous studies have shown the presence of fetal antigens in a variety of neoplasms (Hanna et al. 1971; Lo Gerto et al. 1972; Fritsche and Mach 1975). In liver, one of the most thoroughly studied examples of fetal antigens is that of α-fetoprotein. A number of workers investigating both the rodent and the human have demonstrated the appearance of this antigen in animals and patients bearing one or more hepatic neoplasms (Abelev 1963; Tomasi 1977; Sell and Becker 1978). In addition, Kitagawa and others (Kitagawa et al. 1972) have demonstrated the relatively early appearance of α-fetoprotein in serum of animals treated with several hepatocarcinogens well prior to the appearance of frank neoplasms.

B. Fetal Enzymes and Isozymes in Hepatomas and Hepatocarcinogenesis

A number of investigations have now demonstrated the presence of fetal enzymes and isozymic forms of enzymes in a wide variety of transplanted

Table 1. Examples of fetal enzymes (isozymes) in several transplanted rat hepatomas

Hepatoma	Enzyme	References
Morris hepatoma 7316A	Pyruvate kinase M_2 (fetal)	Tanaka et al. (1972)
	Hexokinases I, II, III (fetal)	Sato et al. (1969)
	Phosphofructokinase IV (adult and fetal)	Kurata et al. (1972)
	Branched chain amino acid transaminase I (fetal)	Ogawa and Ichihara (1972)
	Glutaminase (fetal/kidney isozyme)	Katunuma et al. (1972)
3924A	Phosphofructokinase L, M, N (fetal)	Dunaway et al. (1974)
	Glutaminase (fetal/kidney isozyme)	Katunuma et al. (1972)
	Hexokinases I, II	Shatton et al. (1969)
	Pyruvate kinase III (fetal)	Farina et al. (1974)
5123C	Hexokinases I, III	Shatton et al. (1969)
Yoshida ascites hepatomas	Hexokinases I, II, III	Sato et al. (1969)
	Phosphofructokinases II, III (fetal)	Kurata et al. (1972)
	Branched chain amino acid transaminase I (fetal) and III	Ogawa and Ichihara (1972)
	Pyruvate kinase III	Farina et al. (1974)
Novikoff hepatoma	Hexokinases I, II, III (fetal)	Shatton et al. (1969)

hepatocellular carcinomas (Potter 1968; Sato et al. 1969; Weinhouse 1973; Dunaway et al. 1974; Hatzfeld et al. 1975). A list of representative examples of fetal enzymes and isozymes found in transplanted hepatomas is seen in Table 1. There is significant variation in the presence of specific isozymes in different tumor lines. On the other hand, some neoplasms, e.g., the Morris hepatomas 7793 and 9618A, resemble the differentiated characteristics of adult liver far more than that of fetal or neonatal liver in the rodent (Katunuma et al. 1972; Potter et al. 1972). The regulation of genetic expression in hepatic neoplasms has been shown to be affected in a variety of ways in all neoplasms thus far studied (Goldfarb and Pitot 1976). This fact also bears a resemblance to the regulation of genetic expression in fetal tissues which is quite distinctive from that of the adult organ (Greengard 1970).

Endo et al. (1972) and Potter et al. (1972) have also demonstrated the appearance of fetal isozymic patterns in liver during the process of carcinogenesis prior to the appearance of gross neoplasms. Thus the appearance of biochemical fetal characteristics as exemplified by fetal enzymes also appears to be characteristic of the natural history of hepatocarcinogenesis in a manner similar to that shown by the appearance of fetal antigens, whose function in the hepatocyte is largely unknown at the present time.

II. Developmental Aspects of Hepatocarcinogenesis

The development of the embryo throughout gestation follows a detailed format characteristic of the species. Characteristic changes in the biochemistry and morphology of the developing embryo occur at specific times after fertilization (Flickinger 1963). Such changes in gene expression have been postulated to be the result of nuclear protein repression and derepression of specific genes (Chytil et al. 1974) or the cytoplasmic stabilization of specific messenger RNA's during development (Kafatos 1972).

Analogous to fetal development is the genesis of a neoplasm. Beginning in many instances with a single cell (Fialkow 1976; Nowell 1976), the progeny of the initiated cell develops through a series of stages until the destruction of the host results or the neoplasms continue to proliferate in cell culture or in a number of hosts following repeated transplantations. Although as Potter has pointed out (Potter et al. 1972) the development of a neoplasm may result from "blocked ontogeny", numerous investigations have characterized specific stages in the development of neoplasia.

Earlier studies in the skin (Berenblum and Shubik 1947) characterized at least two stages of carcinogenesis, that of initiation followed by that of promotion. The characteristics of these two stages in epidermal carcinogenesis have been well described (Boutwell 1974). Recently, however, stages analogous to initiation and promotion in the skin have also been demonstrated in rodent hepatocarcinogenesis (Peraino et al. 1973; Pitot et al. 1978). In particular, Peraino and his associates demonstrated that phenobarbital will promote or enhance tumor production induced by low doses of acetylaminofluorene. In our laboratory we demonstrated that single doses of diethylnitrosamine following a partial hepatectomy resulted in hepatoma formation only if animals were fed phenobarbital continuously for a number of months (Pitot et al. 1978).

In these latter studies it was shown that unlike epidermal carcinogenesis, the immediate progeny of the initiated cells could be identified and characterized to some extent. Earlier studies (Scherer and Hoffman 1971) had indicated both the clonality and the irreversibility of the foci of the initiated cells (Scherer and Emmelot 1975). On the bases of these studies it may now be possible to follow the development of hepatic neoplasms from the earliest progeny of the initiated cells.

A. Gene Expression in Initiated Cell Populations

Early investigations of the phenotypic alterations evidenced during hepatocarcinogenesis in the rodent liver as a whole (Reid 1962) or as anatomically recognizable structures (Goldfarb and Zak 1961) demonstrated a wide variety of alterations. A number of these earlier investigations, especially those carried out after animals had been on the carcinogenic regimen for several months, did not take into account the changing mixed cell populations of the liver resulting from this treatment (Nishizumi et al. 1977).

In order to obviate the difficulties found with a mixed cell population, biochemical studies of liver during the initial 3 to 5 weeks of carcinogen feeding, when new cell populations are in the relative minority (Poirier et al. 1969), and studies on hyperplastic nodules and related structures as well as histochemical investigations were carried out (Becker et al. 1972; Kitagawa and Sugano 1973; Cameron et al. 1976). Enzymatic deficiencies were noted including the easily monitored enzymes, glucose-6-phosphatase and canalicular ATPase. In addition,

Table 2. Number and phenotypes of enzyme altered foci/cm^3 liver

	GP, AP, GT	GT, GT	GT, AP	AP, GT	AP	GT	GT
DEN+P.H.	38±25	31±16	93±53	53±25	291±125	234±78	313±125
	(3.6)	(2.9)	(8.8)	(5.0)	(27.6)	(22.2)	(29.7)
DEN+P.H. +0.05% Pheno-barbital	430±89	89±29	208±74	801±208	801±208	2581±326	593±148
	(7.8)	(1.6)	(3.8)	(14.5)	(14.5)	(46.9)	(10.7)

The values are expressed as the number of foci/cm^3 of liver as calculated by the method of Scherer et al. (1972)±standard error of the mean
GP=glucose-6-phosphatase negative; AP=canalicular ATPase negative; GT=γ-glutamyl transpeptidase positive; DEN=diethylnitrosamine (5–10 mg/kg administered 24 h following P.H.); P.H.=partial hepatectomy.
The numbers in parentheses represent the percent distribution of the foci under that condition.
Six rats given DEN+P.H. alone and eight rats given DEN+P.H.+phenobarbital

the accumulation of iron pigment (ferritin) in hepatic cell populations was also utilized by Williams (Williams 1976) to distinguish between relatively normal liver cell populations and those altered by the carcinogen feeding. Most of these investigations were concerned with a single enzymatic determination on individual foci of altered cells. Recently our laboratory attempted to characterize further the phenotypes of the foci of altered cells by means of serial liver sections and by the use of not one, but three histochemical markers. These included glucose-6-phosphatase, canalicular ATPase, and γ-glutamyltranspeptidase (Pitot et al. 1978). These studies revealed enzyme-altered foci which where identical with those characterized by Friedrich-Freksa, Scherer and others (Friedrich-Freksa et al. 1969; Scherer and Emmelot 1975) as the putative progeny of initiated cells in the liver. Variation in phenotypes of such foci may be seen in Table 2 as an example of such phenotypic heterogeneity. Unlike earlier investigations, which suggested generalized deficiencies or appearance of enzymes based on studies of single variables in histologically and histochemically recognizable foci, studies monitoring several biochemical variables demonstrated that such foci were quite heterogeneous, biochemically. Such heterogeneity, as has been pointed out earlier (Pitot et al. 1978), is not unlike that noted in the extreme biochemical diversity of highly differentiated hepatocellular carcinomas (Pitot et al. 1974).

In addition, however, the loss of glucose-6-phosphatase and canalicular ATPase, as well as the gain of γ-glutamyltranspeptidase in prencoplastic and neoplastic hepatocytes (Pugh and Goldfarb 1978; Pitot et al. 1978), is also characteristic of fetal and neonatal hepatocytes. The expression of genes normally limited to fetal life by normal adult and neoplastic cells in vivo and/or in vitro we have termed "fetal-genic expression". However, as can be seen from Table 2, such fetal-genic expression is not complete in every focus, and the phenotypic alterations are maintained as the focus increases in age and size. Furthermore, the fact that as many as three different phenotypic changes can be seen in a single focus suggests that such alterations are not the result of single genetic mutations, but instead they may be due to alterations in the regulation of genetic expression that lead to a variety of different phenotypes, all of which are the result of a single stable genome whose expression is altered to varying degrees.

III. Fetal-genic Expression in Normal and Neoplastic Cells in Vivo and in Vitro

As evidenced by a variety of studies (vide supra), the incomplete and variable expression of the fetal phenotype by a variety of neoplastic cell populations is not only common but is in all likelihood the general rule. One may even speculate that all neoplasms exhibit some degree of fetal-genic expression in their phenotype. However, just as with many other characteristics of neoplasia, fetal-genic expression is not unique to cells which have undergone the neoplastic transformation.

In a variety of pathologic conditions the production of "fetal" proteins occurs including the production of fetal hemoglobin (Cooper and Hoagland 1972), α-fetoprotein (Keller and Tomasi 1976), and other embryonic proteins (Delwiche et al. 1973). These examples do not include the specific genetic alterations resulting in a maintenance of specific fetal-genic expression such as persistent fetal hemoglobinemia (Forget et al. 1976).

As is the case with hepatocarcinogenesis in vivo, cultured cells transformed in vitro acquire several aspects of the fetal phenotype including fetal antigens in SV 40-transformed cells (Shier and Trotter 1978) and in cells spontaneously transformed in vitro (Ting et al. 1978). In addition, liver cell lines maintained in culture also exhibit fetal-genic expression (Walker et al. 1972). However, examples of normal cells in primary culture exhibiting fetal-genic expression are not so common. Recently Ting et al. (1978) demonstrated the appearance of a fetal antigen in adult lung tissue of young mice. These cells did express fetal antigens in late passages but not in early passages. Later these cells spontaneously transformed in culture and continued to express the fetal antigen. Rosenberg et al. (1978) also described the appearance of fetal antigens in normal mouse and human cells in culture.

The expression of fetal characteristics in primary cell cultures of adult liver has been reported recently by Leffert et al. (1978). In our laboratory we have been able to maintain primary cultures of adult rat liver cells on collagen gels for periods of 3 weeks with continued maintenance of specific hepatocyte functions (Michalopoulos and Pitot 1975). Recently by modification of this technique (Sirica et al. 1979) we have also demonstrated fetal-genic expression in primary cultures of adult hepatocytes. In this study the specific demonstration that the fetal-genic expression occurs directly in cells having all of the morphologic characteristics of adult hepatocytes was presented. In Table 3 may be seen a number of the fetal and adult characteristics originating in adult hepatocytes cultured on collagen gel-nylon meshes. Thus it is apparent that rat hepatocytes placed in primary culture while maintaining a number of their adult functions do acquire, beginning after 3 days in culture, the expression of fetal characteristics in the form of both secretory proteins

Table 3. Adult and fetal phenotypic expression by adult rat hepatocytes maintained on collagen gel-nylon meshes (Sirica et al. 1979)

Age of primary culture	Phenotypic expression by hepatocytes
4–72 h	(a) Synthesize and secrete albumin, transferrin, and α_1-acid glycoprotein in amounts which are proportional to those observed in vivo with normal adult rats
	(b) Exhibit efficient induction of tyrosine aminotransferase by dexamethasone
	(c) Show measurable but low levels of cytochrome P_{450} activity
	(d) Express the adult isozyme form of fructose diphosphate aldolase
	(e) Exhibit negligible levels of DNA synthesis and fetal protein production
72–240 h	(a) Exhibit efficient induction of tyrosine aminotransferase by dexamethasone
	(b) Show a decreased albumin and an increased transferrin and α_1-acid glycoprotein production
	(c) Exhibit a rapid decline in cytochrome P_{450} activity
	(d) Express and exhibit a linear increase in activity of the "fetal" hepatocyte enzyme γ-glutamyltranspeptidase
	(e) Produce α_1-fetoprotein
	(f) Express the fetal isozyme form of fructose diphosphate aldolase
	(g) Exhibit an increased DNA synthesis
	(h) Show a progressive buildup of microfilaments beneath the apical cell surface

and intracellular enzymes. The similarity of this expression to fetal characteristics appearing in hepatic neoplasms and during hepatocarcinogenesis is evident. However, what should be emphasized are the distinctive differences between fetal-genic expression in cultured adult hepatocytes and initiated and fully neoplastic hepatocytes in vivo.

A. Apparent Characteristics of Fetal-genic Expression in Hepatic Cells Transformed in Vivo and Cultured in Vitro

Since fetal-genic expression has been best studied during carcinogenesis and in neoplasia in vivo, such phenotypic alterations in and during the neoplastic transformation should be on the basis of a comparison of fetal-genic expression in neoplasia with any other situation. In neoplasia, fetal-genic expression has two principal characteristics: (1) The fetal phenotypic characteristics of a specific neoplasm or initiated focus appear to be stable during the lifetime of the transformed cellular population in the host and even throughout multiple transplants, and (2) fetal-genic expression in neoplasms in general and especially in neoplasms and initiated foci of the liver is quite variable from one lesion to another.

When one considers the two characteristics of stability of and variation in phenotype, the fetal-genic expression seen in cell cultures, especially that of adult rat liver cells, is distinctly different from that in neoplasia. All cultured hepatocytes appear to exhibit qualitatively the expression of at least one fetal characteristic, that of γ-glutamyltranspeptidase. While our studies have not yet been completed on the determination of the expression of other fetal characteristics, it is clear that the majority of cultured hepatocytes exhibit such fetal-genic expression, thus suggesting that all cultured hepatocytes undergo similar fetal-genic expression in relation to γ-glutamyltranspeptidase, unlike the variations seen in initiated foci (Pitot et al. 1978) and the fully developed malignant hepatomas (Pitot et al. 1974). On the other hand the stability of the fetal phenotypic characteristics seen in primary cultures of adult hepatocytes is a much more difficult question to answer. However, the "fetal" or neonatal characteristic of DNA synthesis seen in hepatocytes cultured on collagen gel-coated nylon meshes (Table 3) can be completely suppressed by the addition of dexamethasone to the culture medium. These two bits of data, while not at all conclusive, do suggest distinctive differences in the fetal-genic expression of hepatocytes transformed in vivo and adult hepatocytes cultured in vitro.

IV. Possible Mechanisms of Fetal-genic Expression in Normal and Neoplastic Cells

While it is possible that fetal-genic expression in neoplasia may result from specific genetic mutation(s), it is more likely that the appearance of specific fetal characteristics in neoplastic cells is the result of alteration in the expression of genetic information in the transforming cellular population. While there is not sufficient space to consider all of the arguments for this statement, the reader is referred to previous discussions supporting this hypothesis (Pitot 1969). Furthermore, one may consider that the alterations seen in the environmental regulation of genetic expression characteristic of all hepatocellular carcinomas are closely related to the fetal-genic expression seen in these cells. As has been pointed out earlier (Pitot 1968), the phenotypic variation of hepatocellular carcinomas and possibly initiated foci may reflect a variety of possible "differentiated" states which the neoplastic cells may assume. Thus the environmental response to hormones will be one component characteristic of a differentiated state. If hormones, at least in part, are a controlling factor in the phenotype of normal adult liver, then the loss of hormonal response, characteristically seen in many transformed hepatocyte populations as well as in other neoplasms, may be a significant factor in fetal-genic expression in neoplastic hepatocytes. Evidence for this statement can be seen from earlier studies in which the phenotype of a variety of hepatocellular carcinomas could be altered by the removal of cortisone via adrenalectomy of the tumor-bearing host (Pitot and Cho 1965).

In adult hepatic cells in culture, essentially no hormones are present except those produced by the cells themselves or the insulin required for their normal maintenance. One may thus postulate the lack of the normal hormonal environment of the adult liver to be a major factor in fetal-genic expresssion of adult hepatocytes in culture. The fact that dexamethasone inhibits one component of this expression is in support of this hypothesis.

Therefore, one may propose that the fetal-genic expression seen in transformed hepatocytes and in adult hepatocytes in culture is based on the same mechanism, the lack or relative lack of hormonal repression and derepression of specific components of the genome. The difference lies, however, in the intrinsic loss of a normal responsiveness to hormones by neoplastic hepatocytes in vivo, in contrast to the absence of critical hormones in the environment of the normal adult hepatocyte in culture.

References

Abelev, G.: Study of the Antigenic Structure of Tumors. Acta Unio. Int. Cancer 19, 80-82 (1963)

Becker, F.F., Klein, K.M., Asofsky, R.: Plasma protein synthesis by N-2-fluorenylacetamide-induced primary hepatocellular carcinomas and hepatic nodules: Cancer Res. 32, 914-920 (1972)

Berenblum, I., Shubik, P.: The role of croton oil applications associated with a single painting of a carcinogen in tumor induction of the mouse's skin. Br. J. Cancer 1, 379-391 (1947)

Boutwell, R.K.: The function and mechanism of promoters of carcinogenesis. Crit. Rev. Toxicol. 2, 419-443 (1974)

Cameron, R., Sweeney, G.D., Jones, K., Lee, G., Farber, E.: A relative deficiency of cytochrome p-450 and aryl hydrocarbon [benzo(a)pyrene]hydroxylase in hyperplastic nodules induced by 2-acetylaminofluorene in rat liver. Cancer Res. 36, 3888-3893 (1976)

Chytil, F., Glasser, S.R., Spelsberg, T.C.: Alterations in liver chromatin during perinatal development of the rat. Dev. Biol. 37, 295-305 (1974)

Cohnheim, J.: Lectures on General Pathology, 3 Vols. London: New Syndenham Society 1889

Cooper, H.A., Hoagland, H.C.: Fetal hemoglobin. Mayo Clin. Proc. 47, 402-414 (1972)

Delwiche, R., Zamcheck, N., Marcon, N.: Carcinoembryonic antigen in pancreatitis. Cancer 31, 328-330 (1973)

Dunaway, G.A., Jr., Morris, H.P., Weber, G.: A comparative study of rat liver, muscle, and hepatoma 3924A phosphofructokinase isozymes. Cancer Res. 34, 2209-2216 (1974)

Endo, H., Eguchi, N., Yanagi, S., Torisu, T., Ikehara, Y., Kamiya, T.: Biochemical studies on the preneoplastic state. Gann Monogr. 13, 235-250 (1972)

Farina, F.A., Shatton, J.B., Morris, H.P., Weinhouse, S.: Isozymes of pyruvate kinase in liver and hepatomas of the rat. Cancer Res. 34, 1439-1446 (1974)

Fialkow, P.J.: Clonal origin of human tumors. Biochim. Biophys. Acta 458, 283-321 (1976)

Flickinger, R.A.: Cell differentiation: Some aspects of the problem. Science 141, 608-614 (1963)

Forget, B.G., Hillman, D.G., Lazarus, H., Barell, F., Benz, E.J., Caskey, C.T., Huisman, T.H.J., Schroeder, W.A., Housman, D.: Absence of messenger RNA and gene DNA for β-globin changes in hereditary persistence of fetal hemoglobin. Cell 7, 323-329 (1976)

Friedrich-Freksa, H., Gossner, W., Bonner, P.: Histochemische Untersuchungen der Cancerogenese in der Rattenleber nach Dauergaben von Diathylnitrosamin. Z. Krebsforsch. 72, 226-241 (1969)

Fritsche, R., Mach, J.-P.: Identification of a new oncofoetal antigen associated with several types of human carcinomas. Nature (London) 258, 734-737 (1975)

Goldfarb, S., Pitot, H.C.: Enzymology of highly differentiated hepatocellular carcinomas. In: Frontiers of Gastrointestinal Research, Vol. 2 (ed. L. van der Reis), pp. 194-242. Basel: S. Karger 1976

Goldfarb, S., Zak, F.G.: Role of injury and hyperplasia in the induction of hepatocellular carcinoma. J. Am. Med. Assoc. 178, 729-731 (1961)

Greengard, O.: The developmental formation of enzymes in rat liver. In: Biochemical Actions of Hormones, Vol. I (ed. G. Litwack), pp. 53-87. New York-London: Academic Press 1970

Hanna, M.G., Tennant, R.W., Coggin, J.H.: Suppressive effect of immunization with mouse fetal antigens on growth of cells infected with Rauscher leukemia virus and on plasma cell tumors. Proc. Natl. Acad. Sci. USA 68, 1748-1752 (1971)

Hatzfeld, A., Weber, A., Schapira, F.: Biochemical and immunological studies of some carcinofetal enzymes. Ann. N.Y. Acad. Sci. 259, 287-297 (1975)

Kafatos, F.C.: mRNA stability and cellular differentiation. Karolinska Symp. Res. Methods Reprod. Endocrinol. 5, 319-343 (1972)

Katunuma, N., Kuroda, Y., Yoshida, T., Sanada, Y., Morris, H.P.: Relationship between degree of differentiation and growth rate of minimal deviation hepatomas and kidney cortex tumors studied with glutaminase isozymes. Gann Monogr. 13, 143-151 (1972)

Keller, R.H., Tomasi, T.B.: Alpha-fetoprotein synthesis by murine lymphoid cells in allogenic reaction. J. Exp. Med. 143, 1140-1153 (1976)

Kitagawa, T., Sugano, H.: Combined enzyme histochemical and radiographic studies on areas of hyperplasia in the liver of rats fed N-2-fluorenylacetamide. Cancer Res. 33, 2993-3001 (1973)

Kitagawa, Y., Yokocbi, T., Sugano, H.: α-fetoprotein and hepatocarcinogene sis in rats fed 3'-methyl-4-(dimethylamino)azobenzene or N-2-fluorenylacetamide. Int. J. Cancer 10, 368-381 (1972)

Kurata, N., Matsuchima, T., Sugimura, T.: Multiple forms of phosphofructokinase in rat tissues and rat tumors. Biochem. Biophys. Res. Commun. 48, 473-479 (1972)

Leffert, H., Moran, T., Sell, S., Skelly, H., Ibsen, K., Mueller, M., Arias, I.: Growth state-dependent phenotypes of adult hepatocytes in primary monolayer cultures. Proc. Natl. Acad. Sci. USA 75, 1834-1838 (1978)

Lo Gerto, P., Herter, F.P., Barker, H.G., Bennett, S.: Immunologic tests for detection of gastrointestinal cancers. Surg. Clin. M. A. 52, 829-837 (1972)

Michalopoulos, G., Pitot, H.C.: Primary culture of parenchymal liver cells on collagen membrane: Morphological and biochemical observations. Exp. Cell Res. 94, 70-81 (1975)

Nishizumi, M., Albert, R.E., Burns, F.J., Bilger, L.: Hepatic cell loss and proliferation induced by N-2-fluorenylacetamide, diethylnitrosamine, and aflatoxin B_1 in relation to hepatoma induction. Br. J. Cancer 36, 192-197 (1977)

Nowell, P.C.: The clonal evolution of tumor cell populations. Science 194, 23-28 (1976)

Ogawa, K., Ichihara, A.: Isozyme patterns of branched-chain amino acid transaminase in various rat he patomas. Cancer Res. 32, 1257-1263 (1972)

Peraino, C., Fry, R.J.M., Staffeldt, E., Kisieleski, W.E.: Effects of varying the exposure to phenobarbital on its enhancement of 2-acetylaminofluorene-induced hepatic tumorigenesis in rat. Cancer Res. 33, 2701-2705 (1973)

Pitot, H.C.: Some aspects of the developmental biology of neoplasia. Cancer Res. 28, 1880-1887 (1968)

Pitot, H.C.: The endoplasmic reticulum and phenotypic variability in normal and neoplastic liver. A.M. A. Arch. Pathol. 87, 202-222 (1969)

Pitot, H.C., Cho, Y.S.: Control mechanisms in the normal and neoplastic cell. Prog. Exp. Tumor Res. 7, 158-223 (1965)

Pitot, H.C., Shires, T.K., Moyer, G., Garrett, C.T.: Phenotypic variability as a manifestation of translational control. In: The Molecular Biology of Cancer (ed. Harris, Busch), pp. 523-534. New York-London: Academic Press 1974

Pitot, H.C., Barsness, L., Goldsworthy, T., Kitagawa, T.: Biochemical characterization of stages of hepatocarcinogenesis after a single dose of diethylnitrosamine. Nature (London) 271, 456-458 (1978)

Poirier, L.A., Poirier, M.C., Pitot, H.: Dietary induction of some enzymes of carbohydrate metabolism during 2-acetylaminofluorene feeding. Cancer Res. 29, 470-474 (1969)

Potter, V.R.: Recent trends in cancer biochemistry: The importance of studies on fetal tissue. Can. Cancer Conf. 8, 9-30 (1968)

Potter, V.R., Walker, P.R., Goodman, J.I.: Survey of current studies on oncogeny as blocked ontogeny: Isozyme changes in livers of rats fed 3'-methyl-4-dimethylaminoazobenzene with collateral studies on DNA stability. Gann. Monogr. 13, 121-134 (1972)

Pugh, T.D., Goldfarb, S.: Quantitative histochemical and autoradiographic studies of hepatocarcinogenesis in rats fed 2-acetylaminofluorene followed by phenobarbital. Cancer Res. 38, 4450-4457 (1978)

Reid, E., Significant biochemical effects of hepatocarcinogenesis in the rat: A review. Cancer Res. 22, 398-430 (1962)

Rosenberg, S.A., Parker, G.A., Thorpe, W.P.: Expression of oncofetal antigens by murine and human normal cells in tissue culture. Isr. J. Med. Sci. 14, 98-104 (1978)

Sato, S., Matsushima, T., Sugimura, T.: Hexokinase isozyme patterns of experimental hepatomas of rats. Cancer Res. 29, 1437-1446 (1969)

Scherer, E., Emmelot, P.: Kinetics of induction and growth of precancerous liver-cell foci, and liver tumor formation by diethylnitrosamine in the rat. Eur. J. Cancer 11, 689-696 (1975)

Scherer, E., Hoffman, M.: Probable clonal genesis of cellular islands induced in rat liver by diethylnitrosamine. Eur. J. Cancer 7, 369-371 (1971)

Scherer, E., Hoffman, M., Emmelot, P., Friedrich-Freksa, H.: Quantitative study of foci of altered liver cells induced in the rat by a single dose of diethylnitrosamine and partial hepatectomy. J. Natl. Cancer Inst. 49, 93-106 (1972)

Sell, S., Becker, F.F.: Alpha-fetoprotein. J. Natl. Cancer Inst. 60, 19-26 (1978)

Shatton, J.B., Morris, H.P., Weinhouse, S.: Kinetic, electrophoretic and chromatographic studies on glucose-ATP phosphotransferases in rat hepatomas. Cancer Res. 29, 1161-1172 (1969)

Shier, W.T., Trotter, J.T.: Oncogenic transformation of 3T3 cells is associated with conversion from an "adult" to an "embryonic" esterase isoenzyme pattern. Exp. Cell Res. 111, 285-294 (1978)

Sirica, A.E., Richards, W., Tsukada, Y., Sattler, C.A., Pitot, H.C.: Fetal-Phenotypic expression by adult rat hepatocytes on collagen gel-nylon meshes. Proc. Natl. Acad. Sci. USA 76, 283-287 (1979)

Tanaka, T., Imamura, K., Ann, T., Taniuchi, K.: Multimolecular forms of pyruvate kinase and phosphofructokinase in normal and cancer tissues. Gann Monogr. 13, 219-234 (1972)

Ting, C.-C., Sanford, K.K., Price, F.M.: Expression of fetal antigens in fetal and adult cells during long-term culture. In Vitro 14, 207-211 (1978)

Tomasi, T.B.: Structure and function of alphafetoprotein. Ann. Rev. Med. 28, 453-465 (1977)

Walker, P.R., Bonney, R.J., Becker, J.E., Potter, V.R.: Pyruvate kinase, hexokinase, and aldolase isoenzymes in rat liver cells in culture. In Vitro 8, 107-114 (1972)

Weinhouse, S.: Metabolism and isozyme alterations in experimental hepatomas. Fed. Proc. 32, 2162-2167 (1973)

Williams, G.M.: Functional markers and growth behavior of preneoplastic hepatocytes. Cancer Res. 36, 2540-2543 (1976)

Cell Differentiation and Its Relation to Promotion and Prevention of Bladder Cancer

R.M. Hicks

School of Pathology, Middlesex Hospital Medical School
London, W1P 7LD, England

I. Introduction

There are many ways of considering the biogenesis of cancer, partly because it is not a single entity but a family of diseases all of which are characterized by uncontrolled, atypical cell growth. While it is generally accepted that the neoplastic cell reflects an altered function of the genome, the initial injury to the genetic apparatus may be quite different in different cancers. Some neoplasms may be associated with gross morphological alterations to the chromosomes, for example familial retinoblastoma; in others, neoplastic growth follows incorporation of a viral genome into the cell which suppresses but does not eliminate the genes controlling normal differentiation, as is seen in the transformation of human cells in culture by SV 40 virus. Indeed, there may be no permanent gross alterations of the genome in a tumor cell but instead only altered selection of those genes which are expressed, as indicated by the production of normal chimeric mice from teratocarcinoma cells demonstrated by Dr Mintz and Dr Illmensee (for review of recent work see paper by Illmensee in this volume). In all instances, however, the final result of neoplastic transformation is an altered phenotype, such that the cancer cells express certain properties which differ from those of the normal cells from which they were derived. Thus cancer maybe regarded as a disease not necessarily of differentiation, but certainly a disease in which some concomitant phenotypic change in differentiation is obligatory.

II. Differentiation of the Normal Urothelium

The normal mammalian urinary bladder is lined by highly differentiated epithelium known as the urothelium. The changes which occur in the urinary bladder during carcinogenesis illustrate concomitant changes in differentiation, not all of which are necessarily a part of the neoplastic growth syndrome. The normal differentiation of the adult mammalian bladder has been described and illustrated in detail elsewhere (Hicks et al. 1974; Hicks 1975; Severs and Hicks 1977). The markers for normal differentiation include development of highly specialized, polyploid surface cells, with a uniquely structured surface membrane.

The particular form of transitional cell differentiation seen in the normal adult mammalian urothelium is in a state of balance which is very easily disturbed. Furthermore, the urothelium is a multipotential tissue (Hicks 1975) and has the option of other forms of differentiation. Thus, if subjected to regular irritation as from a bladder calculus, the normal differentiation is disturbed and switches from transitional cell differentiation to epidermalization with the synthesis of gross keratin plaques. Similarly, epidermalization is also produced if the animal is made vitamin A-deficient, when again, the genome for normal transitional differentiation is apparently switched off, and that for differentiation to squamous metaplasia is switched on or amplified (Hicks 1968, 1969, 1975, 1976). The frequency with which epidermalization or squamous metaplasia of the urothelium is seen, suggests that the genome responsible for keratinization, though normally repressed, must be in a readily inducible state and that the controlling area must be readily accessible and easily derepressed. An alternative type of differentiation which is also seen in the urothelium is mucous metaplasia and this frequently develops in mild pathological conditions, for example in chronic cystitis. Again, this form of differentiation is seen sufficiently frequently to suggest that the area of the genome concerned with mucous production, though normally repressed, is in a position or state where it is readily inducible.

These three forms of differentiation may be seen in the adult mammalian urothelium either in normal or in only mildly pathological conditions. They are all, however, morphologically different from the more primitive differentiation of the urothelium found in the fetus (Firth and Hicks 1970). These gross manifestations of differentiation observed in the normal or nearly normal urothelium must represent minor variations at the level of transcription in the control of the genome in the normal urothelial cell. If it is assumed that a balanced supply of various inducers and repressors to the nucleus controls which genes in the noncondensed part of the DNA are actually transcribed, then continuous and regulated synthesis of these modulators is necessary to maintain the status quo. In the normal bladder, the balance of these modulators clearly allows transitional cell differentiation. In vitamin A-deficiency, the balance changes and keratinization is switched on. In other circumstances the supply of modulators is modified yet again and mucous metaplasia results. These are all relatively trivial alterations in the phenotype of the urothelial cells and are not related in any way to malignant transformation or those alterations in the DNA which permit uncontrolled growth and multiplication of the urothelial cancer cell.

III. Carcinogenesis: Initiation and Promotion

Biogenesis of urothelial cancer, though it occasionally occurs as a result of exposure to a high dose of a complete carcinogen, more usually is the outcome of a multistage process of initiation, promotion and propagation (Hicks and Chowaniec 1978; Hicks et al. 1978). In this sequence of events, the initiator, which in man is probably a small amount only of the activated form of a urine-borne chemical carcinogen, reacts with and produces some permanent alteration to the DNA so that the DNA carries altered information by comparison with normal. Interestingly, as shown by Yuspa and his colleagues (Yuspa et al. 1969) initiators do not preferentially react with replicating rather than nonreplicating DNA, and therefore, since most of the genes are in a repressed state, the modified areas of the DNA are more likely to be in the unexpressed than in the transcribing areas of the genome. Therefore no difference in phenotypic expression will necessarily be seen in the urothelial cell after an initiating event. However, even if the initiating event is in an unexpressed area of the genome, it will be permanent and leave some cells harboring misinformation even though they are phenotypically identical to their unaffected neighbours as judged by their functional proteins and membrane structure. Thus, after treatment of the bladder with a low, initiating dose of a carcinogen like N-methyl-nitrosourea (MNU), after the initial response of the urothelium to the toxic effects of the drug, differentiation will return to normal and the animal may survive with a normally differentiated urothelium for the rest of its life (Wakefield and Hicks 1973). Nevertheless, such an initiating dose of MNU selectively introduces methyl groups into the O_6 guanine position of rat bladder DNA, and these persist for long periods of time, certainly many weeks after the introduced (Bird 1978). Hence methylated O guanine may react abnormally with the normal protein/DNA interaction at the point where the methyl group has been introduced (Bird 1978). Hence methylated O_6 guanine may react abnormally with any incoming inducer or repressor proteins and also with the histones of the nucleosomes; since it is the interaction of proteins with the DNA which determines which genes are repressed and which are inducible, alkylation of the bases by a carcinogen has the potential to modify the transcription of the genome. (For reviews on the structure of transcribing and nontranscribing DNA and the interaction of DNA with nuclear proteins see Franke and Scheer and articles by other authors in The Philosophical Transactions of the Royal Society of London, Vol. 283, Structure of Eukaryotic Chromosomes and Chromatin, 1978).

After the initiating event, although the cell may carry altered information, so long as the altered area remains unexpressed no change in phenotype will occur even when the cell divides in the normal course of events. Before a change in phenotype can be seen, a promoting event is required which will allow the altered area of genome to become inducible, e.g., the area carrying a methylated base must be morphologically altered into a transcribing rather than a nontranscribing form, and the balance between inducer and repressor proteins must be altered to allow new genes to be expressed. These are the very changes which known skin promoters are able to bring about; they aid or stimulate the evolution of the initiated site to a latent tumor cell by altering gene expression. Promoters of skin cancer are also known to promote hyperplasia by increasing the rate of DNA synthesis and the rate of turnover of the basal cells, so giving any altered transcribable area of the genome a better chance to be expresssed and to appear as an altered phenotype. Thus promotors stimulate the altered cell to grow and divide and, in effect propagate tumor growth (see review by Scribner and Suss 1978).

In the biogenesis of bladder cancer, vitamin A-deficiency clearly acts as a promoting stimulus. This was observed nearly 20 years ago by Angrist and his co-workers though at the time it was not interpreted as initiation and promotion (Capurro et al. 1960). Updoubtedly vitamin A-deficiency causes an increase in DNA synthesis and increased mitotic activity in the basal cells of the urothelium which results in hyperplasia. Coincident with this rather nonspecific stimulation of cell division and tumor promotion, vitamin A-deficiency also specifically affects the genes controlling the differentiation of the cell so that bladder tumors in vitamin A-deficient animals are also heavily keratinized. The same is true where cancer growth is associated with mechanical irritation, as in the presence of a urinary calculus or in a bilharzial bladder in which the wall is filled with calcified ova, for as mentioned above, irritation predisposes to squamous metaplasia. In both these instances keratinization is concomitant with but not an essential part of the biogenesis of the neoplasm. The keratinization demonstrates a disturbance in the normal control of transcription introduced by vitamin A-deficiency or irritation; if these factors also encourage random transcription of the genome they will automatically increase the chance of transcribing potentially carcinogenic changes in the genome introduced by previous exposure to an initiator. Recently we have demonstrated that saccharin also acts as a promotor of the growth of bladder cancer in animals previously subjected to an initiating dose of the carcinogen MNU (Hicks et al. 1975; Hicks and Chowaniec 1977). These observations have been confirmed by Friedell and his co-workers using an initiating dose of another bladder carcinogen, the nitrofuran FANFT (Cohen et al. 1978). There is no explanation, as yet, as to how the saccharin acts as a promotor in this system. Undoubtedly tumors promoted by saccharin show a wide variety of differentiation patterns both within the tumor and in adjacent, non-neoplastic areas of urothelium (Chowaniec and Hicks, unpublished observations). This suggests that saccharin in some way disturbs the normal control of differentiation and, by implication, suggests that it disturbs the balance of regulator molecules arriving at the nucleus. Like vitamin A-deficiency it may encourage random gene transcription and so increase the chance of expressing any neoplastic modification of the growth controlling genes introduced by pretreatment with an initiator. Furthermore, in the absence of any initiating treatment saccarin undoubtedly causes urothelial hyperplasia in some but not all animals, and increased numbers of cells enter mitosis (Chowaniec and Hicks, unpublished observations). Thus saccharin possesses two properties characteristic of skin promotors, namely it appears to increase the chance of random transcription of the genome and at the same time it encourages cell division thus promoting and propagating tumor growth.

A third property of promoters is the inhibition of terminal differentiation of the tissue. A number of examples are now published where the phorbol ester series of skin promotors maintain tissues in a relatively undifferentiated state and delay the terminal differentiation of cell lines in culture. It is a matter of common observation that the aggressive, dividing, malignant cells in solid tumors are not usually those undergoing terminal differentiation, but are the relatively undifferentiated stem cells which remained unspecialized or anaplastic in highly invasive cancers. For example, in the bladder, the neoplastic urothelial cells are small, predominantly diploid cells which do not reflect any of the obvious adult differentiation patterns seen in the superficial cells of the normal urothelium (Hicks 1976; Hicks and Chowaniec 1978). The developing bladder cancer cells characteristically fail to develop the markers previously described for normal differentiation in the urothelium, but instead may develop other patterns of differentiation as described above which are not strictly linked to the neoplastic process. Fortuitously, the developing bladder cancer cells also develop a new phenotype which we have not observed in either the normal fetal or adult bladder. The cells at the free surface of developing urothelial tumors develop bizarre, glycocalyx-coated microvilli, which have been described in detail elsewhere (Hicks and Wakefield 1976; Newman and Hicks 1977; Hicks and Chowaniec 1978). These changes occur so regularly at the free surface of developing mammalian tumors, both in man and in experimental animals, that it suggests the gene responsible for the surface changes must be closely linked to the gene responsible for the uncontrolled growth of the neoplastic cell. These characteristic surface changes are thus convenient morphological markers for neoplastic transformation in the urinary bladder.

In summary, promotors increase the opportunity for abnormal phenotypic expression by accelerating the rate of mitosis; this also helps to propagate the tumor. In addition, by randomly altering gene expression, possibly by disturbing the balance of inducers and repressors arriving at the nucleus, they increase the opportunity of expressing any altered information previously introduced into the genomes by exposure to an initiator. Finally, by inhibiting terminal differentiation of the tissue they increase the time required before a cell will pass into the "differentiate and die" pathway and retain the cells for longer in the cell cycle where they can grow and divide.

IV. Differentiation and Anticarcinogenesis

Over many years the action of a number of anticarcinogenic agents, including actinomycin D, dexamethasone and the family of vitamin A analogues known as the retinoids has been described in various experimental systems. Recently, renewed interest in the retinoids has been stimulated by Sporn and his colleagues (Sporn 1976; Sporn et al. 1976). These compounds have been shown to be anticarcinogenic for several different epithelial cell systems. For example, feeding 13-cis-retinoic acid to rats after a complete carcinogenic treatment with either MNU or another bladder carcinogen, butylbutanol nitrosamine (BBN), significantly inhibits the development of both preneoplastic and neoplastic growth of the urothelium (Grubbs et al. 1977; Sporn et al. 1977; Squire et al. 1977). This retinoid diminishes the number of cancers which develop and also their severity judged in terms of their invasiveness. Not surprisingly, since it is a vitamin A analogue, the retinoid also decreases very significantly the degree of squamous metaplasia observed in the bladders of carcinogen-treated, retinoid-fed animals by comparison with the carcinogen-treated animals on a control diet. However, in the experimental conditions used so far both for the bladder and other organ systems, the retinoid treatment, though it significantly reduces the growth of neoplasms, still permits some tumors to develop in carcinogen treated animals.

It is not yet known exactly how retinoids exert their anticarcinogenic effect, but in the context of this discussion some suggestions can be made. Retinoids, like some other anticarcinogenic agents, inhibit DNA synthesis and thus retard cell division. They will therefore reduce the opportunity for phenotypic expression of any change in the genome introduced by previous exposure to a carcinogen and at the same time will antagonize tumor propagation. This antihyperplastic effect is clearly nonspecific, but is the converse of the effect produced by promoting agents which encourage hyperplasia and tumor propagation. Both dexamethasone and the retinoids are also known to oppose tumor promotion and to antagonize the action of such established skin promotors as 12-o-tetradecanoylphorbol 13-acetate (TPA) (Bollag 1972; Scribner and Slaga 1973). If, as suggested earlier, promotors act by modifying the balance of inducer and repressor molecules available to the nucleus, thus permitting random transcription and the production of altered phenotypes, it seems possible that antipromoting agents may act by restoring or maintaining the normal balance of inducers and repressors either directly or indirectly. Recently Boutwell and his colleagues have shown synthetic retinoids prevent the synthesis of gene-modulating polyamines in the skin by inhibiting the burst of ornithine decarboxylase synthesis which normally follows exposure of the cells to a promoting agent; furthermore, the efficacy with which they inhibit ornithine decarboxylase activity is roughly proportional to their tumor inhibiting activity in skin (Verma et al. 1978).

As well as acting as antipromoting agents, retinoids reinforce normal differentiation patterns and maintain the normal morphology of epithelial tissues (Moore 1967). They stimulate the production of differentiated cell lines from pluripotent embryonal carcinoma cell lines (Jetten and Jetten 1978) and reduce the degree of abnormal differentiation in experimental bladder cancers (Sporn et al. 1976).

Thus, in a multistage system of initiation and promotion, an anticarcinogenic agent such as one of the retinoids may prevent inbalance in gene modulating molecules by directly antagonizing the biochemical action of promotors, and may also oppose tumor propagation by reducing the rate of cell turnover. Since at the same time retinoids reinforce normal differentiation of tissues, they presumably reduce the chance of random transcription and reduce the chance expression of previously introduced genetic modifications. These effects could apply to any multistage process involving step functions of initiation, promotion, and propagation spread over a temporal sequence of days or months. Unfortunately, if tissues are exposed to high enough doses of powerful complete carcinogens, the carcinogen, rather than just initiating a few changes in the DNA, may produce both initiation and promoting effects and cause irreparable damage to sets of regulator genes located in the normally transcribing regions of the DNA. Under these circumstances, a dominant mutation may be produced which is not susceptible to modification by outside influences which can lead directly to neoplastic growth. In that situation, neither promotors such as saccharin nor anticarcinogenic agents such as retinoids are likely to affect the final outcome. When confronted by a dominant, neoplastic lesion in the transcribing area of the genome, compounds such as retinoids may modulate some aspects of cell differentiation and possibly reduce the rate of tumor growth, but are unlikely to prevent indefinitely the growth of clones of tumor cells arising from the catastrophic damage produced by the carcinogen. Fortunately, our environment is such that we are seldom exposed to high doses of complete carcinogens, though this has occurred before now, e.g., the industrial exposure of workers in the chemical and rubber industries to 2-naphthylamine. Our environment undoubtedly contains low levels of numerous carcinogens of differing potency capable of initiating random damage to the genome of a few cells in multifocal sites in disparate tissues. In this situation, agents which, like the retinoids, reinforce normal cell differentiation and oppose tumor promotion may have a very valuable role to play in reducing both the morbidity and mortality currently attributable to cancer.

V. Conclusions

This short discussion is both speculative and incomplete, for it is not possible to summarize in a short article all published work which bears either upon the problems of cancer promotion and prevention, or upon the effect of modulating cell differentiation patterns on neoplastic growth. The hypotheses proposed here are intended simply as a basis for further research and investigation, and will doubtless be modified in the light of new information. It is clear, however, that any major advance in understanding the mechanisms of tumor promotion and prevention in the next decade depends on further basic research into the control of normal and abnormal cell differentiation.

Acknowledgments. The work cited from this laboratory has been supported by generous grants from the Cancer Research Campaign over the past 19 years. I am most grateful to the Campaign and to my numerous colleagues, past and present, at the Middlesex Hospital Medical School for support, encouragement and helpful discussion over this time.

References

Bird, A.P.: The occurrence and transmission of a pattern of DNA methylation in *Xenopus laevis* ribosomal DNA. Philos. Trans. R. Soc. London Ser. B. 283, 325-327 (1978)

Bollag, W.: Prophylaxis of chemically induced papillomas and carcinomas of mouse skin by vitamin A acid. Experientia 28, 1219-1220 (1972)

Capurro, P., Angrist, A., Black, J., Moumgis, B.: Studies in squamous metaplasia in rat bladder. 1. Effects of hypovitaminosis A, foreign bodies and methylcholanthrene. Cancer Res. 20, 563-567 (1960)

Cohen, S.M., Arai, M., Friedell, G.H.: Promoting effect of DL-tryptophan and saccharin in urinary bladder carcinogenesis in the rat. Proc. Amer. Assoc. Cancer Res. 19, 4 (1978)

Cox, R., Murphy, W.M., Irving, C.C.: Alkylation of DNA in rat bladder epithelium by N-methyl-N-nitrosourea (MNU). Proc. Amer. Assoc. Cancer Res. 18, 167 (1977)

Firth, J.A., Hicks, R.M.: Differentiation and cell death in transitional epithelium in the urinary bladder of foetal and suckling rats. J. Anat. 107, 192-194 (1970)

Franke, W.W., Scheer, U.: Morphology of transcriptional units at different states of activity. Philos. Trans. R. Soc. London Ser. B. 283, 333 (1978)

Grubbs, C.J., Moon, R.C., Squire, R.A., Farrow, G.M., Goodman, D.G., Brown, C.C., Sporn, M.B.: 13-cis-retinoic acid: Inhibition of the development of bladder cancer induced in rats by N-butyl-N-(4-hydroxybutyl)nitrosamine. Science 198, 743-744 (1977)

Hicks, R.M.: Hyperplasia and cornification of the transitional epithelium in the vitamin A-deficient rat. J. Ultrastruct. Res. 22, 206-230 (1968)

Hicks, R.M.: Nature of the keratohyalin-like granules in hyperplastic and cornified areas of transitional epithelium in the vitamin A-deficient rat. J. Anat. 104, 327-339 (1969)

Hicks, R.M.: The mammalian urinary bladder; an accommodating organ. Biol. Rev. 50, 215-216 (1975)

Hicks, R.M.: Changes in differentiation in the urinary bladder during benign and neoplastic hyperplasia. In: Progress in Differentiation Research (ed. N. Muller-Berat), pp. 339-353. Amsterdam: North-Holland 1976

Hicks, R.M., Chowaniec, J.: The importance of synergy between weak carcinogens in the induction of bladder cancer in experimental animals and man. Cancer Res. 37, 2943-2947 (1977)

Hicks, R.M., Chowaniec, J.: Experimental induction, histology and ultrastructure of hyperplasia and neoplasia of the urinary bladder epithelium. Int. Rev. Exp. Pathol. 18, 199-280 (1978)

Hicks, R.M., Wakefield, J.St.J.: Membrane changes during urothelial hyperplasia and neoplasia. Cancer Res. 36, 2502-2507 (1976)

Hicks, R.M., Ketterer, B., Warren, R.C.: The ultrastructure and chemistry of the luminal plasma membrane of the mammalian urinary bladder. A structure with low permeability to water and ions. Philos. Trans. Roy. Soc. London Ser. B. 268, 23-38 (1974)

Hicks, R.M., Wakefield, J.St.J., Chowaniec, J.: Evaluation of a new model to detect carcinogens or co-carcinogens; results obtained with saccharin, cyclamate and cyclophosphamide. Chem. Biol. Interact. 11, 225-233 (1975)

Hicks, R.M., Chowaniec, J., Wakefield, J.St.J.: Experimental induction of bladder tumors by a two-stage system. In: Carcinogenesis, Mechanisms of Tumor Promotion and Carcinogenesis, Vol. 2 (eds. T.J. Slaga, A. Sivak, R.K. Boutwell), pp. 475-489. New York: Raven Press 1978

Jetten, A., Jetten, M.E.R.: Stimulation of differentiation of several embryonal carcinoma cell lines by retinoids. Abstract of Poster (1978)

Moore, T.: Effects of vitamin A deficiency in animals: pharmacology and toxicology of vitamin A. In: The Vitamins, 2nd ed., Vol. 1 (eds. W.H. Sebrell, Jr., R.S. Harris), pp. 245-266, 280-294. New York: Academic Press 1967

Newman, J., Hicks, R.M.: Detection of neoplastic and preneoplastic urothelia by combined scanning and transmission electron microscopy of urinary surface of human and rat bladders. Histopathology 1, 125-135 (1977)

Scribner, J.D., Slaga, T.J.: Multiple effects of dexamethasone on protein synthesis and hyperplasia caused by a tumor promotor. Cancer Res. 33, 542-546 (1973)

Scribner, J.D., Suss, R.: Tumor initiation and promotion. Int. Rev. Exp. Pathol. 18, 137-198 (1978)

Severs, N.J., Hicks, R.M.: Frozen-surface replicas of rat bladder luminal membrane. J. Microsc. 111, 125-136 (1977)

Sporn, M.B.: Approaches to prevention of epithelial cancer during the preneoplastic period. Cancer Res. 36, 2699-2702 (1976)

Sporn, M.B., Dunlop, N.M., Newton, D.L., Smith, J.M.: Prevention of chemical carcinogenesis by vitamin A and its synthetic analogs (retinoids). Fed. Proc. 35, 1332-1338 (1976)

Sporn, M.B., Squire, R.A., Brown, C.C., Smith, J.M., Wenk, M.L., Springer, S.: 13-cis-retinoic acid: Inhibition of bladder carcinogenesis in the rat. Science 195, 487-489 (1977)

Squire, R.A., Sporn, M.B., Brown, C.C., Smith, J.M., Wenk, M.L., Springer, S.: Histopathologic evaluation of the inhibition of rat bladder carcinogenesis by 13-cis-retinoic acid. Cancer Res. 37, 2930-2936 (1977)

Verma, A.K., Rice, H.M., Shapas, B.G., Boutwell, R.K.: Inhibition of 12-o-tetradecanoylphorbol-13-acetate-induced ornithine decarboxylase activity in mouse epidermis by vitamin A analogs (retinoids). Cancer Res. 38, 793-801 (1978)

Wakefield, J.St.J., Hicks, R.M.: Bladder cancer and N-methyl-N-nitrosourea, 11. Subcellular changes associated with a single non-carcinogenic dose of MNU. Chem. Biol. Interact. 7, 165-179 (1973)

Yuspa, S.H., Del Sol, A.E., Morgan, D.L., Bates, R.R.: The binding of 7,12-dimethylbenz(a)anthracene to replicating and non-replicating DNA in cell culture. Chem. Biol. Interact. 1, 223-233 (1969)

Drug-Induced Differentiation of Human Neuroblastoma: Transformation into Ganglion Cells with Mitomycin-C

M.N. GOLDSTEIN and S. PLURAD

Department of Anatomy and Neurobiology
Washington University School of Medicine, St. Louis, MO 63110, USA

I. Introduction

A variety of substances have been used to modify gene expression with the purpose of learning about mechanisms which regulate growth and differentiation in normal and malignant cells. Some block cell differentiation, others promote growth and/or cell differentiation and maturation (Silbert and Goldstein 1972; Kram et al. 1973; Kolber et al. 1974; Goldstein and Brodeur 1975; Brodeur and Goldstein 1976; Fibach et al. 1977; Jones and Moscona 1977; Kolber et al. 1978; Marks et al. 1978; Prasad and Sinha 1978; Sachs 1978).

In our studies with human neuroblastomas, hormones, antimetabolites, cytotoxic antibiotics and cyclic nucleotides have been added to established lines to determine if they would modulate gene expression and promote differentiation.

Mouse nerve growth factor (NGF) protein obtained from male mouse submandibular glands stimulated cells of some lines of human neuroblastoma and caused cells to increase in size and develop long neurites but few were transformed into mature neurons. Dibutyryl cyclic adenosine monophosphate increased adherence to the substrates on which cells were grown and changed their morphology and chemistry but did not irreversibly differentiate the cells.

A number of antimetabolites and antibiotics which inhibit DNA synthesis without markedly affecting RNA or protein synthesis were also assayed for their potential to modulate the differentiated state of human neuroblastomas. One of them, mitomycin-C, slowly produced changes in the morphology of immature neuroblastoma cells and caused them to transform into nondividing ganglion cells. Like normal sympathetic neurons, the transforming cells became dependent on NGF for continuous long-term survival in culture. The sequence of morphologic changes from immature neuroblastoma cells to ganglion cells is described in this paper.

II. Material and Methods

Human neuroblastoma line NGP and clones derived from this line, NGP-2 and NGP-20R (resistant to 10^{-5} M ouabain) were used in these studies (Brodeur and Goldstein 1976). Confluent cultures in T-30 plastic flasks were incubated for 1 h at

37.5° C in the growth medium, 20% fetal calf serum, 80% Eagle's MEM and nonessential amino acids, containing 4γ/ml to 20γ/ml of mitomycin-C (Caspersson 1965). The cultures were rinsed thoroughly with control medium and re-fed with control medium or control medium containing 200 nanograms/ml of 2.5 S NGF (Kolber et al. 1978).

The conversion of immature neuroblastoma cells was followed by time lapse photography. Cultures were also fixed at varying times and control and drug-treated cultures were prepared by conventional techniques for study by transmission and scanning electron microscopy.

III. Results

The small fusiform cells of the lines of human neuroblastoma resembled immature neuroblasts (Fig. 1A) and as cultures became confluent many cells developed thin bipolar neurites. NGF markedly stimulated neurite outgrowth and the formation of spiny processes and collateral fibers. Although NGF affected cell morphology and chemistry, careful survey of dense cultures growing for many weeks in media with NGF showed that there were few cells with the morphology of mature neurons.

When confluent cultures were exposed to mitomycin-C, no striking morphologic changes were observed for 3 to 4 days. HeLa cells or rat C-6 glioma cells were completely lysed by the 4th day after exposure to the antibiotic.

Some debris from degenerating neuroblastoma cells appeared in the medium for another week following the 4-day lag. The cells that survived the treatment were larger and many had begun to develop processes which were even longer than control cells growing in media with NGF. It was surprising to find that there were as many surviving cells in cultures treated with 4γ/ml as in those exposed to 10γ or 20γ/ml of antibiotic. By the 14th day the size of the cell soma and neurites were markedly increased and cells had begun to form aggregates instead of growing monolayers (Fig. 1B). The aggregates of cell bodies and their neurites resembled the pattern of growth of normal sympathetic neurons in culture.

The addition of NGF to the growth medium was essential for the long term survival of mitomycin-C treated cultures. With NGF in the medium cultures were maintained for 3 months, while those without NGF degenerated by the 3rd week. As the culture period increased, the cell bodies of transformed neuroblastoma cells continued to increase in size and they developed more neurites (Fig. 1C). Isolated ganglion-like cells were often found enmeshed in a web of neurites. Scanning electron photomicrographs of cultures at this time showed that there were many cells with long multipolar processes and also many dilated terminals which resembled those observed in cultures of normal neurons (Fig. 1C). The scanning electron photomicrographs showed that the organization and surface topography of the cells were completely different from control cells and were similar to scanning electron photomicrographs of normal sympathetic ganglion cells.

Modification in the ultrastructure of the mitomycin-C treated cells provided further evidence for their transformation into ganglion cells. Control cells (Fig. 2A)

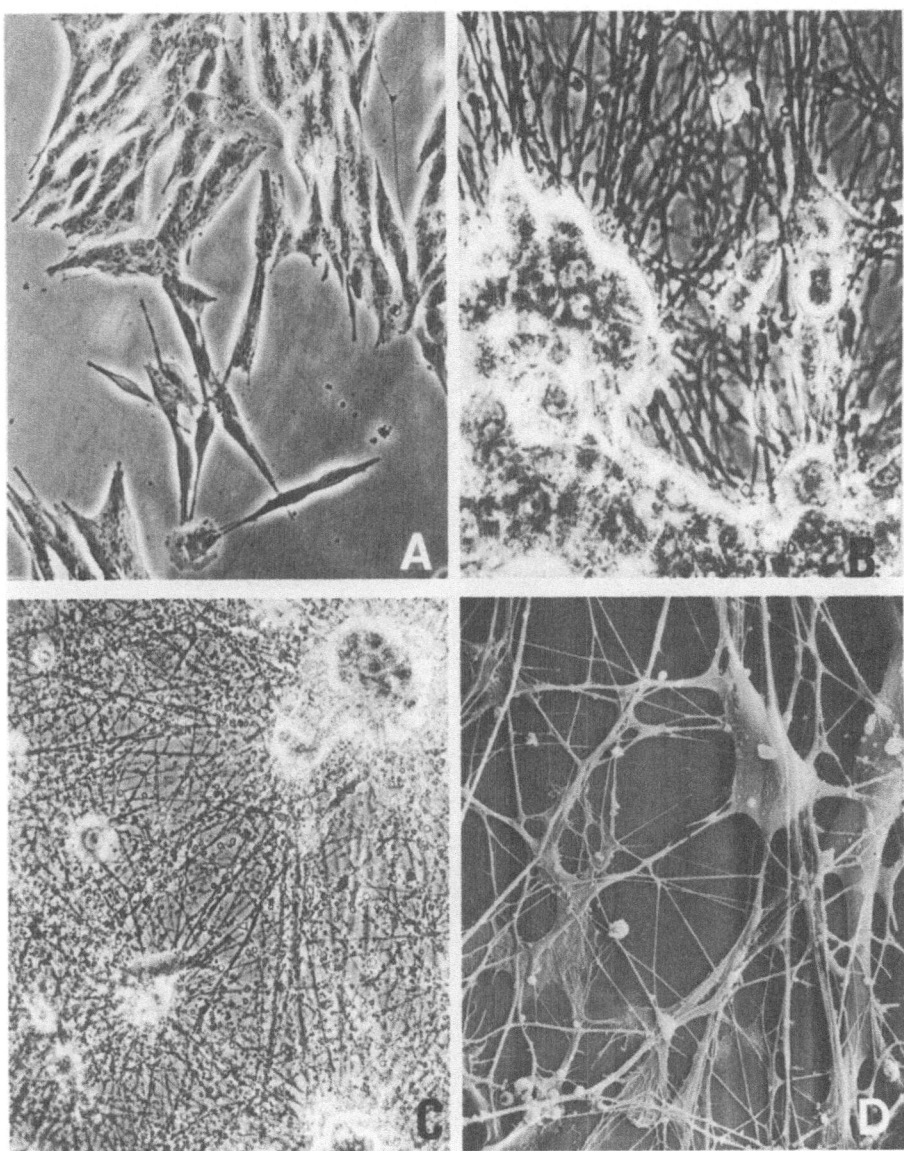

Fig. 1. A Small fusiform cells of cloned line of human neuroblastoma NGP-2 growing in control medium. 4 days after subculture. Positive phase contrast. × 175. **B** NGP-2 cells 19 days after exposure to 5 gamma/ml of mitomycin-C. Medium contained 200 mg/ml of 2.5 S NGF after 4th day. Note the enlargement of the aggregated cells and the extensive outgrowth of neurites and collaterals. Positive phase contrast. × 350. **C** Culture of **B** at 44 days. Note the multipolar ganglion cell. The aggregates of transformed cells had a brownish precipitate adhering to cell bodies. This is characteristic of neurons and pheochromocytes which syntheisize and secrete catecholamines. Positive phase contrast. × 160. **D** S.E.M. of 44 day culture of **C**. Note the ganglion cell and network of interlacing neurites. Many of the endings have formed expanded flat terminals. Coated with gold/paladium. × 1000

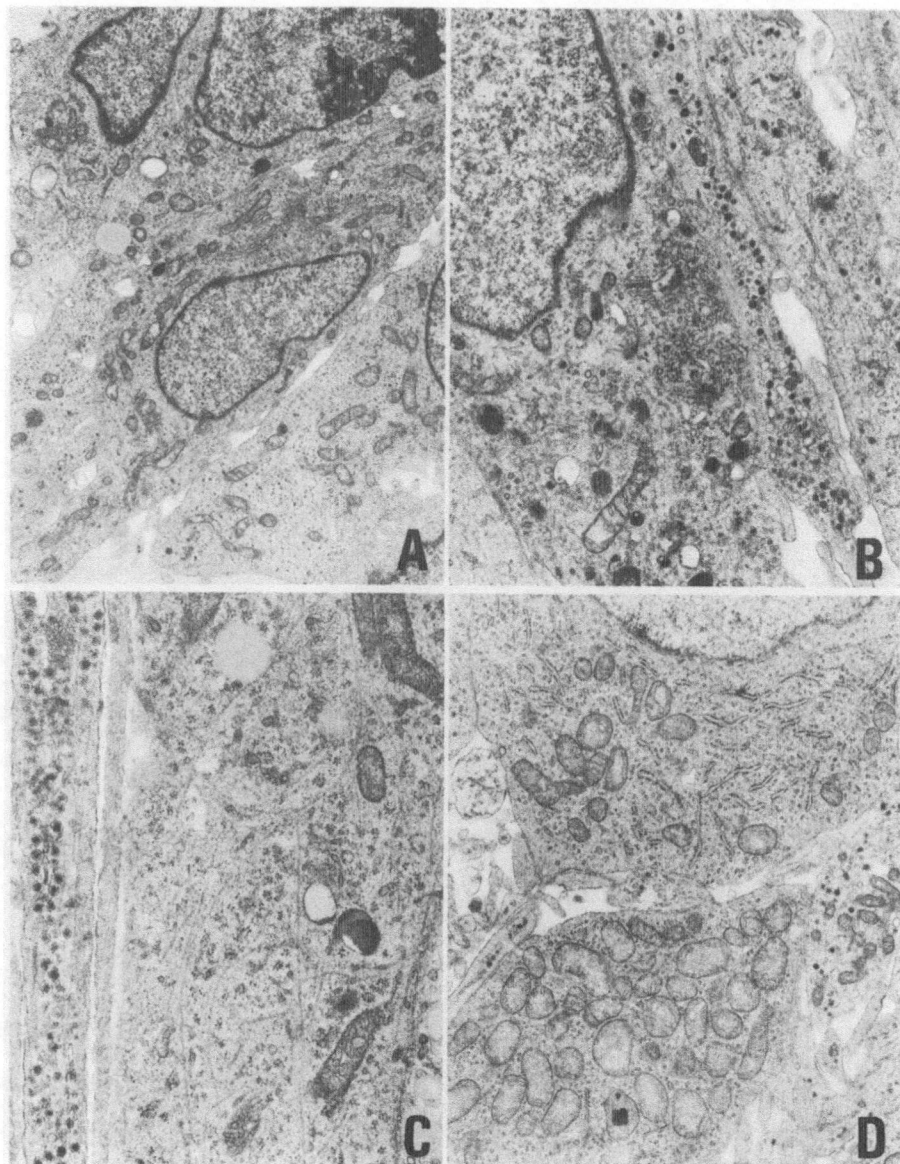

Fig. 2. A T.E.M. of line NGP-2 cells grown for 8 days in standard medium. Note the many small tubular mitochondria, Golgi complex, polysomes and sparse RER. Dense core and empty core secretion granules are in a short process at *lower left.* Control. × 6,200. **B** T.E.M. of sister culture to **A** grown with 200 mg/ml of 2.5 S mouse NGF. Note the increase in the Golgi complex and the many secretion granules derived from Golgi complex. × 7000. **C** T.E.M. of portion of cell and neurite from culture of NGP-2 pulsed with 10 gamma/ml mitomycin-C and grown in medium with NGF for 50 days. The neurite contains prominent secretion granules and small empty granules. Note the many polysomes and mitochondria. × 18,300. **D** T.E.M. of cells from same culture as **C**. Note the arrays of organized RER and many polysomes in the cytoplasm of transformed neuroblastoma cell.
Nucleus is enlarged and cells resemble normal sympathetic neurons. × 10,600

had small nuclei with borders of heterochromatin, and cytoplasm containing an inactive Golgi complex, few polysomes and few dense core secretion granules. Cells grown in media with NGF (Fig. 2B) had larger nuclei but they retained the distribution of heterochromatin as observed in control cells. The cytoplasm showed a marked proliferation of Golgi membranes and many empty and dense core secretion granules. The granules appeared in large numbers in elongated processes.

The mitomycin-C treated cells had the structure of ganglion cells (Goldstein et al. 1964). The chromatin in the nuclei of these cells was not condensed; it was organized as euchromatin. The structure and organization of the cytoplasm was like that of normal neurons. There was an increase in microtubules, in secretion granules, in rough endoplasmic reticulum and now many terminals full of granules were also observed (Fig. 2C-D).

IV. Discussion

The mechanism responsible for the transformation of immature neuroblastoma cells of a line designated NGP and of cells of other neuroblastoma lines not reported on in this paper is unknown. Mitomycin-C is a bifunctional alkylating agent which crosslinks G-C bases in DNA. This action does not explain its potential for inducing cells of some lines of neuroblastoma to become ganglion cells. Hela, C_6 rat glioma and other human and animal cells which were pulsed in the same manner with Mitomycin-C were completely lysed (Lapis and Bernhard 1965).

A surprising number of different agents are able to cause a variety of tumors to differentiate in vitro. It is unlikely that polar compounds, cytotoxic antibiotics, or non-toxic growth factors have a common mechanism of action. However, they appear to have a select effect on a variety of undifferentiated tumors and induce the tumor cells to develop properties of normal cells.

Our studies and those reported by other workers suggest that cancer is a disease of the mechanisms which regulate expression of genes. By modification of the environment of the tumor cells, sufficient pressure can be exerted on the tumor genome so that it can be forced to express in a coordinated manner the cellular information that imparts specificity to the normal cell of origin.

Acknowledgment. This paper is dedicated to Odile Schweisguth, former Chief of Pediatric Oncology at L'Institut Gustave-Roussy. Presented in part at the Annual Meeting of the American Association for Cancer Research, April, 1978.

References

Brodeur, G.M., Goldstein, M.N.: Histochemical demonstration of an increase in acetylcholinesterase in established lines of human and mouse neuroblastoma by nerve growth factor. Cytobios 16, 133-138 (1976)

Caspersson, T., Farber, S., Foley, G.E., Killander, D., Zettenberg, A.: Cytochemical evaluation of metabolic inhibitors in cell culture. Exp. Cell Res. 39, 365-385 (1965)

Fibach, E., Reuber, R.C., Rifkind, R.A., Marks, P.A.: Effect of hexamethylene bisacetamide on the commitment to differentiation of murine erythroleukemia cells. Canc. Res. 37, 440-444 (1977)

Goldstein, M.N., Brodeur, G.M.: Human neuroblastomas in vitro: Activation of the membrane pump for catecholamines by 5-bromodeoxyuridine. In: Proc. XI Int. Cancer Congr. (eds. P. Bucalossi, U. Veronesi, N. Cascinelli), Vol. , pp. 178-182. Amsterdam-New York: Elsevier 1975

Goldstein, M.N., Burdman, J.A., Journey, L.J.: Long term tissue culture of neuroblastomas II. Morphologic evidence for differentiation and maturation. J. Natl. Cancer Inst. 32, 165-199 (1964)

Jones, R.E., Moscona, A.A.: Effects of cytosine arabinoside on differential gene expression in embryonic neural retina. J. Cell Biol. 74, 30-42 (1977)

Kolber, A., Goldstein, M.N., Moore, B.W.: The effect of nerve growth factor on the expression of colchicine binding activity and 14-3-2 protein in an established line of human neuroblastoma. Proc. Natl. Acad. Sci. USA 71, 4203-4207 (1974)

Kolber, A.R., Perumal, A.S., Goldstein, M.N., Moore, B.W.: Drug-induced differentiation of a rat glioma in vitro. II. The expression of S-100, a glial specific protein and steroid sulfatase. Brain Res. 143, 513-520 (1978)

Kram, R., Mamont, P., Tomkins, G.M.: Pleiotypic control by adenosine 3':5' cyclic monophosphate: a model for growth in animal cells. Proc. Natl. Acad. Sci. USA 70, 1432-1436 (1973)

Lapis, K., Bernhard, W.: The effect of mitomycin-C on the nucleolar fine structure of KB cells in culture. Cancer Res. 25, 628-646 (1965)

Marks, P.A., Rifkind, R.A., Bank, A., Tenada, M., Reuben, R., Fibach, E., Nudel, U., Salmon, J., Gazitt, Y.: Induction of differentiation of murine erythroleukemia cells. In: Cell Differentiation and Neoplasia (ed. G.F. Saunders), pp. 453-471. New York: Raven Press 1978

Prasad, K.N., Sinha, P.K.: Regulation of differentiated functions and malignancy in neuroblastoma cells in culture. In: Cell Differentiation and Neoplasia (ed. G.F. Saunders), pp. 111-141. New York: Raven Press 1978

Sachs, L.: Control of normal cell differentiation in leukemic white blood cells. In: Cell Differentiation and Neoplasia (ed. G. Saunders), pp. 223-238. New York: Raven Press 1978

Silbert, S.W., Goldstein, M.N.: Drug induced differentiation of a rat glioma. Cancer Res. 32, 1422-1427 (1972)

Teratocarcinogenesis and Spontaneous Parthenogenesis in Mice

L.C. STEVENS

The Jackson Laboratory, Bar Harbor, ME 04609, USA

I. Introduction

There are several recent reviews on the biology of teratomas (Damjanov and Solter 1974; Martin 1975; Sherman and Solter 1975; Stevens 1975; Graham 1977; Jacob 1977, 1978; Mintz 1978). Most have emphasized in vitro studies of transplantable teratocarcinomas. Here the emphasis will be on studies of spontaneous and experimentally induced teratomas in mice.

Teratomas are tumors that usually occur in the gonads, and unlike other tumors they may be composed of many kinds of normal appearing immature and adult tissues (Stevens 1957; Stevens and Hummel 1957). In the mouse they are usually benign, but occasionally they grow progressively and are malignant. When teratomas contain several types of tissue as well as proliferating pluripotent stem cells (embryonal carcinoma cells), they are called teratocarcinomas. Embryonal carcinoma cells are pluripotent, and upon dividing they may give rise to other cells like themselves as well as to cells that differentiate into almost any kind of cell.

II. Testicular Teratomas

A. Transplantable Teratocarcinomas

Some teratocarcinomas can be maintained as transplantable tumors, and their undifferentiated embryonal stem cells may remain pluripotent or even totipotent for years.

Transplantable teratocarcinomas have been obtained from four sources. They may be obtained from spontaneous testicular (Stevens 1958) and ovarian teratomas (Stevens and Varnum 1974). They may also be derived from teratomas that have been experimentally induced (Stevens 1964). For example, when male genital ridges are grafted to the testes of adults, they develop into testes with teratomas (Stevens 1970a, 1970c). Early embryos when grafted to extrauterine sites will also form teratocarcinomas (Dunn and Stevens 1970; Solter et al. 1970).

Embryonal carcinoma stem cells of transplantable teratocarcinomas are of interest to developmental biologists because they have been shown to resemble normal early embryonic cells morphologically, biochemically, serologically, and in embryonic potency. They are readily available in large numbers for experimental analyses of many aspects of early mammalian differentiation and development.

B. Spontaneous Testicular Teratomas

Testicular teratomas are extremely rare in mice. When they were first observed, about 1% of the males of a subline of strain 129 were affected (Stevens and Little 1954). About 2% of another subline of strain 129 were found to have teratomas. Males in second and later litters have about twice as many teratomas as males in first litters. When the gene steel *(Sl)* was introduced onto strain 129 genetic background to form the congenic strain 129/Sv-*Sl C P* the incidence was increased in *Sl*/+ males to about 5% (Stevens and Mackensen 1961). About a third of the males of a new subline of strain 129/Sv-*ter* had congenital testicular teratomas (Stevens 1973). This high incidence was apparently due to a single gene mutation.

The development of sublines of strain 129 with increased incidence of teratomas made it feasible to perform a developmental study to determine when teratomas arise in the gonad and their cell of origin. Tissues appeared immature in tumors of young mice, and in mice of a given age the tumors appeared to be in about the same stage of development. The first morphological sign of differentiation was observed in mice 2 to 3 days old (Stevens 1959). The tumors were composed of vesicles lined by two different types of epithelium with undifferentiated embryonal cells dispersed between them. One type of epithelium resembled the embryonic ectoderm of embryos of 5 to 6 days of gestation. The other resembled endodermal epithelium like that in the pharynx of 9-day embryos. At no stage was primary endoderm of the type found in 6-day embryos observed. The ectodermal epithelium gave rise to neural tissue and epidermis. The endodermal epithelium gave rise to derivatives of respiratory and alimentary epithelium. The apparently undifferentiated cells between the epithelial vesicles gave rise to cartilage, bone, and muscle.

The early teratomas in fetuses of about 18 days of gestation were composed of clusters of undifferentiated embryonal cells that resembled cells in the inner cell mass of normal blastocysts (Stevens 1962). Cavities developed in the clusters lined by cells that later could be recognized as ectoderm or endoderm.

All of the teratomas in 15- to 16-day fetuses were located within seminiferous tubules. This was the first evidence suggesting that teratomas were derived from germ cells. Seminiferous tubules are lined by only two kinds of cells – those that would become spermatogonia, and those that would become Sertoli cells. The ultrastructure of primordial germ cells bore a striking resemblance to embryonal carcinoma cells (Pierce and Beals 1964; Pierce et al. 1967), and it seemed more likely that all of the many kinds of tissues found in teratomas were derived from totipotent primordial germ cells than from cells that had already become determined to become Sertoli cells. However, the possibility existed that teratomas were derived from cells that originated outside and migrated into the tubules.

C. Experimentally Produced Teratomas

A method was developed of experimentally inducing testicular teratomas from primordial germ cells in genital ridges of several genotypes (Stevens 1964, 1966,

1967, 1970a,c). The earliest spontaneous tumors were recognized in histological sections of 15-day fetuses. The size of teratomas in fetal mice from 15 days of gestation to birth was measured. Judging from the decreasing size in younger fetuses, it was estimated that teratocarcinogenesis was initiated at about 12 days of gestation.

When male genital ridges from 12-day fetuses of some strains were grafted to the testes of adults, they developed into testes, and for some inbred strains most of them had teratomas. Spontaneous testicular teratomas could easily be identified in the hosts at the time of grafting, and there was no problem in identifying induced tumors derived from grafted genital ridges. The induced tumors could be recognized about 6 days after grafting the genital ridges, and at that stage all of them were located within the seminiferous tubules as are early spontaneous teratomas. They appeared to develop as did the spontaneous teratomas.

When genital ridges from older fetuses were grafted to adult testes, they developed into testes without teratomas. Apparently the primordial germ cells undergo a developmental change during the 13th day of gestation that makes them resistant to the process of teratocarcinogenesis.

When 12-day genital ridges were grafted to scrotal sites, including the epididymal fat pads of castrated males, they developed into testes that usually had teratomas. This suggested that testicular hormones were not involved in experimental teratocarcinogenesis. When the genital ridges were grafted to nonscrotal sites, such as the spleen or liver, they developed into testes usually without teratomas. Therefore, temperature seems to be involved in experimental teratocarcinogenesis. All of the evidence we have accumulated supports this hypothesis. For example, if a male receives an intratesticular graft of a genital ridge and then is put in an incubator so that the scrotal contents cannot be cooler than body temperature, the genital ridge develops into a testis without teratomas. Further, if genital ridges are grafted to the pinna where the temperature is presumably cooler than body temperature, they develop into testes with teratomas.

D. Germ Cell Origin of Teratomas

This experimental method of inducing teratomas was used to reinvestigate the origin of testicular teratomas (Stevens 1967). Genital ridges were removed from 12-day fetuses resulting from mating of $Sl/+ \times Sl/+$ strain 129/Sv-Sl^J C P animals. Pieces of dorsal skin were also removed from these fetuses and grafted together with the genital ridge from the same donor. Ten days after grafting, the genotype of the testes that developed from the grafted genital ridge was diagnosed by the color of the hair that grew from the skin grafts. If the hair was pigmented, the genotype was classified as $Sl/+$ or $+/+$. If the hair was white, the genotype was classified as Sl/Sl. Of 129 grafts classified as $Sl/+$ or $+/+$, 85 had many germ cells and teratomas. Of 78 grafts classified as Sl/Sl, none had teratomas. The only difference detected between the $Sl/+$ and $+/+$ testes with teratomas and the Sl/Sl testes without teratomas was the presence of primordial germ cells in the former and their absence in the latter. It was concluded that this genetic evidence supported the morphological evidence that testicular teratomas were derived from primordial germ cells in genital ridges of fetuses after 11 and 12 days of gestation.

III. Ovarian Teratomas

A. Description

About half of the females of the strain LT/Sv have spontaneous ovarian teratomas by 3 months of age (Stevens and Varnum 1974). Usually they were composed of many kinds of differentiated cells and tissues and were benign. A few had proliferating undifferentiated embryonal cells, and we were able to establish them as transplantable teratocarcinomas. Histologically the ovarian tetratomas were similar to the testicular teratomas of strain 129 mice. There were a few notable differences. Nearly all of the testicular tumors contained patches of notochord which was absent in most ovarian teratomas. Nearly all of the ovarian teratomas had trophoblastic giant cells which were missing from the testicular teratomas. Cartilage and bone were more common in the testicular than in the ovarian tumors. Retinal tissue was much more common in the ovarian than in the testicular teratomas. We have no explanation for these histological differences. As was the case for the testicular teratomas, the tumors in the ovaries of young mice contained immature tissues.

B. Origin

The ovaries in LT mice from 1 to 2 months of age contained parthenogenetically cleaving eggs, morulae, blastocysts, and rarely egg cylinder stages. At or before the egg cylinder stage, the parthenogenetic embryos became disorganized and formed masses of undifferentiated embryonal cells like those found in the early stages of development of testicular teratomas. The early stages of development of ovarian teratomas were different from the early stages of testicular teratoma development. The early teratomas of the testis resembled normal embryos of 5 days of gestation. They were composed of ectodermal epithelium surrounding a proamniotic cavity without an envelope of primary endoderm. The egg cylinder stage of ovarian teratoma development was composed of a similar layer of ectoderm lining a proamniotic cavity, but the egg cylinder was surrounded by a single layer of proximal endoderm. Occasionally there was a layer of distal endoderm that secreted Reichert's membrane. This in turn was sometimes surrounded by trophoblastic giant cells which were absent in early testicular teratomas. This difference in the early stages of development of testicular and ovarian teratomas was due to the fact that blastocysts form in the ovarian teratomas. As in normal blastocysts, the ventral layer of inner cell mass cells is exposed to the blastocoel fluid, and in normal development this has been shown to provide the inductive stimulus for endodermal differentiation (see Rossant and Papaioannou 1977). This relationship between inner cell mass cells and blastocoel is missing in early testicular teratoma development. It should be pointed out, however, that some transplantable testicular teratomas in the ascitic form produce thousands of embryoid bodies composed of an outer layer of primary endoderm which encloses undifferentiated embryonal cells. Here the outer layer of cells is exposed to the peritoneal fluid, and it appears that this fluid provides a message for the exposed cells to differentiate into endoderm similar to that provided by the fluid of the blastocoel of normal embryos.

Parthenogenetic development of ovarian eggs is the first stage of teratocarcinogenesis in females. Ovarian teratomas in mice are derived from oocytes that have completed the first meiotic division (Eppig et al. 1977). A recombinant line of mice, LTXBJ, was developed using strains LT and C57BL/6J as progenitors. Nearly all of the females of this line have spontaneous ovarian teratomas. Among the C57BL/6J alleles fixed in this strain was $Gpi-l^b$ at the glucosephosphate isomerase ($Gpi-l$) locus on Chromosome 7. Strain LT is homozygous for the $Gpi-l^a$ allele and (LT × LTXBJ)F_1 hybrids are heterozygous at $Gpi-l$ and express the A, hybrid AB, and B allozymes in a ratio of 1:2:1. Electrophoresis of teratomas from F_1 females revealed a homozygous A or B allozyme banding pattern in most cases. It was concluded that the teratomas in (LT/Sv × LTXBJ)F_1 females originated from parthenogenetically activated oocytes which had completed the first meiotic division. In the few cases where the heterozygous banding pattern was found, the most probable conclusion was that crossover had occurred between the centromere and the $Gpi-l$ locus, so that the $Gpi-l^a$ and the $Gpi-l^b$ alleles were located on two chromatids attached to the same centromere. It has been suggested that ovarian teratomas can be used to determine centromere-to-gene or even gene-to-gene distances, because they are derived from germ cells that have undergone the first meiotic division but fail to complete the second. Another possibility that may be of scientific value is to utilize ovarian teratomas for half-tetrad analysis, because the parthenote, and thus the teratoma, is composed of chromosomes from a half-tetrad (Eicher 1978).

IV. Embryo-Derived Teratomas

A. Method

Embryo-derived teratomas may be produced by grafting embryos of the 2-cell to the 6-day stage to adults in extrauterine sites such as the testis or kidney (Stevens 1968, 1970b). Shortly after grafting, the embryos become disorganized. Their cells continue to proliferate and differentiate and they form teratomatous growths. After a month about half of the growths are composed of chaotic mixtures of many kinds of adult type tissue. The other half are composed of many kinds of immature and adult tissue, and they also contain pluripotent undifferentiated embryonal cells that continue to proliferate. When these growths are retransplanted subcutaneously, most of them stop growing, because all of the embryonic cells differentiate into nonproliferating normal adult cells. Occasionally, they will continue to grow and can be serially retransplanted indefinitely.

B. Origin

Embryo-derived teratocarcinomas resemble gonadal teratomas in morphology and behavior, but their origin is different. They are not derived from fully differentiated primordial germ cells (Dunn and Stevens 1970; Stevens 1970b; Mintz et al. 1978) but arise from grafted embryonal cells, specifically ectodermal cells (Diwan and Stevens 1976). About half of embryo-derived teratomas have a male, and the other half a female karyotype. We have never observed an ovarian teratoma in a strain 129 female, but they have been produced from embryos of this strain (Stevens 1968; 1970b). Similarly we have never been able to produce a testicular teratoma by grafting C3H genital ridges to the testis, but this strain seems to be the easiest with which to produce embryo-derived teratomas.

Mintz and co-workers showed that teratocarcinomas could be produced by grafting 6-day W/W and Sl^J/Sl^J "genetically sterile" embryos to the testes of adults (Mintz et al. 1978). Since W/W and Sl^J/Sl^J have few or no germ cells during the development of the definitive gonad (11–12 days of gestation), these results were interpreted by Mintz et al. (1978) to mean that mutant-derived terato-carcinomas were descended from somatic cells and not from germ cells. Although primordial germ cells of the 11–12 day gonad are defective in these mutants, there is no evidence to indicate that germ cell primordia of 6-day W/W and Sl^J/Sl^J are aberrant. To the contrary, there is evidence that these mutant germ cell primordia are normal at 6-days. "...The evidence at 8 days demonstrates that the mutant genes do not impair initial formation of primordial germ cells..." (Mintz and Russell 1957). Thus the work of Mintz et al. (1978) does not unequivocally demonstrate that teratocarcinomas are derived from somatic as opposed to germ cells, and our long-standing assertion that primordial germ cells are direct precursors of teratomas remains valid.

V. Embryoid Bodies

When some transplantable teratocarcinomas are homogenized and injected intraperitoneally they attach to the peritoneal lining and stimulate it to secrete a fluid. Clumps of tumor cells float around in the fluid. In some cases, these clumps resemble normal 5- to 6-day embryos. They are composed of a central group of cells that is sometimes arranged like the ectoderm of normal embryos. A layer of cells that resembles primary endoderm of normal embryos surrounds this clump. Because of their resemblance to normal embryos, they are called "embryoid bodies". For the first few transplant generations, embryoid bodies may vary markedly in size. Some are small and are composed of a disorganized clump of undifferentiated cells surrounded by an endodermal envelope. In others, the internal cells may resemble the ectoderm of a 5- or 6-day embryo. Still others may be large balloon-like cysts composed of an outer layer of yolk sac with both endodermal and mesodermal components that may have ectodermal and mesodermal derived elements inside (Pierce and Dixon 1959a,b).

After several transplant generations of a transplantable teratocarcinoma designated OTT 6050, all of the embryoid bodies were about the same size. When they grew to a certain size they split in half and formed two daughter embryoid bodies. They were all composed of a core of undifferentiated embryonal cells surrounded by endoderm. The core cells were undifferentiated for many years. However, they remained pluripotent. When single embryoid bodies were grafted to the anterior chamber of the eye or to the testis, they formed large growths composed of many kinds of well differentiated tissues (Stevens 1960).

VI. Embryonic Potency of Embryonal Carcinoma Cells

Kleinsmith and Pierce (1964) transplanted single embryonal carcinoma cells subcutaneously into adult mice. The cells grew and gave rise to teratomas with several tissue types. This was the first direct evidence that single embryonal carcinoma cells were pluripotent, i.e., they were able to give rise to more cells like themselves and also to a wide range of differentiated nonmalignant cells.

Brinster (1974) grafted small groups of embryonal cells from a transplantable teratocarcinoma of a pigmented strain into a blastocyst of a random bred albino strain. He transferred the blastocyst into the uterus of a pseudopregnant mouse where it developed normally and grew into a white mouse with pigmented patches. This suggested that the injected embryonal carcinoma cell had lost its neoplastic nature and had participated in normal development.

Mintz and Illmensee (1975), and Illmensee and Mintz (1976) injected stem cells from the core of a strain 129 embryoid body into blastocysts of other inbred strains with many genetic markers. The tumor that produced the embryoid bodies had been produced by grafting a 6-day embryo to the testis of an adult and it had been malignant for more than 8 years. The injected cells were taken from the inner cells of embryoid bodies. About 40 chimeric mice were obtained. Single injected teratocarcinoma cells contributed clonally to all major tissues of adult mice. In two cases teratocarcinoma cell progeny formed sperm in germ line chimeras which fertilized eggs and gave rise to healthy offspring. Illmensee (1978) injected single embryonal carcinoma cells from a transplantable ovarian teratocarcinoma into blastocysts and they contributed to the formation of all major organs. One female gave rise to offspring derived from eggs from both donor embryonal carcinoma cell and from normal host cell origin. This again demonstrated that embryonal carcinoma cells may be totipotent and in certain environmental conditions can contribute to normal development producing a wide range of differentiated cell types. This remarkable ability of embryonal carcinoma cells to undergo a reversal of their former cancerous expression and contribute to normal embryogenesis is discussed by Illmensee and Stevens (1979).

VII. Parthenogenesis

Ovarian teratocarcinomas in strain LT mice are derived from parthenogenetically activated ovarian eggs that have completed the first meiotic division (Stevens and Varnum 1974; Eppig et al. 1977). Parthenogenesis is also common in strain LT mice after eggs are ovulated. The oocytes cleave, form blastocysts which implant in the uterus, and they may develop to the primitive streak stage. After egg cylinder formation, however, the parthenogenetic embryos die and are aborted. Parthenogenetic egg cylinders were dissected from the uterus and grafted to the testis. They survived and formed teratomatous growths. Teratoma cells in the ovary and grafted parthenogenetic embryos are able to survive, but parthenogenetic embryos in utero are not.

We attempted to "rescue" parthenogenetic cells by aggregating 8-cell parthenogenetic embryos from the pigmented strain LT with 8-cell strain 129 albino embryos (Stevens et al. 1977; Stevens 1978). The aggregates were transferred to the uteri of pseudopregnant females. Four male and two female parthenote normal chimeras with white coats and pigmented patches were produced. All were mated to strain 129 albinos. One female produced five litters of 29 albino offspring. Her sixth and last litter contained two albino and two (one male and one female) pigmented offspring. The pigmented offspring could only have been derived from ova of parthenogenetic origin (strain LT) fertilized by normal albino (strain 129) sperm.

We were able to identify the parthenogenetic origin (LT) of the ova by a genetic marker other than C, which determined the pigmented coats of the offspring. Strain LT is Hbb^s/Hbb^s, which produces an electrophoretically single hemoglobin, and strain 129 is Hbb^d/Hbb^d, which produces a diffuse electrophoretic pattern. Examination of the blood of the two pigmented offspring revealed that they were heterozygous Hbb^s/Hbb^d. The Hbb^s allele could only have been derived from ova of parthenogenetic LT origin.

Even though pure parthenogenetic embryos will not survive in utero, these results demonstrate that their cells are capable of participating in normal development. Offspring with the C and Hbb^s alleles from an aggregation chimera between a parthenogenetic C/C Hbb^s/Hbb^s and a normal c/c Hbb^d/Hbb^d demonstrates that fully functional totipotent ova of parthenogenetic origin were produced.

VIII. Summary

Teratomas are rare in most strains of mice. Testicular teratomas are common in some sublines of inbred strain 129. Ovarian teratomas are common in inbred strain LT. Testicular teratomas are derived from primordial germ cells and can be experimentally produced by grafting 12 ½-day genital ridges to the testes of adults. They develop into testes and for some strains most have teratomas. Ovarian teratomas are derived from parthenogenetically activated ovarian oocytes that have completed the first meiotic division. Teratomas of either sex can be experimentally produced by grafting early embryos to various sites in adults. Embryo-derived teratomas originate directly from undifferentiated embryonal cells. Occasionally teratomas are malignant (teratocarcinomas) and can be maintained as transplantable tumors. Some form embryoid bodies that resemble normal early embryos. When the stem cells of some transplantable teratocarcinomas are injected into blastocysts and transferred to the uteri of pseudopregnant females, they participate in normal development and contribute to the formation of all major tissues including functional sperm and eggs.

Spontaneous parthenogenesis is common in strain LT oocytes after ovulation. The eggs cleave, form blastocysts which implant in the uterus, but after the egg cylinder stage they become disorganized and are aborted. Eight-cell embryos from the pigmented LT strain were aggregated with embryos of albino strain 129 and transferred to the uteri of pseudopregnant females. They participated in development and contributed to the formation of normal chimeric tissues. Offspring from eggs derived from parthenogenetic embryonal cells were produced, demonstrating that parthenogenetic embryonic cells are totipotent. It is still a mystery why parthenogenetic embryos will not survive in utero.

Acknowledgments. This research was supported by a grant CA 02662 from the USNIH. The Jackson Laboratory is fully accredited by the American Association for accreditation of Laboratory Animal Care.

References

Brinster, R.L.: The effect of cells transferred into the mouse blastocyst on subsequent development. J. Exp. Med. 140, 1049-1056 (1974)

Damjanov, I., Solter, D.: Experimental teratoma. Curr. Top. Pathol. 59, 69-130 (1974)

Diwan, S.B., Stevens, L.C.: Development of teratomas from the ectoderm of mouse egg cylinders. J. Natl. Cancer Inst. 57, 937-942 (1976)

Dunn, G.R., Stevens, L.C.: Determination of sex of teratomas derived from early embryos. J. Natl. Cancer Inst. 44, 99-105 (1970)

Eicher, E.: Murine ovarian teratomas and parthenotes as cytogenetic tools. Cytogenet. Cell Genet. 20, 232-239 (1978)

Eppig, J.J., Kozak, L.P., Eicher, E.M., Stevens, L.C.: Ovarian teratomas in mice are derived from oocytes that have completed the first meiotic division. Nature (London) 269, 517-518 (1977)

Graham, C.F.: Teratocarcinoma cells and normal mouse embryogenesis. In: Concepts in Mammalian Embryogenesis (ed. M.I. Sherman). Cambridge, MA-London: MIT Press 1977)

Illmensee, K.: Reversion of malignancy and normalized differentiation of teratocarcinoma cells in chimeric mice. In: Gatlinburg Symposium on Genetic Mosaics and Chimeras in Mammals (ed. L.B. Russell), pp. 3-25. New York: Plenum Press 1978

Illmensee, K., Mintz, B.: Totipotency and normal differentiation of single teratocarcinoma cells cloned by injection into blastocysts. Proc. Natl. Acad. Sci. USA 73, 549-553 (1976)

Illmensee, K., Stevens, L.C.: Teratomas and chimeric mice. Sci. Am. 240, 121-132 (1979)

Jacob, F.: Mouse teratocarcinoma and embryonic antigens. Immunol. Rev. 33, 3-33 (1977)

Jacob, F.: Mouse teratocarcinoma and mouse embryo. The Leeuwenhoek Lecture 1977. Proc. R. Soc. London Ser. B 201, 249-270 (1978)

Kleinsmith, J.J., Pierce, G.B.: Multipotentiality of single embryonal carcinoma cells. Cancer Res. 24, 1544-1552 (1964)

Martin, G.R.: Teratocarcinomas as a model system for the study of embryogenesis and neoplasia. Rev. Cell 5, 229-243 (1975)

Mintz, B.: Gene expression in neoplasia and differentiation. In: The Harvey Lectures. Ser. 71, pp. 193-246. New York-London: Academic Press 1978

Mintz, B., Illmensee, K.: Normal genetically mosaic mice produced from malignant teratocarcinoma cells. Proc. Natl. Acad. Sci. USA 72, 3585-3589 (1975)

Mintz, B., Russell, E.S.: Gene-induced embryological modifications of primordial germ cells in the mouse. J. Exp. Zool. 134, 207-237 (1957)

Mintz, B., Cronmiller, C., Custer, R.P.: Somatic cell origin of teratocarcinomas. Proc. Natl. Acad. Sci. USA 75, 2834-2838 (1978)

Pierce, G.B., Beals, T.F.: The ultrastructure of primordial germ cells of the fetal testis and of embryonal carcinoma cells of mice. Cancer Res. 24, 1553-1567 (1964)

Pierce, G.B., Dixon, F.J.: Testicular teratomas I. Demonstration of teratogenesis by metamorphosis of multipotential cells. Cancer 12, 573-583 (1959a)

Pierce, G.B., Dixon, F.J.: Testicular teratomas II. Teratocarcinoma as an ascitic tumor. Cancer 12, 584-589 (1959b)

Pierce, G.B., Stevens, L.C., Nakane, P.K.: Ultrastructural analysis of the early development of teratocarcinoma. J. Natl. Cancer Inst. 39, 755-773 (1967)

Rossant, J., Papaioannou, V.E.: The biology of embryogenesis. In: Concepts in Mammalian Embryogenesis (ed. M.I. Sherman), pp. 1-36. Cambridge, MA-London: MIT Press 1977

Sherman, M.I., Solter, D.: Teratomas and Differentiation. New York-London: Academic Press 1975

Solter, D., Skreb, N., Damjanov, I.: Extrauterine growth of mouse egg-cylinders results in malignant teratoma. Nature (London) 227, 503-504 (1970)

Stevens, L.C.: Histogenesis in testicular teratomas which occur spontaneously in strain 129 mice. Proc. Am. Assoc. Cancer Res. 2, 252-253, Abstract (1957)

Stevens, L.C.: Studies on transplantable testicular teratomas of strain 129 mice. J. Natl. Cancer Inst. 20, 1257-1275 (1958)

Stevens, L.C.: Embryology of testicular teratomas in strain 129 mice. J. Natl. Cancer Inst. 23, 1249-1295 (1959)

Stevens, L.C.: Embryonic potency of embryoid bodies derived from a transplantable testicular teratoma of the mouse. Dev. Biol. 2, 285-297 (1960)

Stevens, L.C.: Testicular teratomas in fetal mice. J. Natl. Cancer Inst. 28, 247-268 (1962)

Stevens, L.C.: Experimental production of testicular teratomas in mice. Proc. Natl. Acad. Sci. USA 52, 654-661 (1964)

Stevens, L.C.: Development of resistance to teratocarcinogenesis by primordial germ cells in mice. J. Natl. Cancer Inst. 37, 859-868 (1966)

Stevens, L.C.: Origin of testicular teratomas from primordial germ cells in mice. J. Natl. Cancer Inst. 38, 549-552 (1967)

Stevens, L.C.: The development of teratomas from intratesticular grafts of tubal mouse eggs. J. Embryol. Exp. Morphol. 20, 329-341 (1968)

Stevens, L.C.: Environmental influence on experimental teratocarcinogenesis in testes of mice. J. Exp. Zool. 174, 407-413 (1970a)

Stevens, L.C.: The development of transplantable teratocarcinomas from intratesticular grafts of pre- and post implantation embryos. Dev. Biol. 21, 364-382 (1970b)

Stevens, L.C.: Experimental production of testicular teratomas in mice of strains 129, A/He, and their F_1 hybrids. J. Natl. Cancer Inst. 44, 923-929 (1970c)

Stevens, L.C.: A new inbred subline of mice (129/terSv) with a high incidence of spontaneous congenital testicular teratomas. J. Natl. Cancer Inst. 50, 235-242 (1973)

Stevens, L.C.: Teratocarcinogenesis and spontaneous parthenogenesis in mice. In: The Developmental Biology of Reproduction. 33rd Symp. Soc. Dev. Biol. (eds. C.L. Markert, J. Papaconstantinou), pp. 93-106. New York-London: Academic Press 1975

Stevens, L.C.: Totipotent cells of parthenogenetic origin in a chimeric mouse. Nature (London) 276, 266-267 (1978)

Stevens, L.C., Hummel, K.P.: A description of spontaneous congenital testicular teratomas in strain 129 mice. J. Natl. Cancer Inst. 18, 719-748 (1957)

Stevens, L.C., Little, C.C.: Spontaneous testicular teratomas in an inbred strain of mice. Proc. Natl. Acad. Sci. USA 40, 1080-1087 (1954)

Stevens, L.C., Mackensen, J.A.: Genetic and environmental influences on teratocarcinogenesis in mice. J. Natl. Cancer Inst. 27, 443-453 (1961)

Stevens, L.C., Varnum, D.S.: The development of teratomas from parthenogenetically activated ovarian mouse eggs. Dev. Biol. 37, 369-380 (1974)

Stevens, L.C., Varnum, D.S., Eicher, E.M.: Viable chimaeras produced from normal and parthenogenetic mouse embryos. Nature (London) 269, 515-517 (1977)

Teratocarcinoma Cells as Agents for Producing Mutant Mice

M.J. Dewey and B. Mintz

The Institute for Cancer Research, Fox Chase Cancer Center
Philadelphia, PA 19111, USA

I. Introduction

The genetic mechanisms that are involved in the differentiation of a wide spectrum of cell types as well as those affecting higher levels of biological organization remain largely unknown in mammals. In part, this is due to the lack of a ready source of experimentally useful, mutant genes for study. Mammalian genetics, instead, has had to content itself primarily with mutants fortuitously obtained without specific selection and which often have fairly gross phenotypes, without a defined molecular basis, such as coat color changes or behavior defects. On the other hand, there has evolved work on somatic cell genetics in culture that has enabled mutagenesis and selection of specific biochemical mutations to be realized at the cellular level, albeit without appreciable differentiation.

It would thus be desirable to combine in a single experimental system the advantages of the in vivo and in vitro approaches. This depends on the availability of a cell that can be maintained in vitro and there subjected to targeted selection for mutant clones, and that is also capable of differentiating extensively in vivo into virtually an entire mouse. In this way, the full phenotypic potential of a clone bearing a specific mutation could be expressed in the somatic tissue derivatives. Germ cell transmission would ultimately result in a new mutant strain.

A mammalian cell has been found that fulfills these requirements. It is the developmentally totipotent teratocarcinoma stem cell.

II. Developmental Totipotency and Normalization of Mouse Teratocarcinoma Stem Cells

Mouse teratocarcinoma cells are the malignant stem cells of transplantable teratomas (Pierce 1967; Stevens 1967) and are known to be developmentally multipotential (Kleinsmith and Pierce 1964). The solid tumors contain a variety of tissues in a chaotic arrangement. However, some tissues, such as kidney and liver, are consistently missing and others are immature. It was thus not known whether the stem cells were truly totipotential.

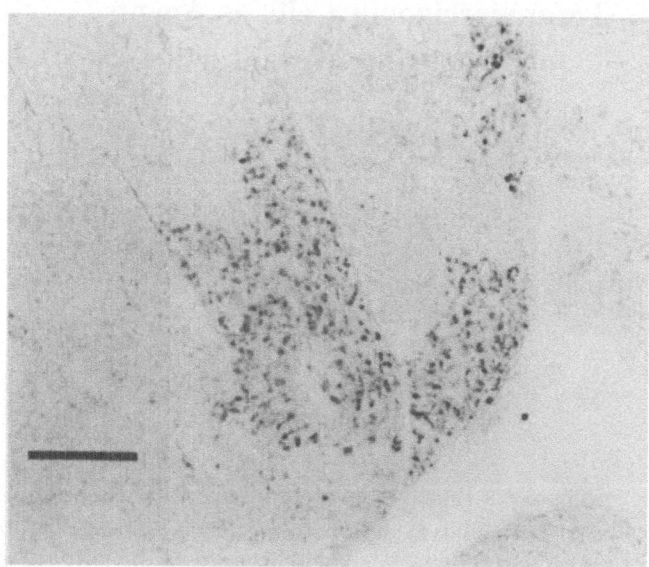

Fig. 1. Direct visualization in mosaic pancreas of differentiated normal cells derived from malignant teratocarcinoma (Dewey and Mintz 1978). The pancreas is from a mouse derived from a blastocyst injected with teratocarcinoma cells. Staining of sections for β-galactosidase allows tumor-strain cells (*dark-staining patch*) to be identified and distinguished from those of the host blastocyst-strain (*light staining*). Bar, 100 microns

The question was decisively resolved recently. Small numbers of euploid (Cronmiller and Mintz 1978) teratocarcinoma cells from an in vivo transplant line were injected into the cavities of genetically marked blastocysts. When these were allowed to develop in the uteri of pseudopregnant females, tumor-free mice were born with normal differentiated cells derived from the injected teratocarcinoma stem cells in virtually all of their somatic tissues (Mintz and Illmensee 1975; Mintz et al. 1975; Illmensee and Mintz 1976). This was evident from the presence of the specific tumor-strain variants of enzymes, and from tissue-specific products such as immunoglobulins, hemoglobins, and liver proteins. The completely normalized tumor-lineage cells were also directly visualized and identified in tissue sections by means of staining for activity of β-galactosidase (Fig. 1; Dewey and Mintz 1978), an enzyme for which there is strain-specific quantitative variation. In addition to the capacity of the stem cells to form normal somatic tissues, an even more striking observation was their ability to contribute to the germ line: The tumor cells were able to give rise to functional gametes from which progeny were obtained.

Thus was demonstrated that teratocarcinoma cells are truly totipotential. In addition, these results provided the first unequivocal example in animals of a nonmutational basis for malignancy and for reversal of malignant cells to normalcy. Whether the tumor stem cells behaved in a normal or a malignant fashion was a matter of change in gene expression rather than gene structure, and the choice was apparently determined by local environmental influences. In the environment of the blastocyst, the teratocarcinoma cells lost their malignancy and behaved as do normal early embryo cells.

III. Teratocarcinoma Cells as Vehicles
for Introducing New Mutations into Mice

That mouse teratocarcinoma cells are developmentally totipotent makes them a source of cells uniquely suited to serve as a means of producing mice with genetic alterations, as proposed elsewhere (Mintz et al. 1975; Mintz 1977; Mintz 1978). The cells readily proliferate in an undifferentiated state in culture; they can there be mutagenized and subjected to procedures that allow specific kinds of mutant cells to be distinguished and plucked from a majority population of nonmutant cells. Cells from a mutant clone are next introduced into a blastocyst of another strain and this blastocyst is allowed to develop in the uterus of a pseudopregnant foster mother. Some mice thus born will be cellular genotypic mosaics; these vary greatly in the proportion of tumor-derived cells and in the tissue distribution of the cells.

The individual variation provides a unique way of identifying the tissue in which primary expression of a mutant gene occurs, hence in which the deviant organismic phenotype originates. This discrimination is arrived at when the mutant phenotype is manifest only when mutant-strain cells are found in a particular tissue.

Animals with tumor-derived gametes are mated to mice of an appropriate marker strain to obtain F_1 progeny heterozygous for the mutation; homozygous mice would be found in the F_2 generation. In this way, a strain of mice bearing a particular predetermined biochemical mutation could be produced.

A. Mice with Cells Bearing a Nuclear Gene Mutation:
Models for Human Genetic Diseases

With such an approach, previously inaccessible genetic analyses are now possible in problems of mammalian metabolism, disease, and differentiation. One area of great promise and practical importance lies in the construction of animal models of certain human genetic diseases. There are many of these diseases for which the specific biochemical lesion has been described, yet in very few cases has the same lesion been found in a laboratory animal where it might be experimentally studied and cures attempted.

An example is the Lesch-Nyhan syndrome (Seegmiller 1976). This X-linked recessive disorder results from a severe deficiency of the enzyme hypoxanthine phosphoribosyltransferase (HPRT). Patients exhibit excess purine biosynthesis, spasticity, mental retardation, and self-mutilation.

Teratocarcinoma cells deficient in HPRT were selected from mutagenized cultures by growing them in medium containing 6-thioguanine, a purine base analogue that kills cells that contain the enzyme (Fig. 2; Dewey et al. 1977). When these HPRT mutant cells were introduced into wild-type blastocysts, they ceased to be malignant and proceeded to contribute to normal embryogenesis, as did nonmutant tumor stem cells. Tumor-free mosaic mice were obtained, and in this case some of the animals had HPRT$^-$ tumor-lineage cells in practically all their tissues. The mature status of the tumor-derived cells was evidenced by their production of such tissue-specific proteins as melanins and liver proteins.

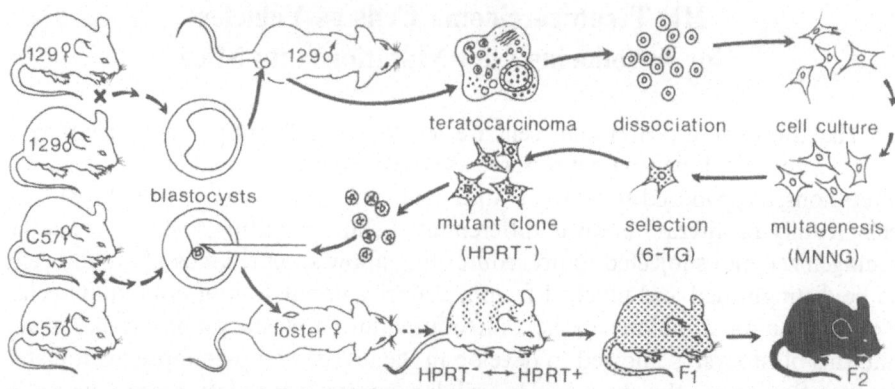

Fig. 2. Experimental scheme for producing mice deficient in HPRT (Dewey et al. 1977). Starting at the *upper left*, a 129-strain blastocyst grafted under the testis capsule of a syngeneic host formed a malignant, transplantable teratocarcinoma. An in vitro culture line of the stem cells was established and from this a clone of 6-thioguanine-resistant (6-TG) cells was derived following mutagenesis with N-methyl-N'-nitro-nitrosoguanidine (MNNG). As expected, the drug-resistant cells were deficient in HPRT. These were injected into blastocysts of another strain (C57). Development of such blastocysts in the uteri of pseudopregnant females resulted in many mice with mosaicism of the somatic tissues. Possible germ line transmission of the injected cells would give rise to a new mutant strain

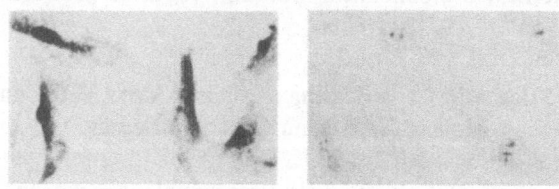

Fig. 3. Autoradiographic visualization of HPRT⁻ colonies in a culture of subcutaneous connective tissue from a mouse derived from a blastocyst injected with HPRT⁻ teratocarcinoma cells. Two kinds of colonies are evident after [³H] hypoxanthine labeling: some with many grains over the cells *(left)* as in HPRT⁺ control cultures, and others with no grains *(right)*. Analysis of the same culture for strain-specific enzyme variants revealed the proportion of tumor-derived cells to be comparable to the number of HPRT⁻ colonies
(Dewey et al. 1977)

    Retention of the HPRT⁻ lesion in the differentiated state was documented in two ways: A marked depression of the HPRT specific activity was found only in those tissues with an appreciable population of tumor-derived cells. In addition, autoradiography of[³H]-hypoxanthine-incubated cultures, from explanted mosaic connective tissue, showed some unlabeled (HPRT⁻) colonies as well as labeled (HPRT⁺) ones (Fig. 3; Dewey et al. 1977).

    While the actual model of the human disease, a "Lesch-Nyhan" male mouse, is not yet on hand, the mosaics themselves have certain unique and useful features not available in animals with the genetic lesion in all of their cells. Certain mosaic individuals had HPRT⁻ cells in only a single tissue, such as brain, and their metabolic and clinical attributes could help to identify the tissue sources of particular aspects of the disease. Although no obvious behavioral defects have yet been seen in the mosaics, there is one striking indication of tissue-specific phenotypic effects manifest in cell selection: HPRT-deficient cells in the blood of mosaic mice are apparently at an especially great selective disadvantage, since the frequency of blood mosaicism was markedly below mosaicism in various other somatic tissues. This result interestingly parallels selective trends in human female HPRT$^+$/HPRT$^-$ heterozygotes, thereby lending credence to the expectation that the mouse would be a suitable animal in which to study the effects of HPRT deficiency.

B. Mice with Cells Bearing Specific Mutant Mitochondrial Genes

An area that is virtually unexplored in mammalian genetics is the role that cytoplasmic determinants play in physiology and development. Cytoplasmic inheritance in eukaryotic cells is known to occur through mitochondria and chloroplasts which contain self-replicating DNA. Although studies of mitochondrial genes in mammalian cells are very limited, evidence from cells in culture has documented the existence of extranuclear inheritance at the cellular level (Bunn et al. 1974). Mouse teratocarcinoma cells with a specific mitochondrial marker would afford a means of investigating the influence of mitochondrial genes, not only at the cellular level but also at the tissue and organismic levels.

Mammalian cells in culture have been selected for resistance to chloramphenicol (CAP), a drug which specifically inhibits mitochondrial protein synthesis. The resistance trait has been found to reside in the cytoplasm (Firkin and Linnane 1968; Bunn et al. 1974).

A stable mutation to CAP resistance (CAP^R) was incorporated into cultured teratocarcinoma cells by the indirect route shown in Fig. 4 (Watanabe et al. 1978). The mutation was first produced in a melanoma cell line by mutagenesis and selection. It was then transferred to CAP-sensitive teratocarcinoma cells by fusing to them enucleated cytoplasts from the melanoma. The fusion clearly resulted in transfer of the CAP^R trait to the teratocarcinoma cells, whereas the nuclear marker (HPRT$^-$) of the melanoma was not transferred. Analysis for strain-specific (nuclear-encoded) electrophoretic enzyme variants in cells of a mutant clone revealed only the enzyme type characteristic of the teratocarcinoma strain. The particular mutation rendering these cells drug-resistant is thus indeed a cytoplasmic trait. Further proof of the probable mitochondrial nature of the CAP^R mutation was seen in the resistance of isolated mitochondria to CAP, in protein synthesis from a labeled amino acid precursor. The CAP^R cybrids, or cytoplasmic hybrids, morphologically resembled the drug-sensitive teratocarcinoma parent, and they formed tumors in vivo that contained a variety of differentiated tissues. Judging from analyses of cells from retransplanted tumors, the lineage was capable of retaining the CAP^R trait for long periods (at least 16 weeks) in vivo in the absence of the selective agent.

Following blastocyst injections with the CAP^R cybrid cells, mosaic animals were obtained. The CAP^R cells participated in the development of various functional specialized tissues in these animals. Therefore, the ability of these teratocarcinoma

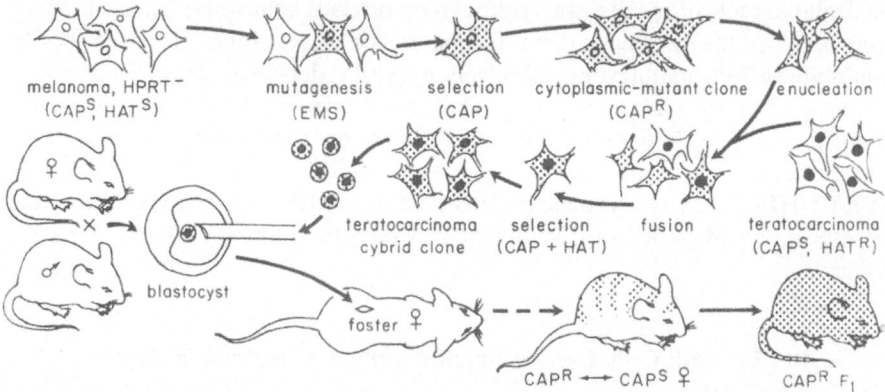

Fig. 4. A diagrammatic representation of the experimental scheme for introducing a specific mitochondrial gene marker into mice. A mutation conferring chloramphenicol resistance (CAP^R) is brought into the teratocarcinoma cells by fusing to them in vitro the cytoplasts of CAP^R melanoma cells. The latter, which also were deficient in HPRT and therefore sensitive to HAT medium, were produced by mutagenesis and selection in medium containing chloramphenicol. Next, the CAP^R melanoma cells were subjected to cytochalasin B enucleation and the resultant cytoplasts fused with Sendai virus to the teratocarcinoma cells; the latter were HPRT$^+$ and able to grow in HAT. Thus, in the fusion only cells with the teratocarcinoma nucleus (HPRT$^+$) and melanoma cytoplasmic mutation (CAP^R) were able to survive the double selection in CAP plus HAT. The CAP^R cybrid cells were microinjected into blastocysts of another strain and these were allowed to develop in the uteri of foster mothers. Mosaic offspring with both cybrid- and blastocyst-derived cells were obtained. According to the expectations based on maternal inheritance, F$_1$ progeny produced by a female with cybrid-derived ova should be CAP^R. (From Watanabe et al. 1978)

stem cells to participate in normal embryogenesis has remained undisturbed, either by introduction of a highly specialized cytoplasm from a melanoma cell, or by the CAP^R mutation itself.

C. Other Prospects

In these experiments, we have demonstrated the unique usefulness of mouse teratocarcinoma stem cells as a means of introducing either specific nuclear or cytoplasmic mutations into mice. The application of this approach can be expected to expand substantially, to include many kinds of mutations, and to be directed at a great diversity of problems.

In the examples discussed above, the isolation of new mutants depended upon the use of selective systems in vitro. An obvious limitation is that relatively few such systems are available. Procedures such as immunoselection may also be utilized, but these exist for a very limited number of genetic lesions. In order to expand the prospects of isolating mutants, other approaches must be developed that would allow screening of colonies for mutant clones in culture. Replica plating has been very effectively employed in bacterial genetics and, in mammalian cell cultures, would also enable deviant colonies to be identified on the replica and then picked from the master plate. Large numbers of clones could be screened in this way by histochemical enzyme staining, autoradiography, etc., to test for certain phenotypes. Once identified, the parent clone could be propagated. Replica plating has been recently successfully applied to hamster cells in culture (Stamato and Jones 1977), although this and other published methods appear not to yield a high replication efficiency with teratocarcinoma cells. We have therefore undertaken, in our laboratory, a series of modifications that have begun to yield very promising results with mouse teratocarcinoma cells (A.J.J. Reuser and B. Mintz, unpublished data). This technique should greatly extend the range and types of mutations available in teratocarcinoma cells and, ultimately, in mice.

Another possibility offered by the unique developmental properties of teratocarcinoma stem cells is the introduction into mice of genetic material from other species. In this way, information could be gained about control mechanisms in gene expression, and how these are modulated as a function of specific differentiation. The foreign material could be relatively crude, such as an entire chromosome or part thereof, or it could be well-defined, such as a bacterial operon with its structural genes and controlling elements, or even a specific piece of molecularly cloned eukaryotic DNA. The possibilities are numerous and, indeed, recent advances in somatic cell genetics and molecular biology now render all of these prospects technically feasible.

Acknowledgments. The work described here was supported by U.S. Public Health Service Grants HD-01646, CA-06927, and RR-05539, and by an appropriation from the Commonwealth of Pennsylvania.

References

Bunn, C.L., Wallace, D.C., Eisenstadt, J.M.: Cytoplasmic inheritance of chloramphenicol resistance in mouse tissue culture cells. Proc. Natl. Acad. Sci. USA 71, 1681-1685 (1974)

Cronmiller, C., Mintz, B.: Karyotypic normalcy and quasi-normalcy of developmentally totipotent mouse teratocarcinoma cells. Dev. Biol. 67, 465-477 (1978)

Dewey, M.J., Mintz, B.: Direct visualization, by β-galactosidase histochemistry, of differentiated normal cells derived from malignant teratocarcinoma in allophenic mice. Dev. Biol. 66, 550-559 (1978)

Dewey, M.J., Martin, D.W., Jr., Martin, G.R., Mintz, B.: Mosaic mice with teratocarcinoma-derived mutant cells deficient in hypoxanthine phosphoribosyl-transferase. Proc. Natl. Acad. Sci. USA 74, 5564-5568 (1977)

Firkin, F.C., Linnane, A.W.: Differential effects of chloramphenicol on the growth and respiration of mammalian cells. Biochem. Biophys. Res. Commun. 32, 398-402 (1968)

Illmensee, K., Mintz, B.: Totipotency and normal differentiation of single teratocarcinoma cells cloned by injection into blastocysts. Proc. Natl. Acad. Sci. USA 73, 549-553 (1976)

Kleinsmith, L.J., Pierce, G.B., Jr.: Multipotentiality of single embryonal carcinoma cells. Cancer Res. 24, 1544-1551 (1964)

Mintz, B.: Teratocarcinoma cells as vehicles for mutant and foreign genes. In: Genetic Interaction and Gene Transfer, Brookhaven Symposia in Biology, Vol. 29 (ed. D.W. Anderson), pp. 82-85. Upton, NY: Brookhaven National Lab. 1977

Mintz, B.: Gene expression in neoplasia and differentiation. Harvey Lect. Ser. 71, pp. 193-245. New York-London: Academic Press 1978

Mintz, B., Illmensee, K.: Normal genetically mosaic mice produced from malignant teratocarcinoma cells. Proc. Natl. Acad. Sci. USA 72, 3585-3589 (1975)

Mintz, B., Illmensee, K., Gearhart, J.D.: Developmental and experimental potentialities of mouse teratocarcinoma cells from embryoid body cores. In: Teratomas and Differentiation (eds. M.I. Sherman, D. Solter), pp. 59-82. New York-London: Academic Press 1975

Pierce, G.B., Jr.: Teratocarcinoma: Model for a developmental concept of cancer. In: Current Topics in Developmental Biology, Vol. 2 (eds. A.A. Moscona, A. Monroy), pp. 223-246. New York-London: Academic Press 1967

Seegmiller, J.E.: Inherited deficiency of hypoxanthine-guanine phosphoribosyl-transferase in X-linked uric aciduria (the Lesch-Nyhan syndrome and its variants). In: Advances in Human Genetics, Vol. 6 (eds. H. Harris, K. Hirschhorn), pp. 75-163. New York-London: Plenum Press 1976

Stamato, T.D., Jones, C.: Isolation of a lactic dehydrogenase-A-deficient CHO-K1 mutant by nylon cloth replica plating. Somat. Cell Genet. 3, 639-647 (1977)

Stevens, L.C.: The biology of teratomas. In: Advances in Morphogenesis, Vol. 6 (eds. M. Abercrombie, J. Brachet), pp. 1-31. New York-London: Academic Press 1967

Watanabe, T., Dewey, M.J., Mintz, B.: Teratocarcinoma cells as vehicles for introducing specific mutant mitochondrial genes into mice. Proc. Natl. Acad. Sci. USA 75, 5113-5117 (1978)

Development of Embryo-Derived Teratomas in Vitro

N. Škreb and V. Crnek

Department of Biology, Faculty of Medicine, Zagreb, Yugoslavia

I. Introduction

Since the relevant data obtained in studies of teratomas have been extensively reviewed (Pierce 1967; Stevens 1967; Damjanov and Solter 1974; Martin 1975; Solter et al. 1975; Jacob 1978), it is unnecessary to mention numerous original papers describing the advantage of this model system for various studies of embryogenesis and neoplasia. Experimental teratomas derived from mouse embryonic shields emphasize the close relationship between early embryo cells and tumor stem cells which can give rise either to embryonal carcinoma cells or to various differentiated cell types. The individual embryonal carcinoma cells probably have the full range of developmental potentialities, but at present it is impossible to single out those factors indispensable for teratocarcinogenesis or those indispensable for differentiation (Illmensee and Mintz 1976).

The study of teratogenesis in vivo has already provided much interesting data (Škreb et al. 1971; Škreb and Švajger 1975). Nevertheless, this technique does not allow a precise analysis of a variety of factors involved in the extrinsic control of tissue differentiation and the influence of host tissues cannot be excluded.

In order to avoid the complex situation following the transfer of the embryonic shields to ectopic sites, we have tried to elaborate an in vitro method in which the isolated rodent embryonic shields are allowed to realize their developmental potentialities and in which the extrinsic factors necessary for the differentiation can be analyzed. The general goals of our various experiments in vitro are: (1) to obtain teratoma-like structures with many differentiated tissues, (2) to increase the appearance of some tissues and to improve the degree of differentiation, and (3) to impede the process of differentiation without significantly influencing the growth.

In this report we shall deal with the following topics relevant to the aims stated above: (1) the growth of rat embryonic shields in vitro, (2) effect of various extrinsic factors on differentiation, and (3) embryonal carcinoma cells after in vitro culture of mouse embryonic shields.

II. Material and Methods

Female rats of the Fischer strain were killed after 9 days and female mice of the C3H strain after 7 days of pregnancy. The egg cylinders at the primitive streak stage were isolated and the extra-embryonic part cut off. The shields were put on lens paper supported by a stainless steel grid placed in an embryological watch glass as described previously (Škreb and Švajger 1973; Škreb and Crnek 1977). The liquid medium consisted of Eagle's minimum essential medium (MEM) supplemented with 40% of rat serum if not otherwise stated. After 14 days the explants were fixed and histological sections were examined.

Some explants were weighed, DNA extracted and determined as usual (Schneider 1945; Burton 1956).

Rat embryonic shields were used more often than mouse shields due to technical problems with mouse embryos, which have not yet been entirely overcome.

III. Results

A. Growth

Although the growth curves were made only on rat explants, the observations of mouse explants in culture allow us to extend the results to the mouse at least in regard to the general behavior in vitro. We measured two diameters of explants and the wet weight as well. Furthermore, in order to assess metabolic activity during in vitro culture, we measured the DNA, RNA, and protein content per explant (unpublished data). All the curves resemble that of DNA (Fig. 1) showing the growth of explant in the first week of culture. In the second week we observed a decline in volume, weight, and in DNA, RNA and protein content, which obliged us to conclude that cell necrosis had taken place. On the other hand, under the kidney capsule even well differentiated embryo-derived teratomas seem to continue

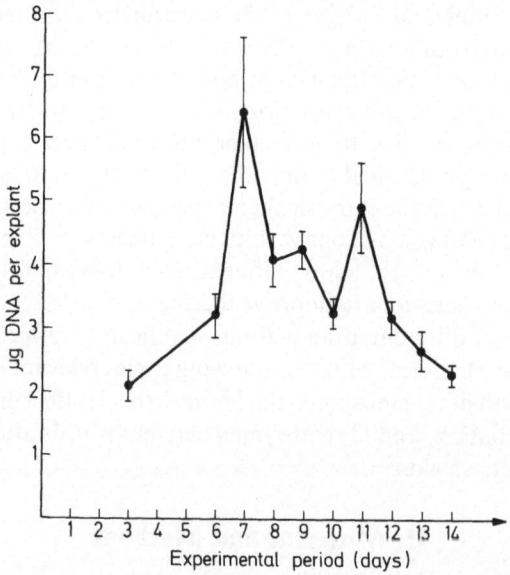

Fig. 1. DNA content of rat embryo explants. The *vertical bars* represent the standard error of the mean of at least 9 explants

growing 6 months after transfer (Škreb et al. 1971). There seems to be a relatively long lag period in both systems as far as overt differentiation is concerned, succeeded by the second week when many adult tissues appear in a similar sequence.

B. Effect of Various Extrinsic Factors on Differentiation

1. Various Sera

Rat serum, calf serum or human serum combined with either Tyrode's saline or Eagle's MEM were used as culture media. All three sera combined with 50–80% Eagle's MEM allowed the same incidence of various differentiated tissues after two weeks of culture. In combination with Tyrode's saline only rat serum gave rise to the same percentage of differentiated tissues as previously, while in the calf and human serum the rat explants were differentiated very poorly. Moreover, the quality of differentiation was better in the rat than in heterologous sera, even if the sera were combined with Eagle's MEM. One can recognize the series cultivated in rat sera by better differentiation of mesenchymal derivatives, especially cartilage (Figs. 2 and 3).

For the time being we have only rarely succeeded in obtaining well differentiated mesenchymal derivatives from mouse embryonic shields if they were

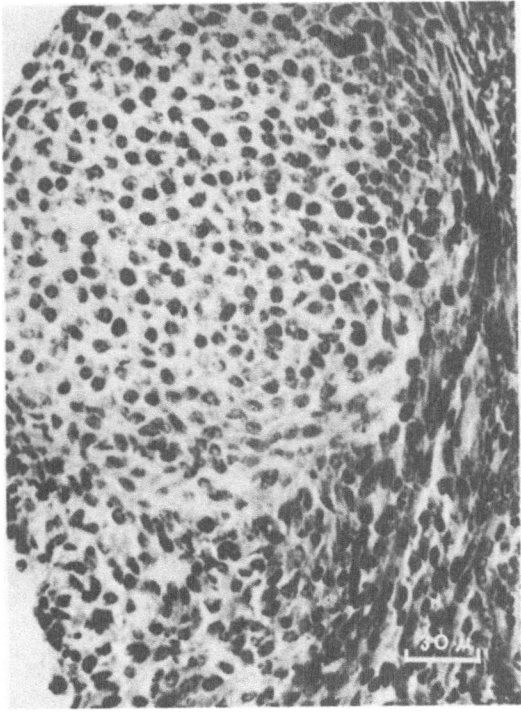

Fig. 2. Detail of a teratoma derived from a rat embryonic shield explanted for 14 days in rat serum + MEM. Note the well differentiated cartilage and some myotubes.

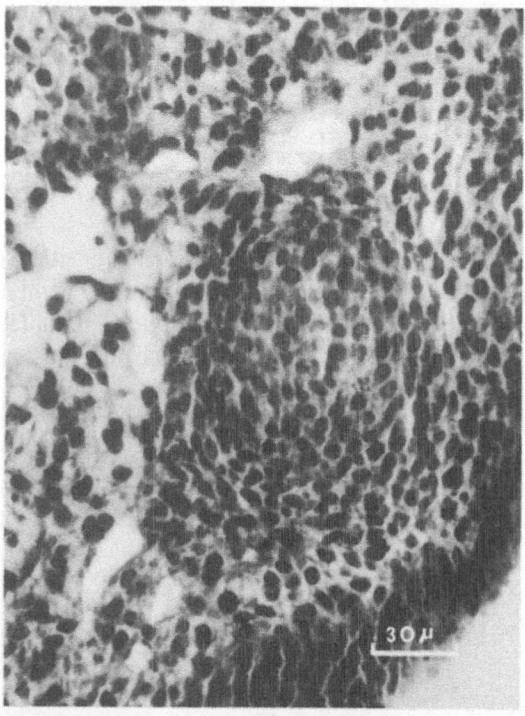

Fig. 3. Detail of a teratoma derived from a rat embryonic shield explanted for 14 days in calf serum + MEM. Note the poor differentiation of cartilage

cultivated in mouse serum (from the strain 129SV) or in fetal calf serum. Experiments dealing with this problem are now in progress.

2. The Metal Grid

Very few data are available concerning the necessity of the metal grid in connection with the in vitro histogenesis. To be able to specify its role in the differentiation, one series of explants was cultivated on the metal grid as described previously, and another without it, floating in the same liquid medium. The 26 explants floating free in the medium survived and even showed a substantial growth, but in the histological sections very few, if any, differentiated tissues could be found. In the majority of cases various epithelia formed large cysts (see Škreb and Švajger 1975; Fig. 6). Sometimes they looked like immature gut, morphologically close to the endoderm, and sometimes like multilayered squamous epithelium. The control series on the metal grid had the same teratomalike structure with many terminally differentiated tissues.

3. The Change of the Medium

A further question was whether we could change the phenotypical expression of the rat embryonic shields by varying the time interval between the changes of the

Table 1. Tissues found in explants after 14 days in vitro

9-day rat embryonic shields		
Change of liquid medium	$8 \times$	$1 \times$
No. explanted	40	30
No. survived	31	18
Kind of tissue		
Epidermis	27	5[a]
Neural tissue	21	11[a]
Gut epithelium	22	10[a]
Gland	2	—
Smooth muscle	18	2[a]
Skeletal muscle	4	—
Cartilage	21	2[a]
Trophoblast	—	1
Hematopoietic	—	4

[a] Immature tissue

liquid medium. If we changed the medium only once during the 2 week period in contradistinction to the usual 8 times, differentiated tissues are very rare, especially as far as mesenchymal derivatives are concerned (Table 1).

4. Addition of Dibutyryl Cyclic Adenosinmonophosphate

As cyclic nucleotides are known to control the growth and differrentiation of various embryonic cells (Friedman 1976), we added N^6O^2 dibutyryl adenosine 3,5-cyclic monophosphate (db-cAMP), theophylline, or both together to the liquid medium in a concentration of 10^{-3} M. In comparison with the control series, in the series treated with db-cAMP and with db-cAMP and theophylline skeletal muscle appeared more frequently (Škreb and Hofman 1977). Theophylline alone seemed to be ineffective. The same holds true for the wet weights of explants which were heavier after db-cAMP than in the control series. Butyric acid alone had either a toxic effect (10^{-2} M) or had no effect whatsoever (10^{-3} M). Thus, we cannot exclude the possibility of the effect of the purine nucleotide on the differentiation.

C. Embryonal Carcinoma Cells After Cultivation of Mouse Shields in Vitro

As mouse embryonic shields cannot so far compete with rat shields as regards differentiation of tissues after the appropriate in vitro culture, the question arises whether embryonal carcinoma cells can be found in mouse explants. The best method to assess the progress of cytodifferentiation in the mouse explants is to transfer them to the kidney capsule after various time periods in vitro. After 2 days in vitro, retransplantable teratocarcinoma was found in kidney grafts (from 38 explants). Therefore, we can conclude that some embryonic cells do not change at least for 2 days in culture.

IV. Discussion and Conclusions

As pointed out in the introduction, the principal aim of the experiments described was to analyze the factors necessary to obtain teratoma-like structures in an in vitro system. As far as rat embryonic shields are concerned, the results obtained in vitro are satisfactory, although the level of differentiation is slightly inferior to that seen in the kidney grafts. The growth of the explants resembles rather that of an organ culture than the growth of the same embryonic shields under the kidney capsule (Deakin 1975). The sequence of appearance of differentiated tissues is about the same as under the kidney capsule. The advantage of homologous serum for the cultivation of rat embryos has already been stressed by New (1967) who studied the development of whole rat embryos in circulating medium. Recently the same results were obtained for the human pancreatic islets in culture (Goldman and Colle 1976). These data seem to explain our failure to obtain differentiated tissues in mouse embryonic shields. The majority of cultures of mouse embryos were performed in rat or fetal calf serum. Good results were only rarely obtained in mouse serum (from the strain 129 SV) or in a special batch of commercial fetal calf serum. The serum from our strain of mice (C3H) seems to be toxic for the embryonic shields.

It was easy to obtain the appearance of fewer tissues than usual by omitting the metal grid or by changing the medium only once during the cultivation period. Nevertheless, in none of these cases did the cells remain at the same level of differentiation as at the moment of explantation. They already showed an irreversible commitment which was seen on the histological sections as stroma-like cells or various slightly immature epithelia.

The proof that such culture conditions can only rarely and not for a long time propagate cells without differentiation was obtained in mouse kidney grafts. If the mouse explants were transferred under the kidney capsule after 2 days in culture in rat serum, some very rare clusters of embryonal carcinoma cells were seen.

The improvement of the differentiation was so far obtained only with db-cAMP which is known to accelerate or induce several differentiated functions in some other embryonic systems (Freedman 1976).

References

Burton, K.: A study of the conditions and mechanism of the diphenylamine reaction for the colorimetric estimation of desoxyribonucleic acid. Biochem. J. 62, 315-322 (1956)
Damjanov, I., Solter, D.: Experimental Teratoma. Curr. Top. Pathol. 59, 69-130 (1974)
Deakin, A.S.: Model for the growth of a solid in vitro tumor. Growth 39, 159-165 (1975)
Friedman, D.L.: Role of cyclic nucleotides on cell growth and differentiation. Physiol. Rev. 56, 652-708 (1976)
Goldman, H., Colle, E.: Human pancreatic islets in culture: Effects of supplementing the medium with homologous and heterologous serum. Science 192, 1014-1016 (1976)
Illmensee, K., Mintz, B.: Totipotency and normal differentiation of single teratocarcinoma cells cloned by injection into blastocyst. Proc. Natl. Acad. Sci. USA 73, 549-553 (1976)
Jacob, F.: Mouse teratocarcinoma and mouse embryo. Proc. R. Soc. London Ser. B 201, 271-284 (1978)

Martin, G.R.: Teratocarcinomas as a model system for the study of embryogenesis and neoplasia. Cell 5, 229-243 (1975)

New, D.A.T.: Development of explanted rat embryo in circulating medium. J. Embryol. Exp. Morphol. 17, 513-525 (1967)

Pierce, G.B.: Teratocarcinoma: model for a developmental concept of cancer. Curr. Top. Dev. Biol. 2, 223-246 (1967)

Schneider, W.C.: Phosphorus compounds in animal tissues I. Extraction and estimation of DNA and RNA. J. Biol. Chem. 161, 293-305 (1945)

Škreb, N., Crnek, V.: Tissue differentiation in ectopic grafts after cultivation of rat embryonic shields in vitro. J. Embryol. Exp. Morphol. 42, 127-134 (1977)

Škreb, N., Hofman, Lj.: Effect of dibutyryl cAMP and theophylline on cultured rat embryonic shields. Experientia 33, 1651 (1977)

Škreb, N., Švajger, A.: Histogenetic capacity of rat and mouse embryonic shields cultivated in vitro. Wilhelm Roux' Arch. Entwicklungsmech. Org. 173, 228-238 (1973)

Škreb, N., Švajger, A.: Experimental teratomas in rats. In: Teratomas and Differentiation (eds. M.I. Sherman, D. Solter), pp. 83-97. New York-London: Academic Press 1975

Škreb, N., Švajger, A., Levak-Švajger, B.: Growth and differentiation of rat egg cylinders under the kidney capsule. J. Embryol. Exp. Morphol. 25, 47-56 (1971)

Solter, D., Adams, N., Damjanov, I., Koprowski, H.: Control of teratocarcinogenesis. In: Teratomas and Differentiation (eds. M.L. Sherman, D. Solter), pp. 139-159. New York-London: Academic Press 1975

Stevens, L.S.: The biology of teratomas. Adv. Morphol. 6, 1-31 (1967)

Loss of Tumorigenicity and Gain of Differentiated Function by Embryonal Carcinoma Cells

E.D. Adamson and Ch.F. Graham

Department of Zoology, University of Oxford, South Parks Road
Oxford, Great Britain

I. Loss of Tumorigenicity

A. Introduction

As the mouse embryo develops, so a variety of cell types are formed: in succession trophectoderm, parietal endoderm and visceral endoderm are derived from the multipotential stem cell population (reviewed Papaioannou et al. 1978a). These cell types are unable to form transplantable tumors when they are injected into syngeneic adult hosts, while the precursors of these cell types are able to form multipotential stem cell tumors (reviewed Diwan and Stevens 1976; Graham 1977). The implication is that the development to these cell types antagonizes the ability to form transplantable tumors.

There have been various attempts to study this antagonism in embryonic systems (see other chapters in this book). Here we are concerned with the multipotential tumors (teratocarcinomas) which have been derived by transplanting embryos to extrauterine sites. The embryonal carcinoma (EC) stem cells of these tumors have been isolated (recent review, Martin 1978), and it has been shown that these cells lose their ability to form tumors in a variety of circumstances. During in vivo differentiation in embryonic (e.g., Illmensee and Mintz 1976; Dewey et al. 1978; Papaioannou et al. 1978b) and adult hosts (e.g., Pierce et al. 1960; Rheinwald and Green 1975), cell types develop which are not tumorigenic. During in vitro differentiation the same loss of tumorigenicity occurs (e.g., Hall et al. 1975; Nicolas et al. 1976). These experiments on the loss of tumorigenicity and differentiation lack detail. Frequently the EC cells were not cloned and they may have been genetically heterogeneous. When cloned EC cells were used they may have been in different developmental states (see later). In both cases the apparent loss of tumorigenicity might be due either to selection against potentially tumorigenic cells or to the loss of tumorigenicity by cell differentiation.

It is certainly the case that cells derived from EC cells can be both tumorigenic and closely resemble (at least in morphology) differentiated cell types such as parietal endoderm (Lehman et al. 1974) and fibroblast (Evans 1972). It is usually thought that these cells arise by some secondary mechanism which "transforms" the cells sometime after they have differentiated. Another possibility is that the capacity for tumorigenic growth lurks in all derivatives of EC cells. It is clearly important to distinguish between these two alternatives by following the kinetics of the loss of tumorigenicity and the gain of differentiated function. We have attempted to do this by the experiments described below.

B. The Transition to END Cells

In culture, some EC cell clones can differentiate into cells which morphologically and histochemically resemble some of the endoderm layers of the embryo (reviewed Evans 1976; Hogan 1977; Martin 1978). This morphological transition may be induced by aggregating EC cells, but it is difficult to observe the behavior of individual cells in these lumps. This difficulty is avoided when the transition is induced in monolayer culture (Burke et al. 1978).

The cells used in our experiments were OC15SI. This is a clone from a single cell (McBurney 1976) which was derived from tumor OTT 6050 passaged in 129J/Sv mice (Stevens 1970). It was maintained with EC cell morphology by subculturing daily and plating at approx. 3.5×10^6 cells in untreated 50 mm diameter tissue culture dishes. The transition was initiated by switching the culture medium from rich (alpha medium, Stanners et al. 1971) to minimal (MEM, Flow Laboratories, Irvine, Scotland), by changing the serum from 10% fetal calf to 10% calf, and by plating at low density on gelatin coated dishes (2 to 4×10^5 cells per 90 mm dish; experimental details in Adamson et al. 1979).

During the 4 to 5 days following the low density plate, most of the small EC cells developed into big flat cells. These cells were called END because they biochemically mimic gene expression in the visceral endoderm of the embryo (see below).

C. Cell Selection During the Transition

The morphological transition was accompanied by some cell death. This was difficult to detect because all the attached cells appeared healthy and continued to divide; the dead cells were only noticed when the tissue culture medium was centrifuged at the end of the experiment. In a typical experiment, 3×10^5 cells were plated and 5 days later these had formed 12×10^5 healthy and attached cells on the bottom of the dish; 1.5×10^5 necrotic cells or cell fragments were found in the medium (cell numbers/90 mm dish).

It was important to discover if this cell death selected out a substantial proportion of the cells which were originally plated, since the figures show that up to half the original cells might have died. An alternative possibility was that cell death occurred at random amongst the derivatives of the cells which were plated. The transition was therefore followed by continuous time lapse recordings using an inverted microscope. Preliminary observations showed that at least 90% of the cells which were originally plated had formed at least one healthy attached cell at the end of the transition (4 days). The conclusion is that the transition does not involve massive selection of the original plate and therefore this was a suitable system for studying the antagonism between differentiation and tumorigenicity.

D. Loss of Tumorigenicity by END Cell Formation

We wished to show that the ability to form tumors in syngeneic hosts changed during the transition. Cells were injected subcutaneously and tumor formation scored after 40 days. The EC cells were highly tumorigenic in contrast to END cells (see Table 1). The capacity for tumor formation was clearly lost during the transition. These experiments do not exclude a role for cell selection in this loss but it is unlikely to be important. Below we show that gene expression changes during the transition and it is likely that it is this developmental change which accounts for the loss of tumorigenicity.

Table 1. Formation of palpable tumors after 40 days by OC15S1 EC and END cells

Cell morphology	Number cells injected	Take at 40 days
Embryonal carcinoma	10,000	6/10
Embryonal carcinoma	5,000	2/10
END	96,000	10/10
END	49,000	0/10*
END	10,000	0/10
END	5,000	0/10

The OC15S1 cells were dissociated to single cells and injected subcutaneously into the neck region of syngeneic 2- to 3-month-old males. The skin was felt every 3 or 4 days to detect tumors. The END cells were tumorigenic when injected in large numbers (96,000) and tumors also appeared in 6/10 animals after 2 to 3 months* when 49,000 END cells were injected into each animal. Two of the latter tumors were sectioned, and the tumors principally contained EC cells. We conclude that a few EC cells do not transit to END during a low density plate for 4 days

II. Gain of Differentiated Function

A. Introduction

The EC cell is small, its cytoplasm is poor in organelles such as mitochondria and lysosomes. It has little endoplasmic reticulum and its ribosomes are largely free. Morphologically, it appears to be a cell with little secretory activity and in general cellular enzymic activities are low. The cell surface of most EC cell lines, however, is marked by the presence of alkaline phosphatase in sufficiently high amounts that it has been the subject of several studies (reviewed Graham 1977) and seems to consist of an isoenzymic form which is unique to teratocarcinoma cells (Wada et al. 1976).

Overtly differentiated cells produced by cloned teratocarcinoma cell lines in culture are formed when dense cultures of EC cells are maintained without subculturing. OC15S1 EC cells rapidly differentiate in vitro, first into nerve; later contracting muscle fibers appear. The production of nerve is accompanied by an increase in the specific activity of acetylcholinesterase, and the appearance of the nerve-specific isoenzyme of aldolase. Muscle tissue is biochemically marked by the production of muscle-specific isoenzymes of creatine phosphokinase and enolase (Adamson 1976; Adamson et al. 1977; Fletcher et al. 1978). The degree of biochemical maturity reached falls short of that of adult tissue and it is clear that culture conditions greatly affect this. However, the conclusion is that the EC cell line used in the current experiments is biochemically multipotential.

B. Heterogeneity Within Cloned EC Cell Lines

Several cell surface antigenic determinants on EC lines have been recognized by immunological methods and some antisera to EC cells have cross-reactivity with early embryonic cells (reviewed Jacob 1977). The best known EC cell antigen, F9, is present on all EC cell lines tested but some cloned EC cell lines contain subpopulations which do not react (Reisner et al. 1977). EC cell antigens disappear when differentiation occurs, and it is possible that one of the reasons why EC lines differ in respect of tumorigenicity, pluripotency or ability to differentiate is that they have become heterogeneous in culture and contain some cells which are covertly differentiated along reversible or irreversible pathways. The C antigen on SIKR cells, for example, (Evans et al. 1978) is not present on a portion of PSA4 EC cells but this seems to be a reversible property.

An antiserum (PG-1) raised against 13th- to 14th-day mouse primordial germ cells (Heath 1978) reacts with 100% of primordial germ cells as tested by indirect immunofluorescence. About 95% of the nullipotent EC line F9 stain, close to 100% of PC13 (related to F9 in origin), but only 30% of OC15S1 and 2% of C145b EC cell lines stain. It has been suggested that the PG-1 antigen may mark those EC cells which can not differentiate or, alternatively, that it marks the cells which have not moved into a transition phase towards differentiation since the antigen is not present on the overtly differentiated cell products (including OC15 END) of these teratocarcinoma cell lines.

The conclusion is that a population of OC15S1 cells, like other multipotential EC cell lines, may have a uniform morphology and also have heterogeneity with respect to antigen expression.

C. Biochemical Changes During Differentiation to END Cells

First, it is important to stress that the first differentiated cells to be formed by the developing inner cell mass of the blastocyst embryo are primitive endoderm cells. These are thought to be the precursors of the parietal endoderm which lines the trophoblastic sac of the embryo, and the visceral endoderm which also finds its end as the outer of two layers of an extra-embryonic membrane, the visceral yolk sac. Below we have summarized the evidence which suggests that END cells mimic the visceral endoderm of the embryo. This conclusion is based on studies of several kinds of gene products synthesized by END cells, some of which have been found to be characteristic products of specific types of endoderm in the embryo.

1. Products Which Are Components of an Extracellular Matrix

a) LETS, fibronectin or cell attachment factor. One of the necessary culture conditions which leads to differentiation of EC cells is their attachment to a substratum. It is therefore appropriate to find that both EC cells and END cells synthesize fibronectin, but only END cells retain it in substantial amounts which are readily detected by indirect immunofluorescence (Wartiovaara et al. 1978a). In the embryo, fibronectin is found in basement membranes underlying the parietal endoderm and the visceral endoderm of the egg cylinder and visceral yolk sac (Wartiovaara et al. 1978b).

b) Collagen. END cells synthesize type I collagen to about 1% of secreted proteins. This has been shown by characterization of radioactive biosynthesized protein and by immunoperoxidase staining using a specific antibody (Adamson et al. 1979). This collagen type is also synthesized by the endoderm of the visceral yolk sac in contrast to the primitive endoderm and parietal endoderm which synthesize type IV collagen. Type I collagen is an integral part of the extracellular environment of many cells and has specific binding sites for fibronectin (Kleinman et al. 1976; Dessau et al. 1978); these properties may combine to promote the flattening out of END cells which occurs during their differentiation. The transition to big flat cells appears to be promoted by the denatured type I swine skin collagen which is used to coat the tissue culture dishes; perhaps the synthesis of type I collagen has an autocatalytic effect on differentiation (for example in corneal differentiation, Meier and Hay 1975). Surprisingly, OC15EC cells also synthesize collagen, but at a much lower rate amounting to 0.2% of secreted products. However only 10–15% of this is type I collagen; 60–70% is type IV or basement membrane collagen. The switch from type IV to type I collagen during differentiation may lead to some clarification of the nature of heterogeneity within and between cloned EC cell lines since the detection of staining amounts of type I collagen may correlate with a transition state of differentiation in subpopulations. The conclusion is that matrix aroound END cells resembles that underlying the visceral endoderm.

2. Other Secreted Products

a) Alpha-fetoprotein (AFP). This glycoprotein is a normal product of the visceral endoderm and liver of the fetus (Dziadek and Adamson 1978). It is also produced by teratocarcinoma cells after they have differentiated into endoderm cells such as are seen as the outer rind of embryoid bodies formed during suspension culture (Adamson et al. 1977). It is just detectable in cultures of OC15 END cells by radio-immunoprecipitation with a specific antibody to mouse AFP. It has not been detected in EC cell cultures.

b)Transferrin. This is also a glycoprotein synthesized by visceral endoderm cells and by fetal and adult liver. In 15th-day embryos it is synthesized in rather smaller amounts than AFP, but END cells synthesize it in larger amounts. It can be readily detected by immunoprecipitation with anti-mouse transferrin after radioactive labeling.

c) Plasminogen activator. When EC lines aggregate and form endoderm there is an increase in plasminogen activator levels (Linney and Levinson 1977), OC15 EC cells synthesize little of this protease although we are not confident of this result because they tend to die during the assay. END cells produce high levels of plasminogen activator and this mimics a similar event in developing embryonic tissues. Visceral yolk sac endoderm as well as parietal endoderm has this property. Visceral endoderm has only recently been shown to be positive in this test and this was found when assays were made at salt levels which did not inhibit the last step of the fibrin overlay assay (Bode and Dziadek 1979; but see Sherman et al. 1976; Strickland et al. 1976).

d) Interferon. END cells differ from EC cells in being interferon inducible and interferon sensitive (Burke et al. 1978). After induction, END cells produce one thousandfold more interferon than EC cells. The cell types which are sensitive to the action of interferon and which are interferon inducible in the embryo are not known.

3. Enzymes

As indicated above, one notable feature of most EC cell lines is the presence of high levels of alkaline phosphatase. This falls to very low values in differentiated cells (Bernstine et al. 1973; Chung et al. 1977). We have not shown this change with OC15S1 but have examined lysosomal acid phosphatase in this cell type. This enzyme activity increases about three or fourfold per cell and as it does so a larger proportion of the fast migrating lysosomal isoenzyme is evident. After the formation of the large amount of cytoplasm in the flattened END cell, numerous (up to 20) acid phosphatase-rich granules or lysosomes appear whereas EC cells have, on average, only one. Other lysosomal enzyme activities also increase during the differentiation to endoderm cells.

III. Conclusions

1. When OC15EC cells differentiate to END cells, they exhibit dramatic changes in gene expression. These include losses of cell surface antigens but increases in the synthesis and/or retention of fibronectin, type I collagen, AFP, transferrin, plasminogen activator, interferon, and acid phosphatase.
2. These biochemical features are also found in the visceral yolk sac endoderm of developing mouse embryo and we suggest therefore that the formation of END cells mimics visceral endoderm formation.
3. In view of the heterogeneity to PG-1 antiserum demonstrated by OC15S1 cells, the question may be raised that the formation of END cells was the result of cell selection. However a one thousandfold increase in the interferon titer per cell during the differentiation of END cells coupled with our time-lapse video tape observations argues against it.
4. Early embryonic cells and EC cells are both transplantable. Both these types of pluripotent cells will differentiate and in each the process is closely linked to the loss of tumorigenicity.

References

Adamson, E.D.: Isoenzyme transitions of creatine phosphokinase, aldolase and phosphoglycerate mutase in differentiating mouse cells. J. Embryol. Exp. Morphol. 35, 355-367 (1976)

Adamson, E.D., Evans, M.J., Magrane, G.G.: Biochemical markers of the progress of differentiation in cloned teratocarcinoma cell lines. Eur. J. Biochem. 79, 607-615 (1977)

Adamson, E.D., Gaunt, S.J., Graham, C.F.: The differentiation of teratocarcinoma cells is marked by the types of collagen which are synthesized. Cell 17, 469-476 (1979)

Bernstine, E.G., Hooper, M.L., Grandchamp, S., Ephrussi, B.: Alkaline phosphatase activity in mouse teratoma. Proc. Natl. Acad. Sci. USA 70, 3899-3903 (1973)

Bode, V.C., Dziadek, M.A.: Plasminogen activator secretion during mouse embryogenesis. Dev. Biol. 73, 272-289 (1979)

Burke, D.C., Graham, C.F., Lehman, J.M.: Appearance of interferon inducibility and sensitivity during differentiation of murine teratocarcinoma cells in vitro. Cell 13, 243-248 (1978)

Chung, A.E., Estes, L.E., Shinozuka, H., Braginski, J., Lorz, C., Chung, C.A.: Morphological and biochemical observations on cells derived from the in vitro differentiation of the embryonal carcinoma cell line PCC4-F. Cancer Res. 37, 2072-2081 (1977)

Dessau, W., Adelmann, B.C., Timpl, R., Martin, G.R.: Identification of the sites in collagen α-chains that bind serum anti-gelatin factor (c-Ig). Biochem. J. 169, 55-59 (1978)

Dewey, M.J., Martin, D.W.M., Martin, G.R., Mintz, B.: Mosaic mice with teratocarcinoma-derived mutant cells deficient in hypoxanthine phosphoribosyl transferase. Proc. Natl. Acad. Sci. USA 74, 5564-5568 (1978)

Diwan, S., Stevens, L.C.: Development of teratomas from the ectoderm of mouse egg cylinders. J. Natl. Cancer Inst. 57, 937-942 (1976)

Dziadek, M., Adamson, E.D.: Localization and synthesis of alphafetoprotein in postimplantation mouse embryos. J. Embryol. Exp. Morphol. 43, 289-313 (1978)

Evans, M.J.: The isolation and properties of a clonal tissue culture strain of pluripotent mouse teratoma cells. J. Embryol. Exp. Morphol. 28, 163-176 (1972)

Evans, M.J.: Totipotency of animal cells. In: The Development of Plants and Animals (eds. C.F. Graham, P.F. Wareing), pp. 64-72. Oxford: Blackwell 1976

Evans, M.J., Lovell-Badge, R., Stern, P.L.: Differentiation of Teratocarcinoma cells. Eur. Terat. Soc. Meet. Symp. Teratology 14, 367 (Abstract) (1976)

Fletcher, L., Rider, C.C., Tayler, C.B., Adamson, E.D., Luke, B.M., Graham, C.F.: Enolase isoenzymes as markers of differentiation in teratocarcinoma cells and normal tissues of mouse. Dev. Biol. 65, 211-224 (1978)

Graham, C.F.: Teratocarcinoma cells and normal mouse embryogenesis. In: Concepts in Mammalian Embryogenesis (ed. M.I. Sherman), pp. 315-394. Cambridge, MA-London: MIT Press 1977

Hall, J.D., Marsden, M., Rifkin, D., Teresky, A.K., Levine, A.J.: The in vitro differentiation of embryoid bodies produced by transplantable teratoma in mice. In: Teratomas and Differentiation (eds. M.I. Sherman, S. Solter), pp. 251-269. New York-London: Academic Press 1975

Heath, J.K.: Characterization of a xenogeneic antiserum raised against the foetal germ cells of the mouse: cross reactivity with embryonal carcinoma. Cell 15, 299-306 (1978)

Hogan, B.L.M.: Teratocarcinoma cells as a model for mammalian development. In: Biochemistry of Cell Differentiation, Vol. XV (ed. J. Paul), pp. 333-376. Baltimore: University Park Press 1977

Hogan, B.L.M.: High molecular weight extracellular proteins synthesized by endoderm cells derived from teratocarcinoma cells and normal extraembryonic membranes. Dev. Biol. 76, 275-285 (1980)

Illmensee, K., Mintz, B.: Totipotency and normal differentiation of single teratocarcinoma cells cloned by injection into blastocysts. Proc. Natl. Acad. Sci. USA 73, 549-553 (1976)

Jacob, F.: Mouse teratocarcinomas and embryonic antigens. Immunol. Rev. 73, 3-32 (1977)

Kleinman, H.K., McGoodwin, E.B., Klebe, R.J.: Localization of the cell attachment region in type I and II collagen. Biochem. Biophys. Res. Commun. 72, 426-432 (1976)

Lehman, J.M., Speers, W.C., Swartzendruber, D.E., Pierce, G.B.: Neoplastic differentiation: characteristics of cell lines derived from a murine teratocarcinoma. J. Cell Physiol. 84, 13-28 (1974)

Linney, E., Levinson, B.B.: Teratocarcinoma differentiation: plasminogen activator activity associated with embryoid body formation. Cell 10, 284-304 (1977)

Martin, G.R.: Advantages and limitations of teratocarcinoma stem cells as models of development. In: Development in Mammals, Vol. 3 (ed. M.H. Johnson), pp. 225-265. Amsterdam-New York: North-Holland 1978

McBurney, M.W.: Clonal lines of teratocarcinoma cells in vitro: differentiation and cytogenetic characteristics. J. Cell Physiol. 89, 441-456 (1976)

Meier, S., Hay, E.D.: Stimulation of corneal differentiation by interaction between cell surface and extracellular matrix. J. Cell Biol. 66, 275-291 (1975)

Nicolas, J.F., Avner, P., Gaillard, J., Guenet, J.L., Jakob, H., Jacob, F.: Cell lines derived from teratocarcinomas. Cancer Res. 36, 4224-4231 (1976)

Papaioannou, V.E., Rossant, J., Gardner, R.L.: Stem cells in early mammalian development. In: Stem Cells and Tissue Homeostasis (eds. B.I. Lord, C.S. Potten, R.J. Cole), pp. 49-69. Oxford: Cambridge University Press 1978a

Papaioannou, V.E., Gardner, R.L., McBurney, M.W., Babinet, C., Evans, M.J.: Participation of cultured teratocarcinoma cells in mouse embryogenesis. J. Embryol. Exp. Morphol. 44, 93-104 (1978b)

Pierce, G.B., Dixon, F.J., Verney, E.L.: Teratocarcinogenic and tissue-forming potentialities of the cell types comprising neoplastic embryoid bodies. Lab. Invest. 9, 583-602 (1960)

Reisner, Y., Gachelin, G., Dubois, P., Nicolas, J.F., Sharon, N., Jacob, F.: Interaction of peanut agglutinin, a lectin specific for non-reducing terminal D-galactosyl residues with embryonal carcinoma cells. Dev. Biol. 61, 20-27 (1977)

Rheinwald, J.G., Green, H.: Formation of keratinizing epithelium in culture by a cloned cell line derived from a teratoma. Cell 6, 317-330 (1975)

Sherman, M.I., Strickland, S., Reich, E.: Differentiation of early mouse embryonic and teratocarcinoma cell in vitro: Plasminogen activator production. Cancer Res. 36, 4208-4216 (1976)

Solter, D., Shevinsky, L., Knowles, B.B., Strickland, S.: The induction of antigenic changes in a teratocarcinoma stem cell line (F9) by retinoic acid. Dev. Biol. 70, 515-521 (1979)

Stanners, C.P., Eliceiri, G., Green, H.: Two types of ribosome in mouse-hamster hybrid cells. Nature New Biol. 230, 52-54 (1971)

Stevens, L.C.: The development of transplantable teratocarcinomas from intertesticular grafts of pre- and post-implantation mouse. Dev. Biol. 21, 364-382 (1970)

Strickland, S., Mahdavi, V.: The induction of differentiation in teratocarcinoma stem cells by retinoic acid. Cell 15, 393-404 (1978)

Strickland, S., Reich, E., Sherman, M.I.: Plasminogen activator in early embryogenesis: enzyme production by trophoblast and parietal endoderm. Cell 9, 231-240 (1976)

Wada, H.G., Vandenberg, S.R., Sussman, H.H., Grove, W.E., Herman, M.M.: Characterization of two different alkaline phosphatases in mouse teratoma: partial purification, electrophoretic and histochemical studies. Cell 9, 37-44 (1976)

Wartiovaara, J., Leivo, I., Virtanen, I., Vaheri, A., Graham, C.F.: Appearance of fibronectin during differentiation of mouse teratocarcinoma in vitro. Nature (London) 272, 355-356 (1978a)

Wartiovaara, J., Leivo, I., Virtanen, I., Vaheri, A., Graham, C.F.: Cell surface and extracellular matrix glycoprotein fibronectin: expression in embryogenesis and in teratocarcinoma differentiation. Ann. N.Y. Acad. Sci. 312, 132-141 (1978b)

Clinical Oncology and Cell Differentiation

B.J. KENNEDY

*Section of Medical Oncology, University of Minnesota Medical School
Minneapolis, MN 55455, USA*

I. Introduction

The patient with cancer is a product of abnormal cell differentiation. To be able to control or cure cancer, fundamental research in cell differentiation and the subsequent understanding of the development of neoplasia can lead to concepts for methods of destruction of abnormal cells.

It is apparent to the clinician that surgical removal of a cancer can only cure a limited number of patients; whereas, radiotherapy can extend the area of control. For disseminated disease, hormonal or chemical therapies provide the ultimate solution. To be able to select the correct chemical, its method of administration, and the proper tumor for its use, developments in basic science are necessary and can lead to cure of these diseases.

The Conference on Cell Differentiation and Neoplasia covered a broad range of subjects including genetic control, chromosome abnormalities, growth regulation of malignant tumor cells, differentiation of stem cells, control of normal cell growth, carcinogenic agents, regulation of differentiation, and many other important subjects. Such a Conference on Cell Differentiation and Neoplasia has many beneficial aspects, many of which are intangible. The meeting of developmental biologists and oncologists can only be a stimulus to solving the problems of neoplasia and from the reports and reviews of this conference new ideas will evolve. Some clinical situations in oncology demonstrate how much of the basic work of the type described in this conference is valuable in the progress against cancer.

The medical oncologists' skill is clinical judgment in matters relating to cancer. It is their task to apply what has been proven and what is available to the immediate situation of a patient ill with cancer. Their weapons for treatment are the product of basic science and the understanding of the biologic nature of cancer.

While the basic scientist can transplant individual cancer cells, culture them and study their progressive growth, the clinical oncologist really deals with advanced cancer. Beginning with a single cancer cell, after 30 doublings of growth, the tumor at 1 cm size finally becomes measurable by the clinician. However, after only five more doublings, the 1 cm lesion attains the size of a football, demonstrating the short time that the clinician has to cope with this disease process.

Clinical oncology has gone through several eras in the past 30 years with respect to the effect of chemotherapy on malignant disease. In the early 40's, the clinician dealt primarily with palliation, such as relief of pain and immediate comfort needs. By the 50's, the availability of hormonal and chemical agents demonstrated that tumor *shrinkage* could occur as a result of such treatments. Symptomatic relief was obtained and in a few instances, long-term disappearance was noted. Hence, in the 60's, with additional agents, it became increasingly apparent that *control* of cancer was possible by hormonal and chemical means. Gross tumors shrank or disappeared and these responses were maintained for long periods. In the 70's, it was clear that some of these patients had never had recurrence of their cancer and that *cure* was obtainable. A few tumor systems such as choriocarcinoma of the uterus, histiocytic lymphoma, and testicular cancer emphasized that since cure was possible in these tumors, it should also be possible in other systems. This is the goal of the oncologist.

As improvement of cancer occurred, the clinician developed new terms: "shrinkage" implying decrease in size of a cancer, "stable disease" implying no evidence of progression but no significant regression, "partial regression" meaning shrinkage of greater than 50% of the tumor for at least two or more months, and "complete regression" with disappearance of all measurable cancer and none detected by any clinical, biochemical, radiologic, or histologic means. It is the complete remission that we now seek in all therapies as the most meaningful response. It is in this group that cure can be appreciated.

Skipper's basic model of cell growth was a major contribution of biologists to the clinician's understanding of the process of cell destruction and goal of cure. As each course of treatment with chemicals was administered, a proportion of tumor cells are destroyed. The greater the proportion and the more frequent the therapy, the more likely is complete destruction of tumor cells possible. A few residual tumor cells might be destroyed by the immuno-competence of the host itself. Anecdotal examples of patients' responses to treatments emphasize what has been accomplished to date.

II. Testis Cancer

Testis cancer is the most common tumor in men between the ages of 20 and 35 (Twito and Kennedy 1975). In patients with recurrence of that disease, 85% of those with embryonal cell cancer were dead in 2 years. Now with intensive chemotherapy involving three or more chemical agents, over 50% of those patients with disseminated cancer are free of disease more than 2 years after their last treatment. They would appear to be cured.

Testis cancer is a disease in which cell differentiation is important. Five major histologic types originate from a common cell: embryonal cell carcinoma, teratocarcinoma, seminoma, choriocarcinoma, and yolk sac tumors being the most common of these germinal cell neoplasms.

The evolution of the chemical therapy of testis cancer demonstrates the advantages of clinicians and basic scientists working together. In 1960, mithramycin was introduced as an agent with anti-tumor potentials. In testis cancer

Table 1. Improvement in the chemical treatment of disseminated testis cancer

Investigation	Year	Drug(s)	Complete regressions (%)
Li et al.	1960	Dactinomycin, Methotrexate, Chlorambucil	13
MacKenzie	1966	Dactinomycin	20
Kennedy	1970	Mithraymcin	23
Samuels	1973	Vinblastine + i.m. bleomycin	32
Samuels	1975	Vinblastine + bleomycin infusions	49
Cvitkovic	1976	VAB III	63
Einhorn	1977	Vinblastine, bleomycin Cis-Platinum	72

striking objective regressions occurred in one-half of the treated patients. However, because of hepatic toxicity and platelet toxicity, a mortality rate of up to 24% was reported (Brown and Kennedy 1965). Although an effective anti-tumor agent, the toxicity prevented adoption of this agent for common use.

In the laboratory the biochemist learned how to measure RNA synthesis and discovered that mithramycin had a dissociated effect on normal cell RNA synthesis and tumor cell RNA synthesis (Yarbo and Kennedy 1967). Following mithramycin injection, RNA synthesis in the mouse liver was decreased, recovering in 25 h. A tumor in that same mouse had depression of RNA synthesis for 48 h. With this dissociation, it became apparent that administration of the drug every other day would avoid accumulative clinical toxicity. This proved to be so and the drug was administered every other day with minimal or controllable toxicity, with no mortality. This was one of the beginning realizations that the time of administration of an agent is important (chronopharmacology of chronotherapy).

Following demonstration of the effectiveness of single drug therapy in testis cancer, resulting in 25% complete remission, additional therapies were developed using combinations of agents (Table 1). Vinblastine plus bleomycin (Samuels et al. 1975) and subsequently, vinblastine, bleomycin and cis-diamminedichloroplatinum (Einhorn and Donahne 1977) resulted in complete regression in 55–60% of patients. In fact, 100% of patients treated have partial or complete regressions demonstrating the remarkable progress being made in this disease.

To aid in the management of testis cancer, the basic scientist has developed the measurement of biologic markers (alphafetoprotein, human serum subunit beta gonadotrophins, lactic-dehydrogenase). These are now valuable in terms of diagnosis, prognosis, monitoring the course of the disease, and assessment of completeness of treatment.

The remarkable progress in the treatment and cure of testis cancer is truly the product of applied basic research.

III. Breast Cancer

Advanced breast cancer can be destroyed by hormonal therapy or chemical therapy. In a post-menopausal woman the administration of estrogenic hormones has destroyed some cancers. Rarely, this had resulted in complete disappearance of the disease and death from other causes. However, only 35% of such treated women respond with objective decreases in the tumor. An explanation for the failures was required.

The clinical revolution in hormone therapy of breast cancer came about with the discovery by the basic scientist of the estrogen receptor protein. With this added biological information as to the nature of the breast cancer, the patient could be selected for specific hormonal therapy and allow the clinician to employ this test as a prognostic factor (Kiang and Kennedy 1977). The biological marker provided not only prognostic implications but therapeutic guidance. This is the essential principle in good tumor therapy, that is, to be able to select the correct treatment in the appropriate patients based on the biological nature of the cancer.

With the measurement of the receptor proteins both in the cytoplasm and nucleus, the biological nature of each tumor can be further defined and explanations developed to explain why the various empirical therapies employed in the past are effective. We now know that small doses of estrogen stimulate the growth of breast cancer in women. Large doses, however, appear to block the action of receptor proteins on DNA synthesis and interfere with cancer growth. Increasing knowledge of this sort will guide us further to specific treatments in selected patients.

IV. Chronobiology

In chronic myelogenous leukemia, it was noted that the white blood cell count varied in cyclic oscillations (Kennedy 1970). These biorhythms also were apparent in the normal white blood cell and represent a switching on and off of a differentiation process. These clinical observations stimulated studies in the timing of the use of drugs, developing different schedules of administration, in an attempt to reduce toxicity and increase anti-tumor effects. Many studies in chronobiology imply that the time of day a drug is administered will have an appreciable effect on toxicity of the drug, if not on the tumor response (Halberg 1976). We are just beginning to see the impact of chronobiology in clinical oncology.

V. Methods of Drug Administration

In the management of cancer patients the method of administration, the timing of administration (chronotherapy), and the amount of drug (dose) became increasing factors in control of tumors. Combinations of therapies are of great value. An agent which may be ineffective intravenously may destroy a tumor effectively by infusion of the drug in the artery leading to that tumor. The addition of hyperthermia with chemotherapy, or chemotherapy and radiotherapy are other modalities that are under investigation. Again, the understanding of the mechanisms of these therapies is a responsibility of the basic scientist.

VI. Adjuvant Chemotherapy

With the success of chemotherapy in advanced malignant diseases, clinicians turned to treatment of earlier disease with the concept that a smaller volume of cancer cells might be destroyed more easily. Following surgery of osteosarcoma, adriamycin or methotrexate have been administered resulting in improving the cure rate from 25% to over 50% (Frei et al. 1975). In premenopausal women with stage II breast cancer (the tumor having spread to the axillary nodes) combination chemotherapy after mastectomy has resulted in a lower rate of recurrence and prolonged survival (Bonnadonna et al. 1978).

VII. Conclusions

Let us not imply that because we are able to obtain complete regression and cure in a few patients with cancer, the problem is solved. Despite the clinicians' efforts, the large majority of patients with advanced cancer fail to be cured. Cancer can be rapidly progressive and destructive.

Breast cancer has not been cured with chemotherapy. Hence, the mortality figures have not changed. Some interpret this as failure of science to make progress in the management of breast cancer. On the other hand, however, the life of patients with breast cancer is prolonged by present therapies, although the disease is not cured.

It is not uncommon to see patients who have undergone all the standard surgical, radiotherapeutic, hormonal, and chemical therapies still faced with disseminated cancer. Following employment of new investigative agents, the disease prevails. Yet, these same patients may be ambulatory, effective and working members of the community; they plead for further treatment. This is the task for the basic scientists. An understanding of cell differentiation, of host resistance to cell growth, of mechanisms of actions of therapeutic agents, and development of new concepts of cancer control are essential to meet the demands of these patients who need help.

With the united efforts of basic scientists and clinicians, we can obtain the cure of cancer. This conference on Cell Differentiation and Neoplasia will have helped in attaining that goal.

Acknowledgment. Supported in part by grant CA-19527 from the National Cancer Institute, United States Public Health Service, the Masonic Hospital Fund, Inc., and the Minnesota Medical Foundation.

References

Bonnadonna, G., Valagussa, P., Rossi, A., Zucali, R., Tancini, G., Bajetta, E., Brambilla, C., De Lena, M., Di Fronzo, G., Banfi, A., Rilke, F., Veronesi, U.: Are surgical adjuvant trials altering the course of breast cancer? Semin. Oncology 5, 450-464 (1978)

Brown, J.H., Kennedy, B.J.: Mithramycin in the treatment of disseminated testicular neoplasms. New Engl. J. Med. 272, 111-119 (1965)

Cvitkovic, E., Hayes, D., Golbey, R.: Primary combination chemotherapy (VAB III) for metastatic or unresectable germ cell tumors. Abstract. Proc. Am. Assoc. Cancer Res. 17, 296 (1976)

Einhorn, L.H., Donahue, J.: Cis-diamminedichloroplatinum, vinblastine, and bleomycin combination chemotherapy in disseminated testicular cancer. Ann. Intern. Med. 87, 293-298 (1977)

Frei, E., III, Jaffe, N., Tattersall, M.H.N., Pitman, S., Parker, L.: New approaches to cancer chemotherapy with Methotrexate. New Engl. J. Med. 292, 846-851 (1975)

Halberg, F.: Chronobiology in 1975. Chronobiologica 3, 1-14 (1976)

Kennedy, B.J.: Cyclic leucocyte oscillation in chronic myelogenous leukemia during hydroxyurea therapy. Blood 35, 751-760 (1970)

Kiang, D.T., Kennedy, B.J.: Estrogen receptor assay in the differential diagnosis of adenocarcinomas. JAMA 238, 32-34 (1977)

Li, M.C., Whitmore, W.F., Golbey, R., Grabstald, H.: Effects of combined drug therapy on metastatic cancer of the testis. JAMA 174, 145-153 (1960)

MacKenzie, A.R.: Chemotherapy of metastatic cancer. Results in 154 patients. Cancer 19, 1369-1376 (1966)

Samuels, M.L., Johnson, D.E., Holoye, P.Y.: The treatment of stage 3 metastatic germinal cell neoplasia of the testis with bleomycin combination chemotherapy. Proc. Am. Assoc. Cancer Res. 14, 23 (1973)

Samuels, M.L., Johnson, D.E., Holoye, P.Y.: Continuous intravenous bleomycin (NSC-125066) therapy with vinblastine (NSC-49842) in stage III testicular neoplasia. Cancer Chemother. Rep. 59, 563-570 (1975)

Twito, D.I., Kennedy, B.J.: Treatment of testicular cancer. Annu. Rev. Med. 26, 235-243 (1975)

Yarbo, J.W., Kennedy, B.J.: A comparison of the rate of recovery from inhibition of RNA synthesis in mouse liver and transplantable glioma. Cancer Res. 27, 1779-1782 (1967)

Subject Index

Results and Problems in Cell Differentiation

Editors: W. Beermann,
W. J. Gehring, I. B. Gurdon,
F. C. Kafatos, J. Reinert

Springer-Verlag
Berlin
Heidelberg
New York